T0191867

Lecture Notes in Physics

Volume 948

The Lecture Notes in Physics

The series Lecture Notes in Physics (LNP), founded in 1969, reports new developments in physics research and teaching-quickly and informally, but with a high quality and the explicit aim to summarize and communicate current knowledge in an accessible way. Books published in this series are conceived as bridging material between advanced graduate textbooks and the forefront of research and to serve three purposes:

- to be a compact and modern up-to-date source of reference on a well-defined topic
- to serve as an accessible introduction to the field to postgraduate students and nonspecialist researchers from related areas
- to be a source of advanced teaching material for specialized seminars, courses and schools

Both monographs and multi-author volumes will be considered for publication. Edited volumes should, however, consist of a very limited number of contributions only. Proceedings will not be considered for LNP.

Volumes published in LNP are disseminated both in print and in electronic formats, the electronic archive being available at springerlink.com. The series content is indexed, abstracted and referenced by many abstracting and information services, bibliographic networks, subscription agencies, library networks, and consortia.

Proposals should be sent to a member of the Editorial Board, or directly to the managing editor at Springer:

Christian Caron
Springer Heidelberg
Physics Editorial Department I
Tiergartenstrasse 17
69121 Heidelberg/Germany
christian.caron@springer.com

More information about this series at http://www.springer.com/series/5304

Christoph Scheidenberger • Marek Pfützner
Editors

The Euroschool on Exotic Beams - Vol. 5

 Springer

Editors
Christoph Scheidenberger
Nuclear Physics Department
GSI Helmholtzzentrum für
Schwerionenforschung GmbH
Darmstadt, Germany

Marek Pfützner
Faculty of Physics
University of Warsaw
Warsaw, Poland

ISSN 0075-8450   ISSN 1616-6361   (electronic)
Lecture Notes in Physics
ISBN 978-3-319-74877-1   ISBN 978-3-319-74878-8   (eBook)
https://doi.org/10.1007/978-3-319-74878-8

Library of Congress Control Number: 2014930224

Printed on acid-free paper

This Springer imprint is published by the registered company Springer International Publishing AG part of Springer Nature.
The registered company address is: Gewerbestrasse 11, 6330 Cham, Switzerland

Preface

This book is the fifth volume of a series of Lecture Notes in Physics, which emerged from the Euroschool on Exotic Beams. This book appears in 2018, when the 25th anniversary of the establishment of this school will take place. This anniversary will be celebrated in August in Leuven (Belgium), where the Euroschool started in 1993. With one exception (in 1999), the Euroschool on Exotic Beams has been held every year, first in Leuven from 1993 to 2000, and then, starting in 2001, it travelled to various places in Europe: Finland (2001, 2011), France (2002, 2007, 2017), Spain (2003, 2010), United Kingdom (2004), Germany (2005, 2016), Italy (2006), Poland (2008), Belgium (2009), Greece (2012), Russia (2013), Italy (2014), and Croatia (2015). Based on lectures given at these Euroschool events, the Lecture Notes provide an introduction, for graduate students and young researchers, to novel and exciting fields of physics with radioactive ion beams and their applications. The fifth volume in this series covers topics that were presented in Euroschool lectures between 2006 and 2015 and comprises recent updates.

Current research in nuclear physics aims at a comprehensive understanding and description of atomic nuclei and their properties, based on their fundamental degrees of freedom, protons and neutrons, and their interaction. The field has advanced tremendously with the advent of radioactive ion-beam facilities and intense stable beams, which allow the production and study of exotic nuclei, i.e., short-lived nuclei far-off stability, and superheavy elements. These studies opened the pathway to the *terra incognita* of the nuclear chart, leading, for instance, to the discovery of new chemical elements, novel magic numbers in nuclei with a large neutron excess, and new phenomena, such as neutron skins and collective excitations, in which these neutron skins oscillate against a rigid nuclear core. Laboratory access to exotic nuclei is also essential for the understanding of many astrophysical objects whose dynamics and associated nucleosynthesis are driven by short-lived nuclei. A new degree of freedom to study the strong interaction in nuclei has been opened up by hypernuclei, which involve hyperons, i.e., baryons that contain at least one strange quark. Studies of the baryon-baryon interactions in hypernuclei are also an essential approach for the understanding of extreme astrophysical objects with high densities, such as the cores of neutron stars, where hyperons are predicted to play important roles in the equation of state. This wide field, including applications in material research, biology, medicine, imaging techniques, security, and other areas, is explored on theoretical grounds and in experiments with exotic beams. These experiments are now being carried out at existing facilities and will be performed at future facilities. It is the goal of the Euroschool lectures and the present Lecture Notes to fill the gap between classical university education and research life in laboratories, and in this way to contribute to the education of the next generation of scientists, who will explore the *terra incognita*. Also, the Nuclear Physics European Collaboration Committee (NuPECC) states, in its Long Range Plan 2017, that activities such as the Euroschool on Exotic Beams are vital for training highly qualified researchers in nuclear science and are a very important element for maintaining development in the field. In the very first original funding request for the Euroschool on Exotic Beams, which was submitted to the European Commission in 1992, we read: "The school forms an ideal basis for exchange of technical and scientific know-how, and for mobility between different research institutes and universities". Now, 25 years later, this statement is still valid, and the Euroschool's Board of Directors, which organizes the Euroschool every year, is indebted to this mission.

Clearly, the present book cannot cover all the topics and methods in the broad field of radioactive beams. Therefore, with this fifth volume, we follow the previous examples and have selected topics from the traditional core of the field of exotic nuclei (fission, alpha decays, giant resonances) and included new directions (hyper-nuclei and nucleon resonances) and applied areas (laser acceleration and dating methods). None of these topics has been treated before in this series; therefore, the present volume complements the previous editions. This is an indicator of the breadth and prosperity of an active field. Owing to the engagement of Euroschool lecturers who are world class experts in their domains, the Euroschool Lecture Notes are a valuable asset for the high-level education of present and next-generation scientists. We hope that this volume will be as useful and as successful as the previous ones.

It is our pleasure to thank the sponsors for their support over many years; this support makes the Euroschool events possible and contributes to the education of next-generation scientists. The sponsors, to whom we are indebted, are:

- ADS; Arenberg Doctoral School (Belgium)
- CEA; Commissariat à l'énergie atomique et aux énergies alternatives (France)
- CNRS; Centre national de la recherche scientifique (France)
- Demokritos; National Center for Scientific Research, Athens (Greece)
- ECT*, European Centre for Theoretical Studies in Nuclear Physics and Related Areas, Trento (Italy)
- GANIL; Grand Accelerateur National d'Ions Lourds, Caen (France)
- Gobierno de España; Ministerio de Economia y Competitividad FANUC Network and CPAN Ingenio 2010, Madrid (Spain)
- GSI: Helmholtz Centre for Heavy Ion Research, Darmstadt (Germany)
- HIC-4-FAIR; Helmholtz International Center for FAIR, Darmstadt (Germany)

- IFIC-CSIC; Instituto de Fisica Corpuscular, Consejo Superior de Investigaciones Cientificas, Madrid (Spain)
- INFN; Istituto Nazionale di Fisica Nucleare (Italy)
- INFN-LNL; Laboratori Nazionali di Legnaro (Italy)
- IRB; Institut Ruder Boskovic (Croatia)
- ISOLDE-CERN, Geneva (Switzerland)
- JINR; Joint Institute for Nuclear Research, Dubna (Russia)
- JYFL; University of Jyväskylä (Finland)
- KU Leuven; Instituut voor Kern- en Stralingsfysica, Leuven (Belgium)
- KVI-CART; Center for Advanced Radiation Technology, Groningen (The Netherlands)
- UCL; Centre de Ressources du Cyclotron, Louvain-la-Neuve (Belgium)
- Università degli Studi di Padova (Italy)
- University of Warsaw (Poland)
- University of Zagreb (Croatia)
- USC; University of Santiago de Compostela (Spain).

At this point, we would like to thank all those who have contributed to this volume in various ways. We owe the lecturers the largest debt of gratitude, for their efforts in preparing and giving the excellent lectures for Euroschool participants, and for their dedication of much time and effort to the preparation of the contributions to this book: very warm thanks for writing these educational and understandable pieces for our students! Next, we would like to thank all the members of the Board of Directors of the Euroschool on Exotic Beams, who inspired the development of this book with their many ideas. Finally, it is, once more, our pleasure to thank Dr. Chris Caron and his colleagues at Springer Verlag for their encouragement and continuous support in a fruitful collaboration.

Warsaw, Poland   Marek Pfützner
Darmstadt/Giessen, Germany   Christoph Scheidenberger
November 2017

Contents

Chapter 1
Applications of ^{14}C, the Most Versatile Radionuclide to Explore Our World

Walter Kutschera

Abstract Carbon consists of the two stable isotopes (^{12}C and ^{13}C) often accompanied with minute traces of the long-lived radioisotope ^{14}C (half-life = 5700 years). This allows one to use isotope-sensitive methods to trace carbon throughout the environment at large on Earth. In particular, ^{14}C can be used for dating during the past 50,000 years. It has thus revolutionized archaeology, but also many fields of geophysics. Although ^{14}C is a cosmogenic radionuclide produced primarily by cosmic-ray interaction in the atmosphere, the dramatic increase of ^{14}C by the atmospheric nuclear weapons testing period generated a characteristic spike (^{14}C bomb peak) in the early 1960s, which can be used for a variety of unique applications. After describing the basic properties of ^{14}C, the current review focusses on applications of both cosmogenic and anthropogenic ^{14}C, emphasizing the versatility of this extraordinary radionuclide.

1.1 Introduction

The development of the ^{14}C dating technique in the late 1940s by Willard Libby at the University of Chicago [1–3] earned him rightfully the 1960 Nobel Prize in Chemistry. The fact that life as we know it is based on carbon, and ^{14}C is a naturally occurring radioactive carbon isotope with a long half-life (5700 years), makes it a unique tracer. From a nuclear physics point of view, such a long half-life of ^{14}C is not expected for the ß decay of ^{14}C to ^{14}N. With a normal strength of this so-called Gamov-Teller ß transition, observed in neighboring nuclei, the half-life should be on the order of hours, completely useless for archaeological dating. The extreme hindrance of the ^{14}C decay cannot be easily explained by nuclear models [4–5], and might perhaps be called a 'gift of nature' to archaeology.

W. Kutschera (✉)
Vienna Environmental Research Accelerator (VERA), Faculty of Physics – Isotope Research and Nuclear Physics, University of Vienna, Wien, Austria
e-mail: walter.kutschera@univie.ac.at

© Springer International Publishing AG, part of Springer Nature 2018
C. Scheidenberger, M. Pfützner (eds.), *The Euroschool on Exotic Beams - Vol. 5*,
Lecture Notes in Physics 948, https://doi.org/10.1007/978-3-319-74878-8_1

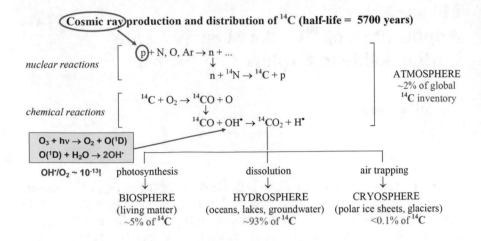

Fig. 1.1 Schematic presentation of the processes leading to the production and distribution of ^{14}C on Earth [6]. The oxidation of carbon is a two-step process, where the very reactive OH radical plays an important role to convert CO into CO_2 [7, 8]. The approximate fractions of the global ^{14}C inventory for the different archives is given in percent

Libby assumed that the continuous cosmic-ray production of ^{14}C through the $^{14}N(n,p)^{14}C$ reaction in the atmosphere and its global distribution through the CO_2 cycle resulted in the same $^{14}C/^{12}C$ isotope ratio in all living matter through all times. Figure 1.1 shows in a schematic way the processes which govern the production and distribution of ^{14}C on Earth, as we know it now [6].

In 1949, the measurement of ^{14}C in several objects of known age [3] gave a reasonably good confirmation that ^{14}C dating works for objects of the last 5000 years if their organic carbon all had the same $^{14}C/^{12}C$ isotopic ratios at the moment of death (Fig. 1.2). However, the assumption of the temporal constancy of the $^{14}C/^{12}C$ isotope ratio turned out to be not strictly fulfilled [9] and caused problems when ^{14}C dating started to challenge the time scale of archaeologists [10].

The situation improved when a ^{14}C calibration was established by measuring ^{14}C in objects of known age. The ideal natural archive for this are the yearly growth rings of trees, whose cellulose content reflects the $^{14}C/^{12}C$ ratio of atmospheric CO_2 for the respective year [11]. With the help of dendrochronology it was possible to establish a continuous series of absolutely dated tree rings covering

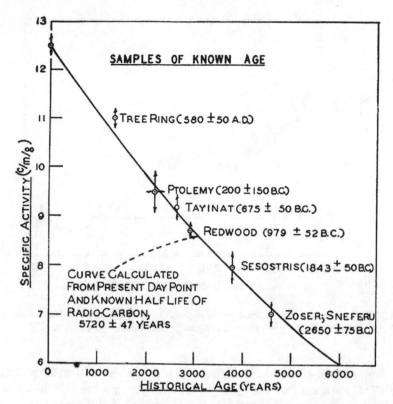

Fig. 1.2 First verification of ^{14}C dating with samples of known age. The figure is reproduced from ref. [3]. All samples used were wood dated by accepted methods [3]. Tree ring counting was used for two samples from North America (Tree Ring and Redwood). The ages of the three samples from Egypt, including specimens from a coffin (Ptolemy), a funerary boat (Sesostris), and pyramid tombs (Zoser; Sneferu), were determined through their Egypt-historical context. One sample (Tayinat) came from a large palace of the Syro-Hittite period in northwestern Syria. Samples of 8 g carbon were measured by ß counting and the specific activity (counts per minute per g of carbon) is plotted against the known age of the samples. The solid curve describes the decrease of the initial ^{14}C activity ^{14}C$_0$ with age t, following the exponential decay law: ^{14}C$_t$ — ^{14}C$_0$e$^{-\lambda t}$. Here, λ is the decay constant of ^{14}C, related to the half-life by $\lambda = \ln2/t_{1/2}$

the last 12,000 years, resulting in a detailed ^{14}C record of the atmosphere showing considerable temporal variations (Fig. 1.3).

The overall trend of atmospheric ^{14}C is due to variations of Earth's magnetic field, which influences the intensity of cosmic rays and therefore the production rate of ^{14}C. The smaller and more rapid ^{14}C fluctuations are caused by solar activity variations and its inter-connected solar magnetic field. This, again, leads to a varying shielding effect of cosmic rays, particularly of the galactic cosmic ray component [13]. In addition, atmospheric ^{14}C/^{12}C ratio can also change, when old CO$_2$ from the deep ocean, which is depleted in ^{14}C, is exhaled for some reason more strongly than usual. The resulting rather 'wiggly' calibration curve limits the precision of

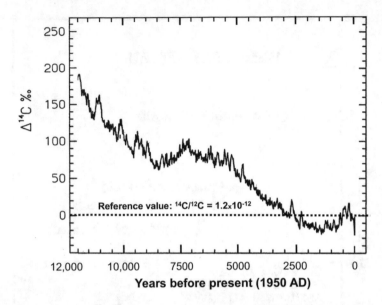

Fig. 1.3 Deviation of the $^{14}C/^{12}C$ ratio in tree rings of known age from a constant reference value of 1.2×10^{-12} for the last 12,000 years. The figure is based on a graph shown in [12]

obtaining absolute ages with ^{14}C dating, which is often lower than the precision of the $^{14}C/^{12}C$ measurement itself [14]. It is, however, important that the calibration curve is globally applicable, because atmospheric CO_2 in the troposphere—which matters for the uptake of CO_2 in plants—is a well-mixed reservoir. Beyond the tree-ring range, other archives of known age are being used for calibration (e.g. varved lake sediments [15]), allowing one to extend the calibration back to the full dating range of ^{14}C, i.e. back to about 50,000 years [16].

It is worthwhile to mention that with the existence of a calibration curve ^{14}C dating became independent of knowing the accurate half-life value of ^{14}C. Because the age of an object is essentially determined by comparing its $^{14}C/^{12}C$ ratio with a tree ring of known age having the same $^{14}C/^{12}C$ ratio. Although this assumption is well fulfilled for many terrestrial objects, small deviations may occur where seasonal ^{14}C variations matter [17].

It should also be mentioned that ^{14}C dating in its present form is not really an absolute dating tool *per se*, since it needs the calibration curve. Absolute dating with ^{14}C would require to measure both ^{14}C and its stable decay product $^{14}N^*$ [18]. Here, $^{14}N^*$ means that it is a radiogenic product. The age t of an unknown object is then determined from the relation $t = 1/\lambda \times \ln(1 + \ ^{14}N^*/^{14}C)$. The hitherto unsolved problem is the measurement of the tiny radiogenic $^{14}N^*$, in the presence of an overwhelming background of ^{14}N (e.g. ^{14}N comprises 78% of the atmosphere!). Besides solving this seemingly unsurmountable problem, one also has to know an accurate half-life value of ^{14}C. On the other hand, absolute ^{14}C dating would be able to get rid of the wiggly calibration curve, which limits the ^{14}C dating precision as mentioned above [14]. There is, however, hope that pronounced yearly spikes in the tree-ring calibration curve [19, 20] will substantially increase the precision of ^{14}C dating.

The use of ^{14}C as a versatile tool for many applications benefitted greatly from the development of accelerator mass spectrometry (AMS) in the late 1970s [21–24]. With AMS, ^{14}C atoms are counted directly rather than detecting it by the classical ß-decay method. The decay of ^{14}C is governed by the radioactive decay law: $d^{14}C/dt = -\lambda \times {}^{14}C$. Here, $d^{14}C/dt$ is the decay rate for a given number of ^{14}C atoms, and λ is the decay constant, which is inversely proportional to the half-life: $\lambda = \ln2/t_{1/2}$. With a typical $^{14}C/^{12}C$ isotopic ratio of 1.2×10^{-12} in living organic matter, 1 mg of carbon contains 6.0×10^{7} ^{14}C atoms. With $t_{1/2} = 5700$ years $= 5.0 \times 10^{7}$ h, the resulting ß-decay rate of ^{14}C is 0.83 decays per hour. In contrast, with AMS approximately 1% of the ^{14}C atoms in the 1-mg sample, i.e. 6×10^{5} ^{14}C atoms, can be detected in 1 h. This led to an enormous gain in detection sensitivity (about one million!), and allows one to reduce the amount of carbon needed for a ^{14}C measurement from grams to milligrams, and even micrograms [25], and the counting time from days to an hour. This clearly revolutionized the use of ^{14}C, both with respect to technical developments [26] and applications [6].

Recently, two methods to measure $^{14}CO_2$ at natural abundances by highly selective laser-induced excitations of $^{14}CO_2$ molecules have been developed: Intracavity optogalvanic spectroscopy (ICOGS) [27] and saturated-absorption cavity ringdown spectroscopy (SCAR) [28]. While several groups could not reproduce the claimed sensitivity of ICOGS [29–31], SCAR looks more promising [32–35]. SCAR is promoted as a "smaller, faster, and cheaper detection of radiocarbon" [36] than AMS. One has to see whether this promise holds up in the years to come. An important factor for 'routine' measurements is also the robustness of the method, which can only be demonstrated by measuring many samples. Table-top size facilities have also been developed for AMS measurements of ^{14}C [37], and recently research is being conducted for mass-spectrometric detection of ^{14}C, even without an accelerator stage [38].

1.2 The Unique Properties of ^{14}C

Since cosmogenic ^{14}C ends up as $^{14}CO_2$ in the atmosphere, it penetrates any object on Earth which is in exchange with the global CO_2 cycle. This includes both plants and animals, and humans as well. Since a human body of 70 kg contains about 16 kg of carbon—mainly in the form of organic molecules—an isotopic ratio of $^{14}C/^{12}C = 1.2 \times 10^{-12}$, leads to 3700 ^{14}C decays per second (Bq)—all life long. When the famous Iceman Ötzi was found in a glacier of the Ötztal Alps at the Austrian-Italian border, this number had decreased to about one half, because he had perished ~5200 years ago [39, 40].

Without carbon, life as we know it would not have developed on Earth. It has been noted that only small changes in the strength of the Coulomb force and the nuclear force governing the nucleosynthesis of carbon in stars could have prevented the existence of carbon and life in the Universe altogether [41]. Besides being so ubiquitous in the living world, several physical and chemical properties of ^{14}C make it a radioactive tracer of extraordinary properties. These properties are summarized in Table 1.1.

Table 1.1 The unique properties of ^{14}C (updated from [6])

Property	Comment
Favorable production process	Spallation reactions of cosmic rays in the atmosphere liberate neutrons, which in turn produce ^{14}C through the ^{14}N(n,p)^{14}C reaction. The high nitrogen content of the atmosphere (78% N_2) results in a production rate of about $\sim$2 ^{14}C atoms cm^{-2} s^{-1} [13], which is substantially higher than that of other long-lived radionuclides [42]
Uniform atmospheric distribution	^{14}C oxidizes to $^{14}CO_2$ (Fig. 1.1) and becomes part of the well-mixed tropospheric CO_2 content. Since the production of neutrons by cosmic rays in the atmosphere is altitude and latitude dependent [43], atmospheric mixing is important to provide a nearly uniform $^{14}CO_2$ concentration in the troposphere around the globe
Photosynthesis of plants	Plants take up atmospheric $^{14}CO_2$ through photosynthesis, which is the main path of ^{14}C to enter the biosphere
High solubility of CO_2	The solubility of CO_2 in water is 0.04 mol/L H_2O at NTP. This is 30 and 60 times higher than that of O_2 and N_2, respectively, and leads to a relatively high $^{14}CO_2$ content of the oceans
Radiocarbon dating	After death, organic material stops the uptake of ^{14}C and the long half-life ($t_{1/2}$ = 5700 years) is ideally suited as a radioactive clock for the last 50,000 years, covering the decisive period for the development of modern humans
Strongly hindered beta decay	The beta decay of ^{14}C to ^{14}N (a so called Gamov-Teller transition) is unusually hindered due to nuclear structure effects [4, 5]. A normal transition strength, as it is observed in neighboring nuclides, would lead to a half-life on the order of hours making it completely useless for archaeological dating. The actual half-life is (5700 $\pm$ 30) years [6]
No build-up of ^{14}C after death	Due to the low isotopic abundance of ^{13}C (1.1%), and the small cross section for the ^{13}C(n,γ)^{14}C reaction (σ = 1.4 mbarn), ^{14}C build-up by environmental neutrons in organic matter after death is negligible
Calibration of ^{14}C	The determination of an absolute age with ^{14}C requires to know the atmospheric ^{14}C content as a function of time. This has been established by ^{14}C measurements in objects of known age such as tree rings, corals, stalagmites [16]. The calibration beyond the range of tree rings ($\sim$12,000 years) has been improved with an extensive analysis of ^{14}C in varved lake sediments [15]

Table 1.1 (continued)

Property	Comment
Mass fractionation correction	Because carbon has two stable isotopes, mass fractionation both in natural and in instrumental processes can be determined through ^{13}C/^{12}C ratio measurements, and the effect on ^{14}C can be corrected correspondingly [44, 45]
The ideal AMS nuclide	Negative ion currents of around 50 μA ^{12}C$^-$ can easily be produced from solid graphite in a cesium-beam sputter source. With a modern ^{14}C/^{12}C ratio of 1.2×10^{-12}, about 400 ^{14}C$^-$ ions per second are then emitted from a carbon sample of 1 mg, and half of it ends up in the final particle detector. Even for a 50,000-year old sample, one will still detect about 12 ^{14}C atoms per minute. Most important is the suppression of the stable isobar ^{14}N in the ion source due to the instability of N$^-$ [22]. In addition, cross contamination between adjacent graphite samples in the ion source is less than 10^{-5}, allowing one to measure the true ^{14}C content of very old samples without the risk of adding ^{14}C from samples with much higher ^{14}C content (e.g. calibration samples)

1.3   Applications of ^{14}C

In the current review, ^{14}C application will be described by using the several domains of our environment at large as a guiding principle: Atmosphere, Hydrosphere, Biosphere, Cryosphere, Lithosphere. There exist, of course, interactions and/or exchanges between these domains (e.g. the CO_2 cycle) which can be well traced by ^{14}C. Besides applications with natural ^{14}C, various uses of anthropogenic ^{14}C will also be described. This includes (1) biomedical studies of new drugs labeled with enriched ^{14}C, (2) the amazing multitude of applications using the so called ^{14}C bomb peak produced by atmospheric nuclear weapons testing between 1952 and 1963, and (3) the effect of releasing 'dead' CO_2 (zero ^{14}C) from fossil fuel burning into the atmosphere, which lowers the ^{14}C/^{12}C ratio of atmospheric CO_2. The last two effects suggest that anthropogenic ^{14}C may be used as one of the markers for a new geological time period, the so called Anthropocene [46], originally promoted by Nobel Laureate Paul Crutzen [47].

1.3.1   Atmosphere

Cosmic Rays were discovered by Victor Hess in 1912, when he found a significant increase of the ionization of air in balloon flights up to an altitude of 5350 m [48]. It took more than 30 years before Willard Libby figured out that ^{14}C must be produced by the interaction of cosmic rays with the air [1] (see Fig. 1.1). Not long after this first idea it became possible to measure natural ^{14}C in organic material on Earth by

Fig. 1.4 Picture of the first hydrogen bomb test ("Mike") of the USA on Eniwetok Atoll of the Marshall Islands in the Pacific Ocean on 1 November 1952. The explosive power was estimated to 10.4 Megatons of TNT, about 700 times the one of the Hiroshima bomb. The simplified schematic indicates that neutrons from the bomb tests convert ^{14}N into ^{14}C, just like the neutrons emerging from the spallation of atmospheric nuclei with high-energy protons from cosmic rays. The Figure is reproduced from [55]

ß-counting [2], and the ^{14}C dating method was developed [3]. In the second half of the 1950s, the first anthropogenic effects altering the modern ^{14}C content were noticed [49, 50]. On the one hand, a few percent reduction of the $^{14}C/^{12}C$ ratio was measured in modern wood due to the atmospheric addition of 'dead' CO_2 from fossil fuel burning depleted in ^{14}C [49]. On the other hand, an increase of a few percent was measured in atmospheric $^{14}CO_2$ in 1957 due to the effect of nuclear weapons testing in the atmosphere [50], and this increase was also found 2 years later in human beings as well [51]. In the ensuing years nuclear weapons testing continued with ever more powerful nuclear explosions (Fig. 1.4), eventually leading to an increase of atmospheric $^{14}CO_2$ in 1963 by almost 100%.

In 1963 the limited Nuclear Test Ban Treaty (NTBT) was signed by the USA, the Soviet Union and Great Britain, essentially putting a halt to above-ground testing. Since that time the $^{14}CO_2$ content decreased due to the exchange with the biosphere and the ocean and by now is only a few percent above the reference value of $^{14}C/^{12}C = 1.2 \times 10^{-12}$. The resulting rise and fall of the ^{14}C content in the atmosphere is usually referred to as '^{14}C bomb peak' [52, 53], and constitutes a rapidly changing ^{14}C-signal for the second half of the twentieth century. Since the emission of anthropogenic CO_2 is likely to continue, ^{14}C will fall well below the reference value, possibly confusing future archaeologists when they apply ^{14}C dating to objects affected by the reduced ^{14}C values [54].

1.3.1.1   Measurements of $^{14}CO_2$ Concentration

Measurements of ^{14}C in atmospheric CO_2 started well before the advent of the AMS technique [56], and to this day can be performed by beta counting if large air samples ($\sim$10,000 L) are available. These measurements already revealed a clear picture of the ^{14}C excess produced by the atmospheric nuclear weapons testing period between 1950 and 1963 [56]. However, often there is much less material available, and AMS measurements of high accuracy have been developed for air samples of a few litres [57]. High-precision monthly measurements of $^{14}CO_2$ at high northern (Point Barrow, the most northern point of Alaska) and southern latitude (South Pole) revealed lower values for the former, which are probably caused by changes in the air-sea ^{14}C flux in the Southern Ocean. Seasonal cycles indicate a strong influence of stratosphere-troposphere air exchanges [58].

1.3.1.2   Measurements of ^{14}CO Concentration

As depicted in Fig. 1.1, ^{14}CO is mainly oxidized to $^{14}CO_2$ by the reaction with the rare but very reactive OH radical in the atmosphere: $^{14}CO + OH \rightarrow {}^{14}CO_2 + H$. Since the production of ^{14}CO is quasi-constant, the measurement of the ^{14}CO concentrations in air has been used as a proxy to monitor seasonal OH concentration [59–61].

In order to measure the ^{14}C production directly in air at higher altitudes, compressed-air cylinders were carried along cruising altitudes of commercial airplanes for 2 years, and the resulting $^{14}CO_2$ and ^{14}CO contents were measured with AMS [62]. The comparison with model calculation showed significant deviation from both the production rate and the $^{14}CO_2/^{14}CO$ ratio. More of these direct measurements would be useful particularly for ^{14}CO, since it is used as a proxy to monitor the hydroxyl radical (OH) concentration in air. OH has been called the 'detergent' of the atmosphere [7], because of its oxidizing power and its important role to remove trace gases [63].

1.3.1.3   Measurements of $^{14}CH_4$ Concentration

Another trace gas of considerable importance in the ongoing debate about global warming is methane (CH_4). Although its concentration in air is about a factor of 200 lower than the one of CO_2 its global warming potential is 20 times higher [64]. At present, the radiative forcing of CH_4 is about a factor of four smaller than the one of CO_2 [65]. $^{14}CH_4$ has been studied with AMS in the atmosphere, e.g. [66–68], and recently also in CH_4 extracted form Greenland ice around the last glacial termination [69].

1.3.2 Hydrosphere

It is well known that our Earth is the only planet in the Solar System which harbors liquid water through most of its 4.5-billion year history. The recent discovery of a temperate terrestrial planet around the red dwarf star Proxima Centauri, the closest star to the Sun (4.22 light-years away) offers hope that liquid water may be present on its surface [70]. Shortly thereafter, seven temperate terrestrial planets were discovered around the ultracold dwarf star TRAPPIST-1(39.5 light-years away) in the constellation *Aquarius* [71]. Even though theses exoplanet systems are out of reach for human travel, it would be very exciting to find out more about their composition, particularly about their atmosphere. This may well be possible with the newest generation of ground-based and/or space-based telescopes.

At the present time on Earth—and probably throughout the Holocene (the last 10,000 years)—the distribution of water on Earth is estimated to be: ocean 97.3%, ice sheets 2.1%, groundwater 0.6% [64]. During the last Ice Age (from 100,000 to 12,000 years ago), the land-locked ice increased to about 5% of the total water. Correspondingly the ocean water was reduced by ∼3%. Since the average depth of the world oceans is now 3800 m, the sea level during the Ice Ages was 100–120 m lower, leading to landbridges between continents (e.g. Bering Strait) and large areas of exposed land. Nowadays, more than two thirds of the surface of the Earth is covered by the oceans, and the transport of heat around the globe by the oceans is roughly of the same magnitude as the one transported by the atmosphere. This clearly has a strong impact on the climate. Therefore the study of ocean currents both on the surface and in the deep ocean in ever more detail is of considerable interest. This is achieved by methods of chemical and physical oceanography, which are briefly described in the following.

1.3.2.1 Chemical Oceanography

In chemical oceanography, ^{14}C and other radionuclides are used as "Tracers in the Sea" [72], allowing a multitude of investigations to study the oceans [73]. One of the most attractive features which came out of these studies, when large water samples (∼250 L each) had to be used for ^{14}C decay counting, is the great ocean conveyor depicted in Fig. 1.5 [74, 75]. This idealized picture was later refined [76], but we are still far away from getting a comprehensive picture of the dynamics of the oceans. A big step forward in analytical capability was AMS, where measurements of ^{14}C in ocean water allowed one to reduce the sample size from 250 to 0.5 L, and well

Fig. 1.5 The great ocean conveyor logo [74]. Illustration by Joe Le Monnier, Natural History Magazine. It is a simplified picture of the main ocean currents transporting heat around the globe. In the northern Atlantic Ocean the warm surface water cools off, gets saltier and sinks into the abyss forming the cold and salty deep current (blue) which moves south and east, eventually surfacing again in the northern Indian and Pacific Oceans

over 13,000 water samples were measured within the World Ocean Circulation experiment (WOCE) project in the 1990s at the National Ocean Science AMS facility (NOSAMS) in Woods Hole [77]. By now the number of ^{14}C measurements in ocean water samples has surely surpassed 20,000. The analysis of this enormous bulk of data is being undertaken by dedicated oceanographers leading to a better understanding of ocean currents in three dimensions [78].

Another cosmogenic radionuclide of great interest for oceanography is ^{39}Ar, which was measured by low level counting as long as large ocean water samples were taken [79]. The half-life of ^{39}Ar (269 years) is well matched to global water movements around the globe (500–2000 years). However, with the advent of AMS only small water samples were taken for ^{14}C measurements (see above), too small for low-level counting of ^{39}Ar. Successful AMS measurements at the ATLAS linac at Argonne National Lab have been undertaken to measure the very low isotope ratios of ^{39}Ar/^{40}Ar ($<8 \times 10^{-16}$) in argon extracted from ocean water samples [80]. But eventually it turned out that these measurements are too sensitive to the stable ^{39}K isobar interference which cannot be consistently controlled to the level required for these ultra-low isotope ratio measurements, despite big efforts [81]. Releases from nuclear reprocessing into the sea provide long-lived anthropogenic radionuclides well suited as quasi-stable tracers for studying ocean circulation. These include ^{129}I ($t_{1/2} = 1.57 \times 10^7$ years) studied in the Norwegian Sea [82], ^{99}Tc (2.11×10^5 years) [83], and ^{237}Np (2.14×10^6 years) [84]. Recently, fall-out of ^{236}U (2.34×10^7 years) from the nuclear weapons testing period is also showing promise to be used as an oceanic tracer [85].

1.3.2.2 Physical Oceanography

In physical oceanography, one attempts to gain knowledge about the dynamics of the oceans by different means as compared to the tracer methods in chemical oceanography. Precise altitude measurements of the ocean surface are performed from satellites called JASON (Joint Altimetry Satellite Oceanographic Network), and temperature, salinity, pressure, velocity and other characteristic parameters of ocean water are measured in the field by more than 3000 floats within the ARGO project [86, 87]. The names of the project where chosen according to Greek mythology, where the Greek hero Jason set out with the Argonauts and the ship Argo to find the Golden Fleece. A recent analysis of the ocean temperature from these measurements can be found in [88].

It is clear that the enormous collection of data in both chemical and physical oceanography will improve our knowledge about the oceans, which in turn should lead to better modeling of the climate.

1.3.2.3 Groundwater Dating

Groundwater is a natural resource of great importance for providing freshwater to many places on Earth. Similar to oil, groundwater is a limited resource, as long as more water is being used than recharged through precipitation. Age determinations of groundwater or objects directly related to groundwater formation allows one to gain information on recharging times. An example for identifying short recharge times was the measurement of the ^{14}C bomb peak in young stalagmites growing in areas of freshwater supply [89]. In the case of the Wombeyan cave near Sydney, the time shift of the ^{14}C bomb peak indicated a recharging time of only 6 years for this important freshwater source. If one employs natural ^{14}C and ^{36}Cl ($t_{1/2} = 3.01 \times 10^5$ years) to ground water dating, more complex hydrologies can be deciphered such as the one in the Palm Valley of central Australia [90].

Of particular interest to groundwater hydrology are the large aquifers in the world, with very old groundwater in the range from several hundred thousand to a million years. In this case ^{14}C cannot be used, but longer-lived cosmogenic radionuclides are available for this purpose. An example is the Great Artisean Basin (GAB) in Australia, which has been first investigated with ^{129}I ($t_{1/2} = 1.57 \times 10^7$ years) [91] and ^{36}Cl [92], indicating groundwater ages of more than 100,000 years. Since dating with both radionuclides is problematic because of unknown underground contributions, groundwater dating in the GAB was also performed with cosmogenic ^{81}Kr ($t_{1/2} = 2.29 \times 10^5$ years), an almost ideal tracer for this purpose. AMS measurements were performed at the National Superconducting Cyclotron Laboratory at Michigan State University, and groundwater ages in the range from 200,000 to 400,000 were established [93]. It is interesting to note, that [81]Kr dating of old groundwater in the range from 0.5 to 1 million year has also been performed with the laser-based ATTA technique (Atom Trap Trace Analysis) in the Nubian aquifer of the Western Egypt Desert [94]. Groundwater dating with the laser method has also been demonstrated for [39]Ar [95]. This would be particularly important for dating younger ground waters, and could be combined with ^{14}C measurements. Noble gas radionuclides cannot be measured with AMS based on tandem accelerators, because they do not form negative ions. AMS at large positive-ion machines is very complex [93], and sometimes marginally possible [80, 81]. They do not allow one to develop a truly applied program of groundwater dating. Therefore, ATTA is the method of choice for measuring noble gas radionuclides in groundwater [96], and very recently also in Antarctic ice [97, 98]. In general, it is relatively straightforward to understand the distribution of cosmogenic noble gas radionuclides into the hydrosphere, because the atmosphere is the main reservoir, and the solubility of noble gases in freshwater and ocean water is well understood.

1.3.3 Biosphere

The main use of cosmogenic ^{14}C in the biosphere is the dating of archaeological objects. In the more than 60 years since its beginning (see Fig. 1.2), countless archaeological projects around the world have benefitted from ^{14}C dating, which sometimes was called the radiocarbon revolution in archaeology. Although limited in precision by the wiggly calibration curve as depicted in Fig. 1.3 [14], ^{14}C dating nevertheless provides an extremely useful tool to establish absolute chronologies in archaeology [99]. Refinements in the calibration process using Bayesian statistics [100] has resulted in a substantial reduction of the uncertainty of calibrated ages, when additional information (so-called prior probability) is available in cases where a sequence of dates with clear chronological ordering exists. In this way good agreement between the historical and ^{14}C-based chronology of dynastic Egypt was established [101]. But not always such an agreement is reached, and in the following we discuss two famous cases were a difference of about 100 years persists since many years, and is still awaiting a solution.

1.3.3.1 Dating the Volcanic Eruption of Santorini

Santorini (ancient Thera) is a Greek island in the Mediterranian Sea, and has experienced several large volcanic eruptions during its history. The most recent one happened about 3500 years ago during the Late Bronze Age. A volcanic eruption is a geophysical phenomenon, but it likely had a large impact on the life of people living on the island and in neighboring areas. In fact, the relatively close-lying island of Crete (~120 km south of Santorini), which had harbored the Minoan

Civilisation for some 2000 years, must have been affected by the eruption too. Some scholars thought that the eruption ended the Minoan Civilisation, and to this day the Santorini eruption is sometimes called the Minoan eruption. Although this was probably not the case to its full extent at the time of the eruption, huge tsunamis may have hit the harbors on Crete [102] causing considerable damage. In any case the eruption constitutes a clear beacon in time, and an accurate date is highly desirable to anchor relative chronologies in archaeology to a reliable absolute date. However, there is a longstanding discrepancy of about 100 years between the date established by archaeological reasoning (1530–1500 BC) and the one determined from ^{14}C dating (1630–1600 BC) [103]. The volcanic eruption may have also produced global climatic effects traceable in tree-ring records, and the deposit of SO_2 emissions and airborn tephra in ice cores are other possible fingerprints of a volcanic eruption. A summary of the situation is shown in Fig. 1.6 [104]. This and the more recent discussion [103] seem to link the 100-year discrepancy to a similar problem which showed up in the chronology of Tell el Dab'a [105].

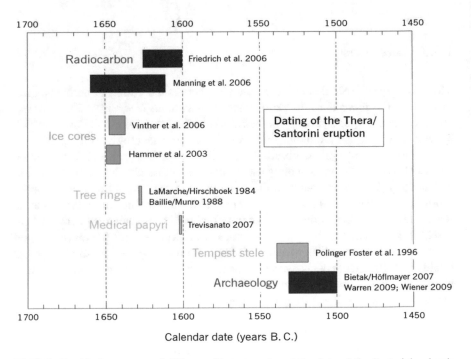

Fig. 1.6 Graphical summary of different efforts to arrive at the date of the Santorini volcanic eruption [104]. The width of the bars indicate the uncertainty of the respective method. While the results from radiocarbon dating, ice core analysis, and tree ring records cluster between 1650 and 1600 BC, most of the archaeology-based results give a date some 100 years later

1.3.3.2 The Chronology of Tell el-Dab'a in the Nile Delta

In simple terms, the political structure of ancient Egypt was a sequence of Kingdoms when powerful kings (pharaos) ruled the entire country, interconnected with so-called Intermediate Periods when several smaller kingdoms—usually in competition which each other—prevailed in parallel for a while. A schematic presentation of The Chronology of Ancient Egypt is shown in Fig. 1.7, indicating also the eruption of Santorini with a current uncertainty of ~100 years, and the end of the Minoan Civilisation which lasted from 3600 BC to 1450 BC. The ^{14}C dating of the dynastic period mentioned above included the Old Kingdom, the Middle Kingdom, and the New Kingdom [101]. No ^{14}C data were assigned for the Intermediate Periods. In contrast, a big effort was undertaken within the SCIEM 2000 project (Synchronisation of Civilisation in the East Mediterranian in the Second Millennium B.C.) [106] to establish the chronology of the Second Intermediate Period at the

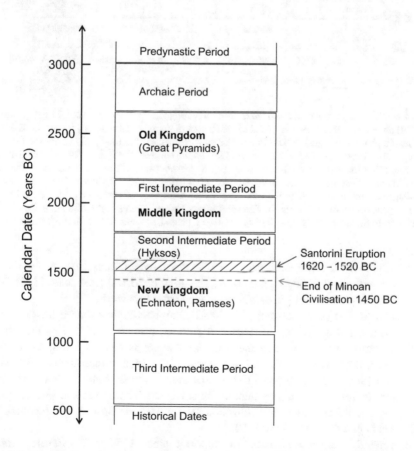

Fig. 1.7 Simplified schematic of the chronology of Ancient Egypt. The crucial events at the transition from the Second Intermediate Period to the New Kingdom around 1500 BC are marked

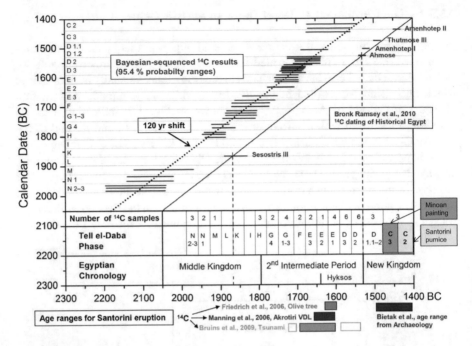

Fig. 1.8 The chronology of Tell el-Dab'a from ^{14}C dating and archaeology [105]. The sequence of the Tell el-Dab'a phases is synchronized to the Historical Egyptian Chronology by two finds linked to specific pharaos (vertical dashed lines). For an agreement of the time scale established by ^{14}C dating and archaeology, the ^{14}C dates should fall on the solid diagonal line. This is obviously not the case, rather there is an overall shift of about 120 years between the two time scales. In contrast, the red bars indicating ^{14}C dates of objects directly related to specific pharaos [101] fall well on the diagonal line, thus agreeing with the Historical Chronology of ancient Egypt. At the bottom the 100-year time shift discussed for Santorini is also shown. Interestingly, Minoan paintings and pumice from Santorini show up in late Tell el-Dab'a phases, incompatible with the ^{14}C results

excavation site of Tell el-Dab'a in the Nile delta by a comparison of ^{14}C dating with archaeological dating linked to the Egyptian historical time scale [105]. The situation is depicted in Fig. 1.8, which is taken from this work.

A shift of 120 years between ^{14}C data and archaeological dating linked to the Egyptian chronology is clearly visible. A detailed discussion of this and a similar shift at Santorini mentioned in the previous chapter is beyond the scope of this review. However, recent work on king Kayan of the Second Intermediate Period at other excavation sites in Egypt [107] opens the possibility that the archaeological assigment at Tell el-Dab'a may need to be corrected. There seems to be also other evidence, that the so-called 'high chronology' for the Late Bronze Age favoured by ^{14}C dates is gaining ground [103, 108].

In general, radiocarbon results are reported since 1959 in the Journal *Radiocarbon* which should be consulted if one is interested in the full breadth of ^{14}C applications.

1.3.3.3   Applications of the ^{14}C Bomb Peak

The production of ^{14}C from neutrons released in nuclear weapons testing is depicted in Fig. 1.4. It resulted in a rapidly changing ^{14}C excess in atmospheric CO_2 in the second half of the twentieth century. Since the above-ground testing was stopped after the limited nuclear test ban treaty (NTBT) in 1963, the rapid rise of $^{14}CO_2$ up to this year was followed by a slower decrease due to the exchange of $^{14}CO_2$ with the biosphere and hydrosphere (Fig. 1.9).

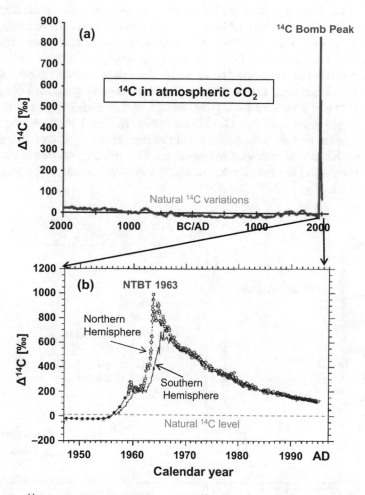

Fig. 1.9 The ^{14}C bomb peak. (**a**) Display of the ^{14}C variations from the reference value of $^{14}C/^{12}C = 1.2 \times 10^{-12}$ in atmospheric CO_2 as measured in tree rings for the last 4000 years [109]. (**b**) Details of the ^{14}C variations for the second half of the twentieth century as measured in atmospheric CO_2 [52]

This created the so-called "^{14}C bomb peak", which has also been called "The mushroom cloud's silver lining" [110]. In the following a few selected examples of using the ^{14}C bomb peak will be described.

Due to the rapid exchange of food on Earth, newly formed living matter will have approximately the same ^{14}C/^{12}C ratio in organic carbon as compared to the one in atmospheric CO_2 of that year. This allows applications in forensic medicine to determine the time of death of a person [111]. It is also possible to determine the year when a person was born by analyzing tooth enamel [112] or eye lens crystallines [113].

The most exciting application of the bomb peak was developed at the Karolinska Institute in Stockholm, where DNA was extracted from brain cells of humans after death, and the time when new brain cells were formed after birth were determined from the ^{14}C content of the DNA [109]. In this work, the authors made a bold conjecture which turned out to be confirmed by the measurements: "Most molecules in a cell are in constant flux, with the unique exception of genomic DNA, which is not exchanged after a cell has gone through its last division. The level of ^{14}C integrated into genomic DNA should thus reflect the level in the atmosphere at any given point, and we hypothesized that determination of ^{14}C levels in genomic DNA could be used to retrospectively establish the birth date of cells in the human body." This method has been applied to different cells in humans, and is graphically summarized in Fig. 1.10.

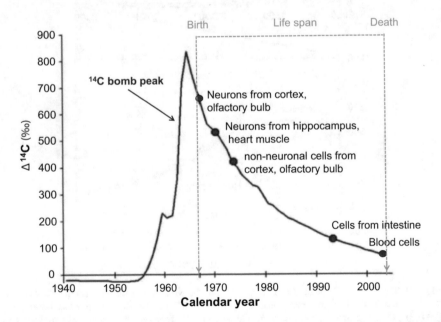

Fig. 1.10 Principle of how to determine the mean birth date of cells from different parts of a human body of known life span by measuring the ^{14}C in the respective DNA extracts [108]

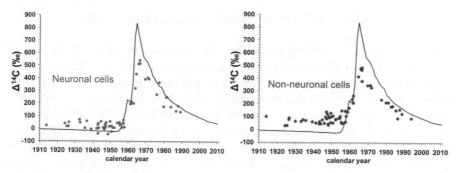

Fig. 1.11 Evidence for neurogenesis in the hippocampus from ^{14}C measurements in DNA of 57 individuals [115]. The ^{14}C results are plotted at the birth date of the respective individuals. Points for individuals born before the rise of the curve indicate a clear evidence of the formation of new cells after birth. The effect is more pronounced for non-neuronal cells

^{14}C Bomb peak dating of human DNA has now been applied to cells of the neocortex in humans [114], providing evidence for (1) neurogenesis in the hippocampus of humans [115], (2) the age of olfactory bulb neurons in humans [116], (3) the age and genomic integrity of neurons after cortical stroke in humans [117], (4) the dynamics of fat cell turnover in humans [118, 119], and (5) the dynamics of cell generation and turnover in the human heart [120]. Particularly important is the observation that neurons are formed after birth in the hippocampus [115], since it is part of the human brain that is involved in memory forming, organizing and storing. Clearly, neurogenesis after birth is good news in this case (Fig. 1.11).

In a different application, the ^{14}C bomb peak has been used for age estimates of white sharks. It has revealed ages up to 73 years [121], longer than previously assumed to be possible for this species. Recently, a ^{14}C study of eye lens crystallines in Greenland sharks resulted in extreme longevity of this species, up to 400 years [122]. This would indeed be the longest-lived vertebrate known.

Since the ^{14}C bomb peak is such a distinct signature in time, it can also be used to discover forgeries and illegal trade. For example, ^{14}C measurement of the canvas of a painting, supposedly produced by the French painter Fernard Legér (1855–1955) at the beginning of the twentieth century, revealed a date between 1959 and 1962, several years after the death of the painter [123]. The illegal trade of ivory has been uncovered by "Carbon-14 measurements on 231 elephant ivory specimens from 14 large ivory seizures ($\geq$0.5 ton) made between 2002 and 2014 showing that most ivory (ca. 90%) was derived from animals that had died less than 3 years before ivory was confiscated" [124]. The poaching of elephants in Africa had also been investigated by genetic assignement of ivory [125].

1.3.3.4   Biomedical Developments with ^{14}C-labelled Compounds

Atom counting of ^{14}C with AMS opened the possibility to reduce the amount of ^{14}C-labelled compounds to such an extent that experiments with negligible radiation doses could be performed on living subjects including humans. Pioneering work in this respect was conducted at the Biomedical Science Division and the AMS facility of Lawrence Livermore National Laboratory [126]. AMS measurements of ^{14}C-labelled carcinogenic compounds in rats and human cancer patients were performed at the VERA Lab in Vienna [127]. The field developed steadily at existing AMS facilities, driven by both biologist [128] and by physicists [129]. When small AMS facilities were developed for this kind of work [130], the pharmaceutical industry got interested in using the method of 'microdosing' in humans [131]. This means that in the pre-clinical and clinical stage of developing new drugs, only 1/100th or less than the pharmacological dose is being administered to humans to study e.g. the metabolic products in detail [132]. This shortens the total development time of a new drug by a year or so, resulting in considerable savings. Several companies specializing on biomedical ^{14}C measurements using small AMS facilities are now on the market, e.g. Exceleron (Germantown, MD), Vitalea Science (Davis, CA), Accium BioSciences (Seattle WA). They are performing their own research but also providing service to pharmaceutical companies.

1.3.4   Study of the Natural Changes of Alpine Glaciers

Climate change and global warming is one of the most discussed issues in our time. It concerns science, economy, politics and the public. One of the difficulties in making prediction about climate change is the fact that it is affected by both anthropogenic and natural influences. One of the striking observations is the ongoing retreat of Alpine glaciers most likely accelerated by anthropogenic greenhouse gases (e.g. CO_2, CH_4, N_2O). On the other hand, large variations of Alpine glaciers happened also at much earlier times, in fact throughout the Holocene (last 10,000 years), when human impact must have been negligible.

The interest in the geophysical phenomenon of glacial movements was greatly enhanced in 1991 with the accidental discovery of the famous Tyrolean Iceman "Ötzi" at a high-altitude mountain pass (3210 m) in the Ötztal Alps near the Austrian-Italian border. The well-preserved body emerged from a shallow ice patch, and ^{14}C measurements quickly established that the Iceman had lived some 5200 years ago [133, 134]. Later, a large number of ^{14}C measurements were performed on equipment and other objects found at the discovery site, essentially confirming the great age of the Iceman [40]. In addition, a detailed study of radiogenic (lead and strontium) and stable (oxygen and carbon) isotopes allowed one to pin down the origin of the Iceman to a particular region in South Tyrol, Italy [135]. It was also revealed from ^{14}C measurements at other locations in the Ötztal Alps that humans were present at high altitudes already some 9000 years ago [40].

Considerable glacial movements have been established by ^{14}C dating of tree logs and other organic remains released from retreating glaciers or buried in morains, sometimes combined with dendrochronology [136–146, 40]. A surprising result was found at the Pasterze Glacier in Austria illustrated in Fig. 1.12 [147]. Several pine tree logs, which surfaced in the forefield of the retreating glacier, were dated by ^{14}C and dendrochronology [136], with the surprising result of very old age (9000–10,000 years). This means that the Pasterze Glacier was essentially absent at that time allowing trees to grow in a region which today is still covered with ice.

These studies were complemented by surface exposure dating of bolders and moraines with ^{10}Be, which is produced by secondary cosmic ray particles (neutrons, muons) in quartz [148–151]. The method of surface exposure dating with ^{10}Be has also been applied on glaciers in the New Zealand Alps, where similar fluctuations were found [152, 153]. Interestingly, some of these fluctuations are out of phase with the one from the European Alps, indicating hemisphere-selective climate changes. In addition to the specific use of studying glacial movements, terrestrial *in situ* cosmogenic nuclides are widely used in lithological studies [154], e.g, to determine incision rates of rivers [155] and erosion rates of landscapes [156].

A graphic summary of the observed movements of glaciers and tree lines in the European Alps during the last 10,000 years (Holocene) is shown in Fig. 1.13 [147]. This allows one to reconstruct temperature fluctuations which are shown in the top curve of the figure. It is quite obvious from this curve that the first half of the Holocene was on average warmer than the second half, becoming progressively colder after the time of the Iceman Ötzi (red line). This may have contributed to the excellent preservation of the Iceman's body, being always covered by snow and ice at the high altitude of his resting place (3210 m) [147].

The fact that considerable natural variations of glaciers and temperatures existed throughout the Holocene complicates predictions about climate change in our time. Even though we are quite sure that human actions do influence the climate, it is difficult to predict the combined influence of anthropogenic and natural causes, as long as we do not fully understand the causes for the natural fluctuations. One of the possibilities for the latter may be small activity changes of the sun [157, 158], which could trigger climatic processes on Earth ultimately leading to the observed phenomenon of glacier fluctuations.

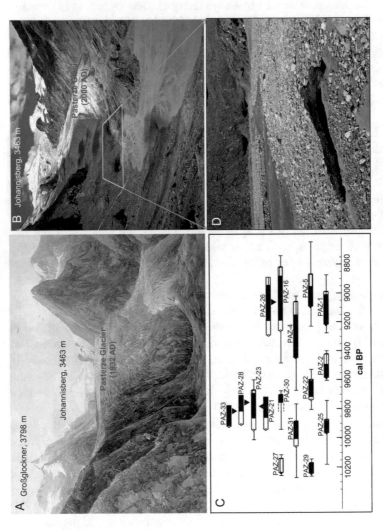

Fig. 1.12 Changes of the Pasterze Glacier in the Austrian Alps [147]. (**a**) The glacier as seen by the painter Thomas Eder in 1832, close to the end of the Little Ice Age. (**b**) Photograph of the strongly reduced glacier in 2000. Today the tongue of the glacier retreats approximately 15 m per year. (**d**) Enlarged area from B showing a log of a stone pine (*Pinus cembra*) found in the forefield of the glacier. (**c**) Age distribution of all wood specimens collected from the glacier forefield [136] by combining dendro dating (filled sections) with ^{14}C measurements (error bars indicate 65% calibrated age ranges). The open sections depict missing inner tree-ring sections, and the hatched sections indicate compressed wood

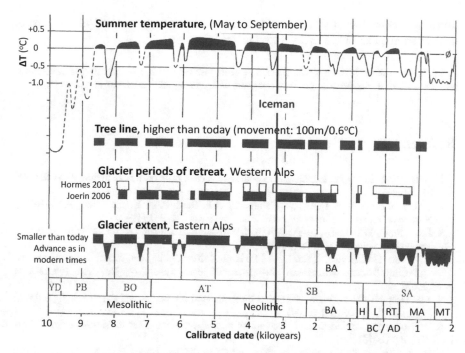

Fig. 1.13 Schematic presentation of glacier and tree-line movements during the Holocene, reproduced from [147]. The periods of smaller glaciers and higher tree lines are indicated with the box symbols. Glacial advances are indicated with filled triangles and curves. The largest advances took place during the Little Ice Age (~AD 1300–1850). The top curve depicts the relative summer temperature variations deduced mainly from the tree-line movement. The mean temperature between AD 1900 and 2000 is used as the zerodegree reference. The red vertical line marks the time of the Iceman [147]. At the bottom of the figure, the paleoclimatic periods (*YD* Younger Dryas, *PB* Preboreal, *BO* Boreal, *AT* Atlantic, *SB* Subboreal, *SA* Subatlantic) and the archaeological periods (*BA* Bronze Age, *H* Hallstatt period, *L* La Tène period, *L+H* Iron Age, *RT* Roman times, *MA* Middle Ages, *MT* modern times) are indicated

1.4   Conclusion

This review could only give a glimpse of the enormous breadth of applications which are possible using ^{14}C as an isotope dating and tracing tool. By selecting a few examples which the author is familiar with, it is nevertheless hoped that the uniqueness of this radioisotope is somehow conveyed. ^{14}C is truly a 'gift of nature', which allows us to explore many domains of our environment at large in a way not possible with any other radioisotope. Worldwide, more than 100,000 ^{14}C measurements are performed each year, allowing us progressively deeper views into the wonders of the world around us. The author remembers well an inspiring lecture by the great astrophysicist and Noble laureate Subrahmanyan Chandrasekhar at Argonne National Laboratory in Chicago, where Chandrasekhar posed the simple question, "How does it all happen?". There is indeed hope that ^{14}C will help us to answer such questions in many different fields.

Acknowledgement This work would not have been possible without the countless collaborations and discussions the author enjoyed over the years with colleagues and friends around the world.

References

1. W.F. Libby, Atmospheric helium three and radiocarbon from cosmic radiation. Phys. Rev. **69**, 671 (1946)
2. E.C. Anderson, W.F. Libby, S. Weinhouse, A.F. Reid, A.D. Kirshenbaum, A.V. Grosse, Natural radiocarbon from cosmic radiation. Phys. Rev. **72**, 931 (1947)
3. J.R. Arnold, W.F. Libby, Age determination by radiocarbon content: checks with samples of known age. Science **110**, 678 (1949)
4. J.W. Holt, G.E. Brown, T.T.S. Kuo, J.D. Holt, R. Machleidt, Shell model description of the ^{14}C dating ß decay with Brown-Rho-scaled NN interactions. Phys. Rev. Lett. **100**, 062501 (2008)
5. J.W. Holt, N. Kaiser, W. Weise, Chiral three-nucleon interaction and the ^{14}C-dating ß decay. Phys. Rev. C **79**, 054331 (2009)
6. W. Kutschera, Applications of accelerator mass spectrometry. Int. J. Mass Spectrom. **349–350**, 203 (2013)
7. P. Crutzen, in *My Life with O₃, NOx and Other YSOxs*, ed. by B.G. Malmström. Nobel Lectures in Chemistry 1991–1995, (World Scientific Publishing, Singapore, 1997), pp. 189–242
8. B.J. Bjork et al., Direct frequency comb measurement of OD + CO → DOCO kinetics. Science **354**, 444 (2016)
9. H. de Vries, Variaton in concentration of radiocarbon with time and location on earth. Proc. Koninkl. Nederl. Akad. Wetenschappen B **61**, 94 (1958)
10. W.F. Libby, Accuracy of radiocarbon dates. Science **140**, 278 (1963)
11. B. Kromer, Radiocarbon and dendrochronology. Dendrochronologia **27**, 15 (2009)
12. M. Stuiver et al., INTCAL98 radiocarbon age calibration, 24,000–0 cal BP. Radiocarbon **40**(3), 1041 (1998)
13. J. Masarik, J. Beer, Simulation of particle fluxes and cosmogenic nuclide production in the Earth's atmosphere. J. Geophys. Res. **104**(D10), 12,099 (1999)
14. T.P. Guilderson, P.J. Reimer, T.A. Brown, The boon and bane of radiocarbon dating. Science **307**, 362 (2005)
15. C. Bronk Ramsey et al., A complete terrestrial radiocarbon record for 11.2 to 52.8 kyr B.P. Science **338**, 370 (2012)
16. P.J. Reimer et al., INTCAL13 and MARINE13 radiocarbon age calibration curves 0-50,000 years cal BP. Radiocarbon **55**, 1869 (2013)
17. M.W. Dee et al., Investigating the likelihood of a reservoir offset in the radiocarbon record for ancient Egypt. J. Archaeol. Sci. **37**, 687 (2010)
18. J. Szabo, I. Carmi, D. Segal, E. Mintz, An attempt at absolute ^{14}C dating. Radiocarbon **40**(1), 77 (1998)
19. F. Miyake, K. Nagaya, K. Masuda, T. Nakamura, A signature of cosmic-ray increase in AD 774-774. Nature **486**, 240 (2012)
20. M.W. Dee, B.J.S. Pope, Anchoring historical sequences using a new source of astro-chronological tie-points. Proc. R. Soc. A **472**, 20160263 (2016)
21. R.A. Muller, Radioisotope dating with a cyclotron. Science **196**, 489 (1977)

22. K.H. Purser, R.B. Liebert, A.E. Litherland, R.P. Beukens, H.E. Gove, C.L. Bennet, H.R. Clover, W.E. Sondheim, An attempt to detect stable N- ions from a sputter ion source and some implications of the results on the design of tandems for ultrasensitive carbon analysis. Rev. Phys. Appl. **12**, 1487 (1977)
23. C.L. Bennet, R.P. Beukens, M.R. Clover, H.E. Gove, R.B. Liebert, A.E. Litherland, K.H. Purser, W.E. Sondheim, Radiocarbon dating using electrostatic accelerators: negative ions provide the key. Science **198**, 508 (1977)
24. D.E. Nelson, R.G. Korteling, W.R. Stott, Carbon-14: direct detection at natural concentrations. Science **198**, 507 (1977)
25. P. Steier et al., Preparation methods of μg carbon samples for ^{14}C measurements. Radiocarbon **59**(3), 803 (2017)
26. H.-A. Synal, Developments in accelerator mass spectrometry. Int. J. Mass Spectrom. **349–350**, 192 (2013)
27. D.E. Murnick, O. Dogru, E. Ilkmen, Intracavity optogalvanic spectroscopy. An analytical technique for ^{14}C analysis with subattomole sensitivity. Anal. Chem. **80**, 4820 (2008)
28. I. Galli, S. Bartalini, S. Borri, P. Cancio, D. Mazzotti, P. De Natale, G. Giusfredi, Molecular gas sensing below parts per trillion: radiocarbon-dioxide optical detection. Phys. Rev. Lett. **107**, 270802 (2011)
29. A. Persson, G. Eilers, L. Rydersors, E. Mukhtar, G. Possnert, M. Salehpour, Evaluation of intracavity optogalvanic spectroscopy for radiocarbon measurements. Anal. Chem. **85**, 6790 (2013)
30. D. Paul, H.A.J. Meijer, Intracavity optogalvanic spectroscopy not suitable for ambient level radiocarbon detection. Anal. Chem. **87**, 9025 (2015)
31. C.G. Carson, M. Stute, Y. Ji, R. Polle, K.S. Lackner, Invalidation of the intracavity optogalvanic method for radiocarbon detection. Radiocarbon **58**(1), 213 (2016)
32. I. Galli et al., Optical detection of radiocarbon dioxide: first results and AMS intercomparison. Radiocarbon **55**(2–3), 213 (2013)
33. A.D. McCartt, T. Ognibene, G. Bench, K. Turteltaub, Measurements of carbon-14 with cavity ring-down spectroscopy. Nucl. Instrum. Methods Phys. Res. B **361**, 277 (2015)
34. G. Genoud, M. Vainio, H. Phillips, J. Dean, M. Merimaa, Radiocarbon dioxide detection based on cavity ring-down spectroscopy and a quantum cascade laser. Opt. Lett. **40**, 1342 (2015)
35. I. Galli et al., Spectroscopic detection of radiocarbon dioxide at parts-per-quadrillion sensitivity. Optica **3**(4), 385 (2016)
36. R.M. Wilson, Smaller, faster, cheaper detection or radiocarbon. Phys. Today **69**(6), 19 (2016)
37. H.-A. Synal, M. Stocker, M. Suter, MICADAS: a new compact radiocarbon AMS system. Nucl. Instrum. Methods Phys. Res. B **259**, 7 (2007)
38. M. Seiler, S. Maxeiner, L. Wacker, H.-A. Synal, Status of mass spectrometric radiocarbon detection at ETHZ. Nucl. Instrum. Methods Phys. Res. B **361**, 245 (2015)
39. W. Kutschera, W. Müller, "Isotope language" of the Alpine Iceman investigated with AMS and MS. Nucl. Instrum. Methods Phys. Res. B **204**, 705 (2003)
40. W. Kutschera et al., Evidence for early human presence at high altitudes in the Ötztal Alps (Austria/Italy). Radiocarbon **56**(3), 923 (2014)
41. H. Oberhummer, A. Csótó, H. Schlattl, Stellar production rates of carbon and its abundance in the universe. Science **289**, 88 (2000)
42. W. Kutschera et al., Long-lived noble gas radionuclides. Nucl. Instrum Methods Phys. Res. B **92**, 241 (1994)
43. J.A. Simpson, Neutrons produced in the atmosphere by cosmic radiation. Phys. Rev. **83**, 1175 (1951)
44. M. Stuiver, H.A. Pollach, Discussion reporting of ^{14}C data. Radiocarbon **19**(3), 355 (1977)
45. S.M. Fahrni, J.R. Southon, G.M. Santos, S.W.L. Palstra, H.A.J. Meijer, X. Xu, Reassessment of the ^{13}C/^{12}C and ^{14}C/^{12}C isotopic fractionation ratio and its impact on high-precision radiocarbon dating. Geochim. Cosmochim. Acta **213**, 330 (2017)

46. C.N. Waters et al., The Anthropocene is functionally and stratigraphically distinct from the Holocene. Science **351**, 137 (2016)
47. P.J. Crutzen, Geology of mankind. Nature **415**, 23 (2002)
48. V.F. Hess, Über Beobachtungen der durchdringenden Strahlung bei sieben Freiballonfahrten. Phys. Z. **13**, 1084 (1912)
49. H.E. Suess, Radiocarbon concentration in modem wood. Science **122**, 415 (1955)
50. T.A. Rafter, G.J. Ferguson, "Atom bomb effect"-recent increase of carbon-14 content of the atmosphere and biosphere. Science **126**, 557 (1957)
51. W.S. Broecker, A. Schulert, E.A. Olson, Bomb carbon-14 in human beings. Science **130**, 331 (1959)
52. I. Levin, V. Hesshaimer, Radiocarbon – a unique tracer of global carbon cycle dynamics. Radiocarbon **42**(1), 69 (2000)
53. Q. Hua, M. Barbetti, A.Z. Rakowski, Atmospheric radiocarbon for the period 1950-2010. Radiocarbon **55**(4), 2059 (2013)
54. H.D. Graven, Impact of fossil fuel emissions on atmospheric radiocarbon and various applications of radiocarbon over this century. Proc. Natl. Acad. Sci. **112**, 9542 (2015)
55. W. Kutschera, Accelerator mass spectrometry: state of the art and perspectives. Adv. Phys. X **1**(4), 570 (2016)
56. I. Levin et al., 25 years of tropospheric ^{14}C observations in Central Europe. Radiocarbon **27**(1), 1 (1985)
57. H.A.J. Meijer, M.H. Pertuisot, J. van der Plicht, High-accuracy ^{14}C measurements for atmospheric CO_2 samples by AMS. Radiocarbon **48**(3), 355 (2006)
58. H.D. Graven, T.P. Guilderson, R.F. Keeling, Observations of radiocarbon in CO_2 at seven global sampling sites in the Scripps flask network: analysis of spatial gradients and seasonal cycles. J. Geophys. Res. **117**, D02303 (2012)
59. C.A.M. Brenninkmeijer et al., Interhemispheric asymmetry in OH abundance inferred from measurements of atmospheric ^{14}CO. Nature **356**, 50 (1992)
60. W. Rom et al., A detailed 2-year record of atmospheric ^{14}CO in the temperate northern hemisphere. Nucl. Instr. Meth. Phys. Res. B **161–163**, 780 (2000)
61. V. Gros et al., Detailed analysis of the isotopic composition of CO and characterization of the air masses arriving at Mount Sonnblick (AustrianAlps). J. Geophys. Res. **106**(D3), 3179 (2001)
62. C. Bronk Ramsey, C.A.M. Brenninkmeijer, P. Jöckel, H. Kjeldsen, J. Masarik, Direct measurements of the radiocarbon production at altitude. Nucl. Instrum. Methods Phys. Res. B **259**, 558 (2007)
63. J. Lelieveld et al., New directions: watching over tropospheric hydroxyl. Atmos. Environ. **40**, 5741 (2006)
64. T.E. Graedel, P.J. Crutzen, *Chemie der Atmosphäre* (Spektrum AkademischerVerlag, Heidelberg, 1994), p. 551
65. J.H. Butler, *The NOAA Annual Greenhouse Gas Index (AGGI)*, (2017) updates are available on the web http://www.esrl.noaa.gov/gmd/aggi/
66. D.C. Lowe, C.A.M. Brenninkmeijer, M.R. Manning, R. Sparks, G. Wallace, Radio-carbon determination of atmospheric methane at Baring Head, New Zealand. Nature **332**, 522 (1988)
67. R. Eisma, K. van der Borg, A.F.M. de Jong, W.M. Kieskamp, A.C. Veltkamp, Measurements of the ^{14}C content of atmospheric methane in The Netherlands to determine the regional emissions of $^{14}CH_4$. Nucl. Instrum. Methods. Phys. Res. B **92**, 410 (1994)
68. P. Quay, J. Stutsman, D. Wilbur, A. Snover, E. Dlugokencky, T. Brown, The isotopic composition of atmospheric methane. Global. Biogeochem. Cycles **13**, 445 (1999)
69. V.V. Petrenko et al., $^{14}CH_4$ measurements in Greenland Ice: investigating last glacial termination CH_4 sources. Science **324**, 506 (2009)
70. G. Anglada-Escude et al., A terrestrial planet candidate in a temperate orbit around Proxima Centauri. Nature **536**, 437 (2016)
71. M. Gillon et al., Seven temperate terrestrial planets around the nearby ultracool dwarf star TRAPPIST-1. Nature **542**, 456 (2017)

72. W.S. Broecker, T. Peng, Z. Beng, *Tracers in the Sea* (Eldigio Press, Palisades, 1982), p. 690
73. W.S. Broecker, A. Mix, M. Andree, H. Oeschger, Radiocarbon measurements on coexisting benthic and planktic foraminifera shells: potential for reconstruc-ting ocean ventilation times over the past 20000 years. Nucl. Instrum. Methods. Phys. Res. B **5**, 331 (1984)
74. W.S. Broecker, The great ocean conveyor. Oceanography **4**, 79 (1991)
75. W.S. Broecker, The carbon cycle and climate change: memoirs of my 60 years in science. Geochem. Perspect. **1**, 221 (2012)
76. S. Rahmstorf, Ocean circulation and climate during the past 120,000 years. Nature **419**, 207 (2002)
77. A.P. Mc Nichol et al., Ten years after—the WOCE AMS radiocarbon program. Nucl. Instrum. Methods. Phys. Res. B **172**, 479 (2000)
78. R.M. Key et al., A global ocean carbon climatology: results from Global Data Analysis Project (GLODAP). Glob. Biogeochem. Cycles **18**, 1 (2004)
79. H.H. Loosli, in *Argon 39: A Tool to Investigate Ocean Water Circulation and Mixing*, eds. by P.J. Fritz, C.H. Fontes. Handbook of Environmental Isotope Geochemistry, vol 3 (Elsevier, Amsterdam, 1989), p. 387
80. P. Collon et al., Development of an AMS method to study oceanic circulation characteristics using cosmogenic ^{39}Ar. Nucl. Instrum. Methods. Phys. Res. B **223–224**, 428 (2004)
81. P. Collon et al., Reducing potassium contamination for AMS detection of ^{39}Ar with an electron-cyclotron-resonance ion source. Nucl. Instrum. Methods. Phys. Res. B **283**, 77 (2012)
82. J.-C. Gascard, G. Raisbeck, S. Sequeira, F. Yiou, K.A. Mork, The Norwegian Atlantic current in the Lofoten basin inferred from hydrological and tracerdata (^{129}I) and its interaction with the Norwegian coastal current. Geophys. Res. Lett. **31**, L01308 (2004)
83. L. Wacker, L.K. Fifield, S.G. Tims, Developments in AMS of ^{99}Tc. Nucl. Instrum. Methods. Phys. Res. B **223–224**, 185 (2004)
84. M.J. Keith-Roach, J.P. Day, L.K. Fifield, F.R. Livens, Measurement of ^{237}Np in environmental water samples by accelerator mass spectrometry. Analyst **126**, 58 (2001)
85. S.R. Winkler, P. Steier, J. Carilli, Bomb fall-out ^{236}U as a global oceanic tracer using an annually resolved coral core. Earth Planet. Sci. Lett. **359–360**, 124 (2012)
86. J. Gould et al., Argo profiling floats bring new era of in situ ocean observations. Eos Trans. AGU **85**, 185 (2004)
87. D. Roemmich et al., Unabated planetary warming and its ocean structure since 2006. Nat. Clim. Chang. **5**, 240 (2015)
88. S. Wijffels et al., Ocean temperatures chronicle the ongoing warming of Earth. Nat. Clim. Chang. **6**, 116 (2016)
89. E. Hodge et al., Using the ^{14}C bomb pulse to date young spaleothems. Radiocarbon **53**(2), 345 (2011)
90. J.D.H. Wischusen, L.K. Fifield, R.G. Cresswell, Hydrology of Palm Valley, central Australia; a Pleistocene flora refuge? J. Hydrol. **293**, 20 (2004)
91. J. Fabryka-Martin, H. Bentley, D. Elmore, P.L. Airey, Natural iodine-129 as an environmental tracer. Geochim. Cosmochim. Acta **49**, 337 (1985)
92. A.J. Love, A.L. Herczeg, L. Sampson, R.G. Cresswell, L.K. Fifield, Sources of chloride and implications for ^{36}Cl dating of old groundwater, southwestern Great Artesian Basin. Water Resour. Res. **36**(6), 1561 (2000)
93. P. Collon et al., ^{81}Kr in the Great Artesian Basin, Australia: a new method for dating very old groundwater. Earth Planet. Sci. Lett. **182**, 103 (2000)
94. N.C. Sturchio et al., One million year old groundwater in the Sahara revealed by krypton-81 and chlorine-36. Geophys. Res. Lett. **31**, L5503 (2004)
95. F. Ritterbusch et al., Groundwater dating with Atom Trap Trace Analysis of ^{39}Ar. Geophys. Res. Lett. **41**, 6758 (2014)
96. W. Jiang et al., An atom counter for measuring ^{81}Kr and ^{85}Kr in environmental samples. Cosmochim. Geochim. Acta **91**, 1 (2012)

97. W. Aeschbach-Hertig, Radiokrypton dating finally takes off. Proc. Natl. Acad. Sci. **111**, 6857 (2014)
98. C. Buizert et al., Radiometric ^{81}Kr dating identifies 120,000-year-old ice at Taylor Glacier, Antarctica. Proc. Natl. Acad. Sci. **111**, 6876 (2014)
99. C. Renfrew, Archaeology introduction. Radiocarbon **51**(1), 121 (2009)
100. C. Bronk Ramsey, Bayesian analysis of radiocarbon dates. Radiocarbon **51**(1), 337 (2009)
101. C. Bronk Ramsey et al., Radiocarbon-based chronology for dynastic Egypt. Science **328**, 1554 (2010)
102. H.J. Bruins, J. van der Plicht, J.A. MacGillivray, The Minoan Santorini eruption and tsunami deposits in Palaikastro. Radiocarbon **51**, 397 (2004)
103. S.W. Manning et al., Dating the Thera (Santorini) eruption: archaeological and scientific evidence supporting a high chronology. Antiquity **88**, 1164 (2014)
104. W. Kutschera, Dating of the Thera/Santorini volcanic eruption, Tagungen des Landesmuseums für Vorgeschichte Halle. Band **9**, 59 (2013)
105. W. Kutschera et al., The chronology of Tell el-Dab'a: a crucial meeting point of ^{14}C dating, archaeology and Egyptology in the 2nd millennium BC. Radiocarbon **54**, 407 (2012)
106. M. Bietak, in *Antagonisms in Historical and Radiocarbon Chronology*, ed. by A.J. Shortland, C. Bronk Ramsey, Radiocarbon and the Chronology of Ancient Egypt, Chapter 8, (Oxbow Books, Oxford, 2013), pp. 76–109
107. N. Moeller, G. Marouard, Discussion of Late Middle Kingdom and Early Second Intermediate Period history and chronology in relation to the Kayan sealings from Tell Edfu. Egypt Levant **21**, 87 (2011)
108. F. Höflmayer et al., New evidence for Middle Bronze Age chronology and synchronisms in the Levant: radiocarbon dates from Tell el-Burak, Tell el-Dab'a, and Tel Ifshar compared. Bull. Am. School Orient. Res. **375**, 53 (2016)
109. K.L. Spalding, R.D. Bhardwaj, B.A. Buchholz, H. Druid, J. Frisén, Retrospective birth dating of cells in humans. Cell **122**, 133 (2005)
110. D. Grimm, The mushroom cloud's silver lining. Science **321**, 1434 (2008)
111. E.M. Wild et al., ^{14}C dating with the bomb peak: an application to forensic medicine. Nucl. Instrum. Methods. Phys. Res. B **172**, 944 (2000)
112. K.L. Spalding, B.A. Buchholz, L.-E. Bergman, H. Druid, J. Frisén, Age written in teeth by nuclear tests. Nature **437**, 333 (2005)
113. N. Lynnerup, H. Kjeldsen, S. Heegaard, C. Jacobsen, J. Heinemeier, Radiocarbon dating of the human eye lens crystallines reveal proteins without carbon turnover throughout life. PLoS One **3**(1), e1529 (2008)
114. R.D. Bhardwaj et al., Neocortical neurogenesis in humans is restricted to development. Proc. Natl. Acad. Sci. **103**, 12564 (2006)
115. K.L. Spalding et al., Dynamics of hippocampal neurogenesis in adult humans. Cell **153**, 1219 (2013)
116. O. Bergmann et al., The age of olfactory bulb neurons in humans. Neuron **74**, 634 (2012)
117. H.B. Huttner et al., The age and genomic integrity of neurons after cortical stroke in humans. Nat Neurosci. **17**, 801 (2014)
118. K.L. Spalding et al., Dynamics of fat cell turnover in humans. Nature **453**, 783 (2008)
119. M.T. Hyvönen, K.L. Spalding, Maintenance of white adipose tissue in man. Int. J. Biochem. Cell Biol. **56**, 123 (2014)
120. O. Bergmann et al., Dynamics of cell generation and turnover in the human heart. Cell **161**, 1566 (2015)
121. L.L. Hamady, L.J. Natanson, G.B. Skomal, S.R. Thorrold, Vertebral bomb radiocarbon suggests extreme longevity in white sharks. PLoS One **9**(1), e84006 (2014)
122. J. Nielsen et al., Eye lens radiocarbon reveals centuries of longevity in the Greenland shark (*Somniosus microcephalus*). Science **353**, 702 (2016)
123. L. Caforio et al., Discovering forgeries of modern art by the ^{14}C bomb peak. Eur. Phys. J. Plus **129**, 6 (2014). https://doi.org/10.1140/epjp/i2014-14006-6

124. T.E. Cerling et al., Radiocarbon dating of seized ivory confirms rapid decline in African elephant populations and provides insight into illegal trade. Proc. Natl. Acad. Sci. **113**, 13330 (2016)

125. S.K. Wasser et al., Genetic assignment of large seizures of elephant ivory reveals Africa's major poaching hotspots. Science **349**, 84 (2015)

126. K.W. Turteltaub et al., Accelerator mass spectrometry in biomedical dosimetry: relationship between low-level exposure and covalent binding of heterocyclic amine-carcinogens to DNA. Proc. Natl. Acad. Sci. **87**, 5288 (1990)

127. R.C. Garner et al., Comparative biotransformation studies of MeIQx and PhIP in animal models and humans. Cancer Lett. **143**, 161 (1999)

128. G. Lappin, R.C. Garner, Big physics, small doses: the use of AMS and PET in human microdosing. Nat. Rev. Drug Disc. **2**(3), 233 (2003)

129. J.S. Vogel, N.M. Palmblad, T. Ognibene, M.M. Kabir, B.A. Buchholz, G. Bench, Biochemical paths in humans and cells: frontiers of AMS bioanalysis. Nucl. Instrum. Methods. Phys. Res. B **259**, 754 (2007)

130. T. Schulze-König, S.R. Dueker, J. Giacomo, M. Suter, J.S. Vogel, H.-A. Synal, BioMI-CADAS: compact next generation AMS system for pharmaceutical science. Nucl. Instrum. Methods. Phys. Res. B **268**, 891 (2010)

131. R.M.J. Ings, Welcome to 'microdosing'. Bioanalysis **2**(3), 371 (2010)

132. R.C. Garner, Practical experience of using human microdosing with AMS analysis to obtain early human drug metabolism and PK data. Bioanalysis **2**(3), 429 (2010)

133. R.E.M. Hedges, R.A. Housley, C.R. Bronk, G.J. van Klinken, Radiocarbon dates from the Oxford AMS system: Archaeometry datelist 15. Archaeometry **34**(2), 337 (1992)

134. G. Bonani, S.D. Ivy, I. Hajdas, T.R. Niklaus, M. Suter, AMS ^{14}C age determination of tissue, bone and grass samples from the Ötztal Iceman. Radiocarbon **36**(2), 247 (1994)

135. W. Müller, H. Fricke, A.N. Halliday, M.T. McCulloch, J.-A. Wartho, Origin and migration of the Alpine Iceman. Science **302**, 862 (2003)

136. K. Nicolussi, G. Patzelt, Discovery of Early-Holocene wood and peat on the forefield of the Pasterze Glacier, Eastern Alps, Austria. The Holocene **10**(2), 191 (2000)

137. A. Hormes, B.U. Müller, C. Schlüchter, The Alps with little ice: evidence for eight Holocene phases of reduced glacier extend in the Central Swiss Alps. The Holocene **11**(3), 255 (2001)

138. M. Magny, J.N. Haas, A major widespread climatic change around 5300 cal. yr BP at the time of the Alpine Iceman. J. Quat. Sci. **19**(5), 423 (2004)

139. C. Schlüchter, U. Joerin, Die Alpen ohne Gletscher (in German). Die Alpen **6**, 34 (2004)

140. H. Holzhauser, M. Magny, H.J. Zumbühl, Glacier and lake-level variations in west-central Europe over the last 3500 years. The Holocene **15**(6), 789 (2005)

141. K. Nicolussi, M. Kaufmann, G. Patzelt, J. van der Plicht, A. Thurner, Holocene tree-line variability in the Kauner Valley, Central Eastern Alps, indicated by dendrochronological analysis of living trees and subfossil logs. Veget. Hist. Archaeobot. **14**, 221 (2005)

142. U.E. Joerin, T.F. Stocker, C. Schlüchter, Multicentury glacier fluctuations in the Swiss Alps during the Holocene. The Holocene **16**(5), 697 (2006)

143. M. Grosjean, P.J. Suter, M. Trachsel, H. Wanner, Ice-borne prehistoric finds in the Swiss Alps reflect Holocene glacier fluctuations. J. Quat. Sci. **22**(3), 203 (2007)

144. U.E. Joerin, K. Nicolussi, A. Fischer, T.F. Stocker, C. Schlü chter, Holocene optimum events inferred from subglacial sediments at Tschierva Glacier, Eastern Swiss Alps. Quat. Sci. Rev. **27**, 337 (2008)

145. K. Nicolussi, C. Schlüchter, The 8.2 ka event—calendar-dated glacier response in the Alps. Geology **40**(9), 819 (2012)

146. M. Le Roy et al., Calendar-dated glacier variations in the western European Alps during the Neoglacial: the Mer de Glace record, Mont Blanc massif. Quat. Sci. Rev. **108**, 1 (2015)

147. W. Kutschera, G. Patzelt, P. Steier, E.M. Maria Wild, The Tyrolean Iceman and his glacial environment during the Holocene. Radiocarbon **59**(2), 395 (2017)

148. B.M. Goehring et al., The Rhone Glacier was smaller than today for most of the Holocene. Geology **39**(7), 679 (2011)

149. I. Schimmelpfennig, J.M. Schaefer, N. Akçar, S. Ivy-Ochs, R.C. Finkel, C. Schlüchter, Holocene glacier culminations in the Western Alps and their hemispheric relevance. Geology **40**(10), 891 (2012)
150. I. Schimmelpfenning et al., A chronology of Holocene and Little Ice Age glacier culminations of the Steingletscher, Central Alps, Switzerland, based on high-sensitivity beryllium-10 moraine dating. Earth Planet. Sci. Lett. **393**, 220 (2014)
151. C. Wirsig, J. Zasadni, S. Ivy-Ochs, M. Christl, F. Kober, C. Schlüchter, A deglaciation model of the Oberhasli, Switzerland. J. Quat. Sci. **31**(1), 46 (2016)
152. J.M. Schaefer et al., High-frequency Holocene glacier fluctuations in New Zealand differ from the northern signature. Science **324**, 622 (2009)
153. A.E. Putnam et al., Regional climate control of glaciers in New Zealand and Europe during the pre-industrial Holocene. Nat. Geosci. **5**, 627 (2012)
154. J.C. Gosse, F.M. Phillips, Terrestrial *in situ* cosmogenic nuclides: theory and application. Quat. Sci. Rev. **20**, 1475 (2001)
155. D.W. Burbank et al., Bedrock incision, rock uplift and threshold hillslopes in the northwestern Himalayas. Nature **379**, 505 (1996)
156. M.K. Pavicevic et al., Erosion rate study at the Allchar deposit (Macedonia) based on radioactive and stable cosmogenic nuclides (^{26}Al, ^{36}Cl, ^{3}He, and ^{21}Ne). Geochem. Geophys. Geosyst. **17**, 410 (2016)
157. F. Steinhilber, J. Beer, C. Fröhlich, Total solar irradiance during the Holocene. Geophys. Res. Lett. **36**, L19704 (2009)
158. J.A. Abreu, J. Beer, F. Steinhilber, M. Christl, P.W. Kubik, 10Be in ice cores and 14C in tree rings: separation of production and climatic effects. Space Sci. Rev. **176**(1), 343 (2013)

Chapter 2
Giant Resonances: Fundamental Modes and Probes of Nuclear Properties

M. N. Harakeh

Abstract To study the properties of nuclear matter, we use nuclear reactions to excite the fundamental modes of the nucleus, which can yield information on the equation of state (EOS) and are also important for understanding nuclear structure aspects of nuclei. Furthermore, it is very important to understand the nuclear processes that precede a supernova event and to understand the properties of nuclear matter in order to explain why stars sometimes explode throwing most of the star material into space leaving a neutron star or a black hole behind.

In the last three decades, the compression modes, the isoscalar giant monopole (ISGMR) and dipole resonances (ISGDR), were extensively studied because of their importance for the determination of the nuclear-matter incompressibility and consequently their implications for the EOS of nuclear matter. Though the nuclear matter incompressibility (K_∞) has been reasonably well determined (240 ± 10 MeV) through comparison of experimental results on several spherical nuclei with microscopic calculations, the asymmetry term was determined with much larger uncertainty. This has been addressed in measurements on a series of stable Sn and Cd isotopes, which resulted in a value of $K_\tau = -550 \pm 100$ MeV for the asymmetry term in the nuclear incompressibility.

Spin-isospin modes, and in particular the Gamow–Teller (GT) transitions, aside from their interest from the nuclear structure point of view, play very important roles in various phenomena in nature. In nucleosynthesis, the β-decay of nuclei along the s- and r-processes determine the paths that these processes follow and the abundances of the elements synthesised. In supernova explosions, GT transitions are of paramount importance in the pre-supernova phase where electron capture occurs on neutron-rich fp-shell nuclei at the high temperatures of giant stars. Electron capture is mediated by GT transitions. Electron capture removes the electron pressure that keeps the star from collapsing precipitating an implosion followed by a cataclysmic explosion throwing much of the star material into space and leaving a neutron star or black hole behind.

M. N. Harakeh (✉)
KVI-CART, University of Groningen, Groningen, The Netherlands
e-mail: m.n.harakeh@kvi.nl

© Springer International Publishing AG, part of Springer Nature 2018
C. Scheidenberger, M. Pfützner (eds.), *The Euroschool on Exotic Beams - Vol. 5*,
Lecture Notes in Physics 948, https://doi.org/10.1007/978-3-319-74878-8_2

2.1 Introduction

The isovector giant dipole resonance (IVGDR) was discovered by Bothe and
Gentner [1]. Later, Migdal, Baldwin, and Klaiber [2, 3] rediscovered the IVGDR,
which was interpreted by Goldhaber and Teller [4] and Steinwedel and Jensen [5] as
dipole oscillations of protons versus neutrons. In the following decades, systematic
studies were performed employing bremsstrahlung radiation produced by electron
accelerators to study the resonant γ absorption in nuclei. In addition, experiments
with quasi-monochromatic photons could be performed by using a tagged-photon
beam or e^+e^- annihilation in flight; see [6] and references therein.

In addition to the IVGDR, there exist other types of giant resonances that can be
described in a hydrodynamical model as the oscillations of a liquid drop composed
of four different types of fluids: protons and neutrons with spins up and down.
The giant resonances can therefore be characterised by their multipolarity, spin
and isospin. An isoscalar mode corresponds to protons and neutrons oscillating in
phase whereas an isovector mode corresponds to protons and neutrons oscillating
out of phase. Similarly, a scalar (electric) mode corresponds to nucleons with spin-
up and spin-down oscillating in phase and a vector (magnetic) mode corresponds
to nucleons with spin-up and spin-down oscillating out of phase. In Fig. 2.1, a
schematic representation is given for the lowest three multipolarities: $\Delta L = 0$
(monopole), $\Delta L = 1$ (dipole) and $\Delta L = 2$ (quadrupole) and for isoscalar electric,
isovector electric, isoscalar magnetic and isovector magnetic, respectively. The
representation for the isoscalar giant dipole resonance (ISGDR) is missing because
in first order it corresponds to a spurious centre-of-mass (c.o.m.) motion, *i.e.*
translational motion of the nucleus as a whole without intrinsic excitation of it. In
higher order, the ISGDR exists as will be discussed below.

The study of giant resonances including the IVGDR has been very intensive in
the last six decades. This started with the discovery of the isoscalar giant quadrupole
resonance (ISGQR) in electron scattering in 1971 [7] and isoscalar giant monopole
resonance (ISGMR) in inelastic α scattering in 1977 [8–10]. Other modes were
discovered and investigated extensively. Among these, the ISGMR and ISGDR are
of special interest because they are compression modes and their excitation energies
are dependent on the compression modulus of the nucleus. Charge-exchange modes
are also very important because of the role they play in nucleosynthesis and other
astrophysical phenomena as we will discuss in more detail below.

In microscopic models, a giant resonance can be described as a coherent
superposition of one-particle-one-hole excitations that are induced by the operation
of one-body operators on the ground state of a nucleus:

$$|\Psi^{\lambda\sigma\tau}\rangle = O^{\lambda\sigma\tau}|\Psi_{g.s.}\rangle$$

where λ refers to the multipolarity of the giant resonance, and σ and τ to its spin and
isospin structure, respectively. In the long-wave length limit ($qr \ll 1$), the multipole
operators are obtained from the first-order terms in the expansion of the spherical

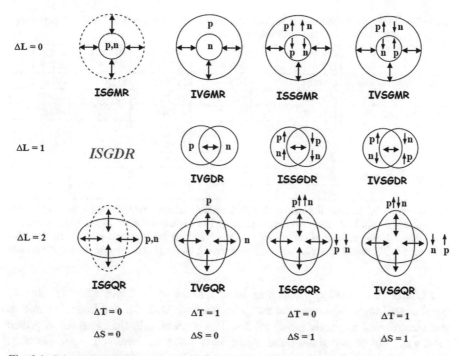

Fig. 2.1 Schematic representation of giant resonances (GRs) for the monopole (M), dipole (D) and quadrupole (Q) modes. Isoscalar (IS) and isovector (IV) modes correspond to neutrons and protons moving in phase and out of phase, respectively. Similarly, electric (scalar) and magnetic (vector; also spin (S)) modes correspond to nucleons with spin-up and spin-down moving in phase and out of phase, respectively. The notation used is self-explanatory

Bessel function j_λ (qr) of order (multipolarity) λ [6]; here, q is the momentum transfer and r is the spherical radial coordinate. This is true except for the isoscalar monopole and dipole operators.

The transition operators for the isoscalar monopole and dipole excitations can be obtained from the expansion of the Bessel function of order $\lambda = 0$ and $\lambda = 1$, respectively. For the monopole, the first-order term is a constant that does not induce any excitations. The second-order term in the expansion is proportional to r^2 and thus can lead to 2 $\hbar\omega$ monopole excitations. For the dipole, the first-order term is proportional to the centre-of-mass (c.o.m.) coordinate and corresponds to the spurious c.o.m. motion.

$$O_\mu^\lambda = \sum_k r_k Y_{1\mu}(\Omega_k) + \frac{1}{2}\sum_k r_k^3 Y_{1\mu}(\Omega_k) + \ldots$$

Intrinsic isoscalar dipole excitations can be induced by the second-order term in the expansion which is proportional to r^3 and thus can lead to 1 $\hbar\omega$ and 3 $\hbar\omega$ excitations. The isoscalar dipole resonance is associated with the 3 $\hbar\omega$ excitation.

Table 2.1 Multipole operators for the first four multipolarities ($\Delta L = 0, 1, 2$ and 3) isoscalar and isovector electric modes and isovector magnetic modes

	$\Delta S = 0$, $\Delta T = 0$	$\Delta S = 0$, $\Delta T = 1$	$\Delta S = 0$, $\Delta T = 1$	$\Delta S = 1$, $\Delta T = 1$	$\Delta S = 1$, $\Delta T = 1$
$\Delta L = 0$: Monopole	ISGMR	IAS	IVGMR	GTR	IVSGMR
	$r^2 Y_0$	$\tau_\pm Y_0$	$\tau r^2 Y_0$	$\tau_\pm \sigma Y_0$	$\tau \sigma r^2 Y_0$
$\Delta L = 1$: Dipole	ISGDR		IVGDR		IVSGDR
	$(r^3 - 5/3 \langle r^2 \rangle r) Y_1$		$\tau r Y_1$		$\tau \sigma r Y_1$
$\Delta L = 2$: Quadrupole	ISGQR		IVGQR		IVSGQR
	$r^2 Y_2$		$\tau r^2 Y_2$		$\tau \sigma r^2 Y_2$
$\Delta L = 3$: Octupole	ISGOR				
	$r^3 Y_3$				

The operators for the isoscalar magnetic modes are not included, since these are expected to be weak and have not been observed experimentally. The operators for the IAS (Fermi) transition and Gamow–Teller resonance (GTR) are also given and do not involve a radial part. Note the extra term in the second-order isoscalar dipole operator, which is to correct for the c.o.m. motion

In Table 2.1, multipole operators for the first four multipolarities: monopole, dipole, quadrupole and octupole are given for both isoscalar and isovector modes, and electric and magnetic (spin) modes. The operators for the isoscalar magnetic modes are not shown since these modes have not been observed experimentally. The operator for the isobaric analogue state (IAS) is given and it consists of an isospin lowering or raising operator that changes a neutron to a proton or a proton to a neutron, respectively, without changing the other quantum numbers, *i.e.* the radial, orbital and spin wave functions remain the same. The operator for the Gamow–Teller (GT) excitation involves in addition the spin-operator and therefore it will induce a spin flip in addition to the isospin flip. It induces thus transitions of neutrons to protons (lowering) or protons to neutrons (raising) without changing the radial and orbital wave functions. However, the spin wave function may change. Thus, transitions between spin-orbit partners are also allowed.

2.2 The Compression Modes and Incompressibility of Nuclear Matter

As stated above the compression modes, ISGMR and ISGDR, provide the possibility to determine the nuclear matter incompressibility experimentally because of the relation between their excitation energies and the incompressibility of nuclei. These two compression modes are depicted in Fig. 2.2. The ISGMR (also called breathing mode) shows a volume oscillation around an equilibrium shape, thereby moving between rarefied and denser situations. The ISGDR (also called squeezing

Fig. 2.2 Oscillation of the
nucleus about an equilibrium
shape shown for ISGMR
(top), ISGDR (middle) and
ISGQR (bottom). The
ISGMR (breathing mode) and
the ISGDR (squeezing mode)
display density variations and
therefore are denoted as
compression modes

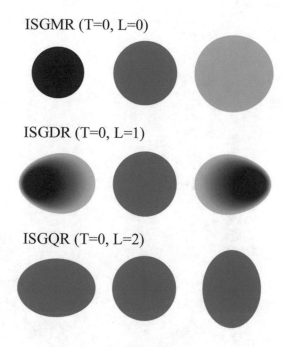

ISGMR (T=0, L=0)

ISGDR (T=0, L=1)

ISGQR (T=0, L=2)

mode) displays oscillations wherein the density increases on one side of the nucleus and decreases on the other keeping the c.o.m. fixed.

The incompressibility of 'infinite' nuclear matter is an important ingredient of the equation of state (EoS) of nuclear matter. The EoS is more complex than for infinite neutral liquids because the nuclear fluid has protons and neutrons with different interactions. In particular, the long-range Coulomb interaction for protons becomes very important for large volumes. The EoS governs: (1) the collapse and explosion of giant stars (supernovae explosions), (2) the formation of neutron stars (mass, radius and crust) and (3) the collisions of heavy ions.

In fluid mechanics, compressibility is a measure of the relative volume change of a fluid as a response to a pressure change:

$$\beta = -\frac{1}{V}\frac{\partial V}{\partial P}$$

where P is the pressure, V is the volume. Incompressibility or bulk modulus (K) is a measure of a substance's resistance to uniform compression and can be formally defined:

$$K = -V\frac{\partial P}{\partial V}$$

The incompressibility is an important ingredient of the nuclear forces that determine the properties of nuclei. For a definition of the incompressibility of 'infinite' nuclear matter, one notes that for the EoS of nuclear matter at saturation nuclear density of

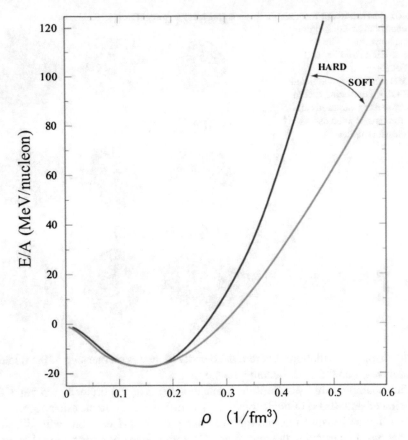

Fig. 2.3 Binding energy per nucleon as function of nuclear density for a soft and a hard EoS. At saturation density (0.16 fm^{-3}) the curves are flat

0.16 fm^{-3} [11] we have (see also Fig. 2.3):

$$\left[\frac{d\,(E/A)}{d\rho}\right]_{\rho=\rho_0} = 0$$

Here, E/A is the binding energy per nucleon, ρ is the nuclear density and ρ_0 is the nuclear density at saturation.

The incompressibility of infinite nuclear matter can then be derived as the second derivative with respect to the nuclear density [11]:

$$K_{nm} = \left[9\,\rho^2\frac{d^2\,(E/A)}{d\rho^2}\right]_{\rho=\rho_0}$$

The excitation energies of the ISGMR and ISGDR are given in the constrained and scaling models as [6, 11, 12]:

$$E_{ISGMR} = \hbar \sqrt{\frac{K_A}{m \langle r^2 \rangle}}$$

and

$$E_{ISGDR} = \hbar \sqrt{\frac{7}{3} \frac{K_A + \frac{27}{25} \varepsilon_F}{m \langle r^2 \rangle}}$$

where m is the nucleon mass, $\langle r^2 \rangle$ is the mean-square radius, ε_F is the Fermi energy and K_A, the finite-nucleus incompressibility, is given by [11]:

$$K_A = \left[r^2 \left(d^2 \left(E/A \right) / dr^2 \right) \right]_{r=R0}$$

where R_0 is the nucleus half-density radius. In the scaling model the nucleus incompressibility can be expanded (Leptodermous (thin-skin) expansion) into several terms, as follows [13]:

$$K_A = K_{vol} + K_{surf} A^{-1/3} + K_{sym}((N - Z)/A)^2 + K_{Coul}Z^2 A^{-4/3}$$

In principle, one can obtain the various terms through a fit to K_A values obtained from the excitation energies of the ISGMR and ISGDR for many nuclei. In this way, K_{vol}, K_{surf} and K_{sym} (also referred to in the literature as K_τ) can be determined. K_{surf} does not play a role in the EoS of infinite nuclear matter. K_{Coul} can be calculated from the well-known Coulomb interaction. K_{vol} can be associated with the incompressibility of nuclear matter, $K_{nm} \equiv K_\infty$. This is, however, not what is done because using the Leptodermous expansion is not considered a reliable method to determine the parameters. Instead, it is considered more correct and reliable to determine the excitation energies of the compression modes through non-relativistic or relativistic calculations with different types of nucleon-nucleon (NN) interactions. The incompressibility of nuclear matter is then taken from the NN interaction that leads to the best reproduction of the isoscalar giant monopole and dipole resonances.

The most important experimental and analysis steps for determining the excitation energies of the ISGMR and ISGDR are discussed briefly in the following. The best reaction to study these isoscalar compression modes is inelastic α scattering at incident energies of   160 MeV or higher and at very forward angles including 0°. At these angles and bombarding energies, the angular distributions are characteristic of the different multipolarities. However, for measuring at and near 0° the use of a magnetic spectrometer is imperative to separate the beam from inelastically scattered α particles. Inelastic α-scattering spectra for different scattering angles are generated by software division of the angular bins. Special techniques have to be used to remove the experimental background. Once the experimental background is removed, a multipole-decomposition analysis (MDA) is performed. In this procedure, excitation-energy bins are defined in the inelastic α-scattering spectra obtained for the different scattering angles measured. The angular distributions for the different excitation-energy bins are determined. These are then fit with the angular distributions of the various possible multipolarities. The fraction of the energy-weighted sum rule (EWSR) strength for each multipolarity exhausted in each excitation-energy bin is deduced in this way. This finally results in spectra for the strength distributions of the different multipoles as a function of excitation energy.

In Fig. 2.4, the results of such an analysis for the strength distributions of the ISGMR and the ISGDR in ^{90}Zr, ^{116}Sn, and ^{208}Pb are shown. The experimental spectra have been obtained in inelastic α scattering at an incident energy of

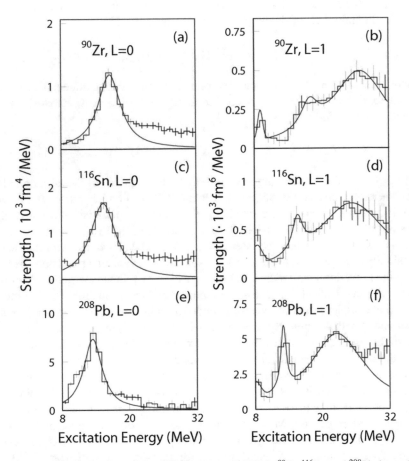

Fig. 2.4 Strength distributions of the ISGMR and the ISGDR in ^{90}Zr, ^{116}Sn, and ^{208}Pb determined from MDA of experimental spectra obtained [14] in inelastic α scattering at an incident energy of 386 MeV. The results of fitting the ISGMR with a Breit-Wigner function are indicated by solid curve. For the ISGDR, the results of fitting with two Breit-Wigner functions for the region of $E_x > 10$ MeV are shown by solid curve

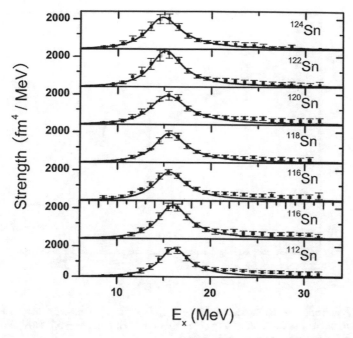

Fig. 2.5 Strength distributions of the ISGMR for all even-A Sn isotopes obtained [16, 17] from inelastic α scattering at an incident energy of around 400 MeV. The results of fitting the ISGMR strength distributions with a Lorentzian function are shown

386 MeV [14]. The results of fits with Breit-Wigner functions are shown. From the ISGMR data on ^{90}Zr and ^{208}Pb, the incompressibility of nuclear matter is determined [15] to be $K_\infty = 240 \pm 10$ MeV. This value seems to be consistent with both ISGMR and ISGDR Data and with non-relativistic and relativistic calculations.

To determine the coefficient of the symmetry term of the incompressibility, inelastic α scattering at around 400 MeV incident energy was studied on all stable even-A Sn isotopes [16, 17]. Isoscalar monopole strength as a function of excitation energy was determined via MDA. These are shown in Fig. 2.5 together with fits with Lorentzian functions. The energies of the ISGMR obtained from these fits are within the uncertainties in good agreement with the values obtained also from the moment ratio $\sqrt{m_1/m_{-1}}$, where the kth moment of the strength distribution is defined as:

$$m_k = \int E_x^{\,k}\, S\,(E_x)\, dE_x$$

where $S(E_x)$ is the EWSR strength function and the integration has been performed in the excitation-energy interval 10.5–20.5 MeV. The moment ratio $\sqrt{m_1/m_{-1}}$ gives the excitation energy of the resonance which has the strength function $S(E_x)$.

The nucleus incompressibility given above can be rewritten:

$$K_A - K_{Coul}Z^2 A^{-4/3} \sim K_{vol}\left(1 + cA^{-1/3}\right) + K_\tau\left[(N-Z)/A\right]^2$$

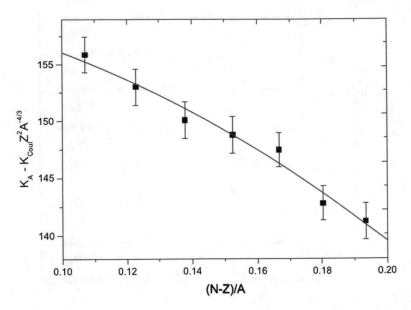

Fig. 2.6 Systematics of the difference $K_A - K_{Coul}Z^2A^{-4/3}$ determined for the Sn isotopes as discussed in the text as function of $\delta = (N - Z)/A$ [16, 17]; $K_{Coul} = -5.2$ MeV [19]. The line represents a least-squares quadratic fit to the data

Here, $c \approx -1$ [18] and K_{Coul} is essentially model independent allowing to calculate the Coulomb term for each of the isotopes. Hence, the difference $K_A - K_{Coul}Z^2A^{-4/3}$ for an isotopic chain can be approximated to be a quadratic function of $\delta = (N - Z)/A$, with K_τ as coefficient. In Fig. 2.6, the difference $K_A - K_{Coul}Z^2A^{-4/3}$ is plotted for all the even Sn isotopes, for which the moment ratio $\sqrt{m_1/m_{-1}}$ was determined as discussed above, as function of the asymmetry parameter, $(N - Z)/A$. The values of K_A were determined using the moment ratios $\sqrt{m_1/m_{-1}}$ for E_{ISGMR}; see above. The data were fitted with a quadratic function. This is shown as a solid line in Fig. 2.6. The fit gives $K_\tau = 550 \pm 40$ MeV; the error is only due to the fitting procedure. Considering all other uncertainties that arise from different sources, a value $K_\tau = 550 \pm 100$ MeV is obtained.

2.3 Spin-Isospin Excitations

Neutrino scattering from nuclei can proceed via neutral currents, *i.e.* elastic and inelastic scattering (ν, ν'), and charged currents, *i.e.* (ν_e, e^-), (antiν_e, e^+). Neutrino scattering proceeds largely through spin-flip and isospin-flip transitions (in charged-current transitions isospin-flip occurs by nature of reaction). These types of transitions are important in various phenomena some of which will be discussed below and are mediated by the weak interaction. Neutral-current transitions of spin-flip and isospin-flip character, in short spin-isospin excitations, can be studied also with inelastic electron scattering and/or inelastic proton scattering at intermediate energies. In this way, the important transition matrix elements, such as for M0, M1 and M2 transitions, can be determined. Charged-current transitions can be studied with charge-exchange reactions such as (p,n) and $(^{3}\text{He},t)$ which induce transitions analogous to β^{-} decay, *i.e.* isospin-lowering transition $\Delta T_{z} = -1$, or as (n,p), $(d,^{2}\text{He})$ and $(t,^{3}\text{He})$ which induce transitions analogous to β^{+} decay, *i.e.* isospin-raising transition $\Delta T_{z} = +1$. These charge-exchange reactions induce excitations of GTR, IVSGMR, IVSGDR, etc. These excitation modes play important roles in nuclear astrophysics, neutrino physics, double-beta decay, determination of neutron-skin thickness, etc. In order to study spin-isospin modes with charge-exchange reactions a few points have to be considered. For example, if one wants to study spin-flip transitions one has to choose an incident energy where the $V_{\sigma\tau}$ spin-isospin part of the nucleon-nucleon (NN) interaction dominates over the V_{τ} isospin part. This is illustrated in Fig. 2.7 where the various central parts of the NN interaction are plotted versus incident energy [20]. In the region between 100 and 500 MeV incident energy, $V_{\sigma\tau}$ dominates strongly over V_{τ}. In this region, also the dominant V_{0} part of the NN interaction has a broad minimum. In the $(^{3}\text{He},t)$ studies that were performed at RCNP, Osaka a bombarding energy of $\sim$150 MeV/u has been used, where the ratio of $V_{\sigma\tau}/V_{\tau} \geq 3$ resulting in a ratio of cross sections of $\sim$10.

Other considerations that come into play are the complexity of the reaction mechanism and the experimental conditions. The first favours light projectiles with no bound states at intermediate incident energies, *i.e.* in the region where the V_{0} part

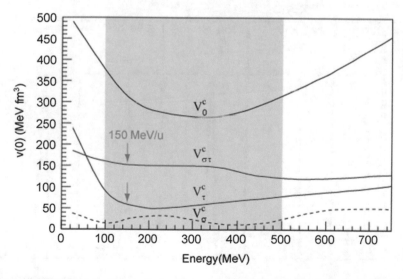

Fig. 2.7 Volume integrals of the central components of the NN interaction are plotted as function of the incident energy [20]. "c" denotes central. Note that whereas $V_{\sigma\tau}$ varies slightly in the energy range 100–500 MeV, V_{τ} goes through a minimum

of the NN interaction has its broad minimum. The second favours certain reactions over others. For example, the (p,n) reaction has a simpler reaction mechanism than the $(^3He,t)$ reaction, but it is more difficult to get high energy resolution because this would require very long flight paths with the time-of-flight method. Long flight paths imply loss of geometrical efficiency. In the case of the $(^3He,t)$ reaction, very high energy resolution can be obtained with the use of a magnetic spectrometer with a detection efficiency approaching 100%. For the (n,p)-type reactions, all of the light-ion reactions (n,p), $(d,^2He)$ and $(t,^3He)$ present experimental problems. The first and third reactions require secondary beams which result in low beam intensities and very low energy resolutions, which is worse for neutron beams than for triton beams. In principle, one can use primary triton beams with very high energy resolution, but this presents other problems with radio protection. The $(d,^2He)$ reaction can be performed with primary beams of high energy resolution but the ^2He ejectile is not bound and the two protons have to be detected with high energy and angular resolutions in order to reconstruct the ^2He in its singlet S state with high resolution.

In Fig. 2.8, the relative dependence of $V_{\sigma\tau}$ and V_τ as function of proton incident energy is demonstrated making use of the $^{14}C(p,n)^{14}N$ reaction at $0°$. The ground and first-excited states of ^{14}N have spin-parity of 1^+ and 0^+, respectively. Transitions from the 0^+ ground state of ^{14}C to the 1^+ and 0^+ levels in ^{14}N are mediated by the $V_{\sigma\tau}$ and V_τ parts of the NN interaction, respectively. The displayed spectra have been measured for proton incident energies ranging from 60 MeV to about 650 MeV. The height of the peak of the 1^+ level at 3.95 MeV excitation energy

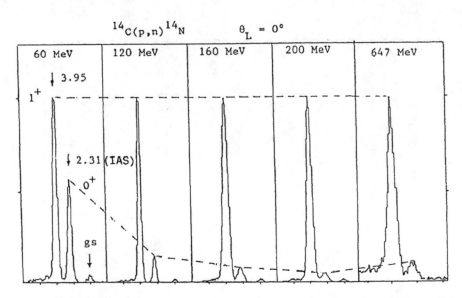

Fig. 2.8 The spectra show the yields for the 1^+ and 0^+ states of ^{14}N populated by the $^{14}C(p,n)^{14}N$ reaction at $0°$ and the indicated bombarding energies. The spectra have been arbitrarily normalised so that the height of the 1^+ peak at 3.95 MeV excitation energy is constant as function of bombarding energy. See [21, 22]

has been arbitrarily normalised to be a constant as function of bombarding energy similar to the dependence of $V_{\sigma\tau}$ in Fig. 2.7. Interestingly, the height of the peak of the 0^+ level follows the same energy dependence as that of the V_τ in Fig. 2.7.

Therefore, to study GT transitions ($\Delta L = 0, \Delta S = 1, \Delta T = 1$) in nuclei it is important to measure at and near $0°$ using a bombarding energy between 100 and 500 MeV/u where strong excitation of levels by GT transitions is expected. For GT$^-$ transitions, the (p,n) and $(^3\text{He},t)$ reactions can be used. If the target ground state has isospin T_0 then three components of the GTR can be populated with: $T_0 - 1, T_0$, and $T_0 + 1$. For GT$^+$ transitions, the (n,p), $(d,^2\text{He})$ and $(t,^3\text{He})$ reactions can be used. If the target ground state has isospin T_0 then in this case only one component of the GTR can be populated with $T_0 + 1$.

2.3.1 GT Strength in fp-Shell Nuclei

Bethe et al. [23] and Fuller, Fowler and Newman (FFN) [24–27] recognised that (GT) transitions in fp-shell nuclei play an important role in determining the weak-interaction rates of processes taking place during the last few days of a heavy star in its pre-supernova stage. This led to renewed interest in determination of the GT strength distributions in fp-shell nuclei, and there were many studies, both experimentally and theoretically, of the GT strength in fp-shell nuclei in the last two to three decades. This is important for a number of reasons. The core of a heavy star with a mass of 10 solar masses and heavier is composed of fp-shell nuclei after having burnt its fuel and reached the pre-supernova stage. These fp-shell nuclei can capture electrons at the high temperatures in which the stars find themselves thus reducing the electron pressure that counteracts gravitational collapse precipitating the collapse that is followed by a cataclysmic supernova explosion. Furthermore, neutrino absorption cross sections by fp-shell nuclei are essential for understanding of nuclear synthesis in supernova explosions. The matrix elements that govern electron capture and neutrino absorption have been difficult to calculate in the shell model because of the open shell structure of fp-shell nuclei though there has been much progress recently with large basis shell-model calculations. Therefore, it is of the utmost importance to measure spin-isospin responses of fp-shell nuclei to gauge theoretical calculations.

For GT$^-$ transitions, the (p,n) and $(^3\text{He},t)$ reactions can be used, and for GT$^+$ transitions, the (n,p), $(d,^2\text{He})$ and $(t,^3\text{He})$ reactions can be used. The cross section can be written at momentum transfer $q = 0$ as [28, 29]:

$$\frac{d\sigma}{d\Omega}(q=0) = \frac{\mu_i \mu_f}{(\pi\hbar^2)^2}\left(\frac{k_f}{k_i}\right)\left(N_D^\tau |J_\tau|^2 B(F) + N_D^{\sigma\tau}|J_{\sigma\tau}|^2 B(GT)\right)$$

Here, μ_i and μ_f are the reduced masses in the initial and final channels, and k_i and k_f the momenta of the incoming and outgoing particles, respectively. N_D is the distortion factor for Fermi (τ) or GT ($\sigma\tau$) transitions, J_τ and $J_{\sigma\tau}$ are the volume integrals of the isospin and spin-isospin components of the central part of the NN interaction and B(F) and B(GT) are the transition strengths for the Fermi and GT parts of the transition, respectively. Note the similarity between this cross section of the (p,n) and $(^3\text{He},t)$ reactions at $q = 0$ and the neutrino absorption cross section given by:

$$\sigma = \frac{1}{\pi \hbar^4 C^3} \left[G_V^2 B(F) + G_A^2 B(GT) \right] \times F\left(Z, E_e\right) p_e E_e$$

Here, G_V and G_A are the vector and axial-vector coupling constants. This similarity indicates the importance of charge-exchange reactions at intermediate energies for determining B(GT) values. In the case of pure GT transitions, the cross-section equation simplifies to:

$$\frac{d\sigma}{d\Omega}(q = 0) = \frac{\mu_i \mu_f}{(\pi \hbar^2)^2} \left(\frac{k_f}{k_i} \right) N_D^{\sigma\tau} |J_{\sigma\tau}|^2 B(GT)$$

In principle, one can determine the B(GT) value directly from measuring the cross section at $0°$ and extrapolating it to $q = 0$ and calculating N_D and $J_{\sigma\tau}$. However, after these steps are taken usually the value is normalised to a calibrated B(GT) value for a known transition from β-decay. In this way, a unit cross section for B(GT) transitions could be defined.

2.3.2 Determination of GT⁻ Strength

In order to demonstrate the power of intermediate energy charge-exchange reactions, we show in the following the results of the $^{176}\text{Yb}(^3\text{He},t)^{176}\text{Lu}$ reaction [30]. Because of the low-lying GT transitions from ^{176}Yb to ^{176}Lu, ^{176}Yb has been considered for solar-neutrino detection to resolve the discrepancy between the observed neutrino flux from the Sun [31, 32] in comparison with the predicted neutrino flux by the Standard Solar Model [33]. The dominant part of the neutrino flux from the Sun is due to the $pp \rightarrow de^+\nu$ with a maximum energy of the neutrinos of 420 keV, which nicely matches with the energy needed for the GT transition to the lowest 1^+ level in ^{176}Lu. The GT transition to the second 1^+ level requires neutrinos with energies above 445 keV, which is beyond the maximum energy of the pp neutrinos; see left panel of Fig. 2.9, which shows the maximum neutrino energies from the different sources as arrows and gives the low-lying levels of ^{176}Lu.

In a (p,n) experiment [34] performed in parallel, the two levels at 195 and 339 keV could not be resolved and a summed B(GT) value of 0.32 ± 0.04 was reported. Since the pp neutrinos can only excite the 195 keV level, it was imperative to resolve these two levels. An experiment was performed [30] with the high-resolution Grand Raiden Spectrometer [35] at RCNP, Osaka with ^3He beam at an incident energy of 150 MeV/u. The two low-lying 1^+ levels could be resolved as

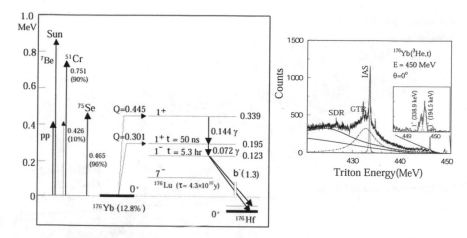

Fig. 2.9 Left panel: Level scheme and γ-ray tags for solar-neutrino detection by ^{176}Yb (12.8% natural abundance). All energies are in MeV. Right Panel: Triton energy spectrum from the ^{176}Yb(^{3}He,t)^{176}Lu reaction taken at 0° and incident energy 150 MeV/u. In addition to structures observed at high excitation energies, the two 1^+ states at low excitation energy are observed and resolved as displayed in the expanded inset; see [30] for details

Table 2.2 Gamow–Teller B(GT) of the levels observed in the ^{176}Yb(^{3}He,t)^{176}Lu [30]

E_x (MeV)	0.195 + 0.339 (p,n)	0.195	0.339 (^{3}He,t)
B(GT)	0.32 ± 0.04	0.20 ± 0.04	0.11 ± 0.02

can be seen in the inset of the right panel of Fig. 2.9. Their B(GT) values could be calibrated with the known ^{164}Dy weak matrix element. The obtained B(GT) values are listed in Table 2.2. The B(GT) of the higher level is found to be about half of that of the lower level, but their sum is in excellent agreement with the result obtained from the (p,n) experiment [34].

Almost an order of magnitude improvement of the resolution of detected scattered particles was obtained when the West-South (WS) beam line was constructed at RCNP, Osaka [36]. This allowed unprecedented high-resolution measurements at a ^{3}He incident energy of ~450 MeV using the momentum-dispersion technique. Beams from the Ring Cyclotron are transported to the target through the high-dispersive WS beam line; see Fig. 2.10. After charge-exchange reactions with target nuclei, the ejectile tritons are measured with Grand Raiden applying dispersion matching. A high resolution of around 30 keV has been achieved with thin targets. This provides excellent new opportunities to study GT strength in fp-shell nuclei in great detail.

To illustrate the excellent potential of this facility, we present here the study of the ^{58}Ni(^{3}He,t)^{58}Cu reaction with the aim to study the structure of the GTR in ^{58}Cu and decompose its isospin components. Starting from ^{58}Ni ground state with isospin $T_0 = 1$, one can, in principle, separate the isospin components by comparison among the spectra obtained by (n,p), (p,p') and (p,n) reactions on

Grand-Raiden
Spectrometer

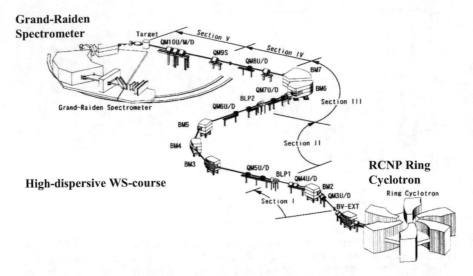

Fig. 2.10 Schematic layout of the experimental setup at RCNP allowing high-resolution measurements making use of dispersion-matching technique. Beams from the Ring Cyclotron are transported to the scattering chamber of the Grand-Raiden spectrometer through the high-dispersive WS beam line [36]

^{58}Ni. This is, however, very difficult in practice because of the high level density for populated states with different isospin. High energy resolution experiments are therefore necessary to be able to perform this decomposition. A successful attempt to decompose the isospin components of the GTR in ^{58}Cu was performed by Fujita et al. [37, 38].

The ^{58}Ni(^{3}He,t) reaction at 0° strongly excites $J^\pi = 1^+$ states with $T = 0$, 1 and 2 in ^{58}Cu starting from the $T_0 = 1$ ^{58}Ni ground state because of the $\Delta T = 1$ transfer in the charge-exchange process. In (p,p') and (e,e'), excited 1^+ ($M1$) states with $T = 1$ or 2 are observed as analogue states in ^{58}Ni. In Fig. 2.11, the various isospin components that can be populated in the (n,p), (p,p') and (p,n) reactions and their analogue relationships and isospin Clebsch-Gordan (CG) coupling coefficients are shown. Since ^{58}Ni ground state has isospin $T_0 = 1$, the square of the CG coefficient of isospin coupling leads to a transition strength ratio of $\sigma_{T=0}:\sigma_{T=1}:\sigma_{T=2} = 2:3:1$ for the $T_0 - 1$, T_0 and $T_0 + 1$ components, respectively; see Fig. 2.11. Because of the nuclear structure of ^{58}Ni, which corresponds to a large extent to 2 neutrons outside a ^{56}Ni core of 28 protons and 28 neutrons filling all shells up to and including $1f_{7/2}$, one can show that in this simple model the expected $J^\pi = 1^+$ B(GT) strength of 19.7 in ^{58}Cu is distributed among the three isospin components $T_0 - 1$, T_0 and $T_0 + 1$ with the ratio of 47%, 41% and 12%, respectively [37, 38].

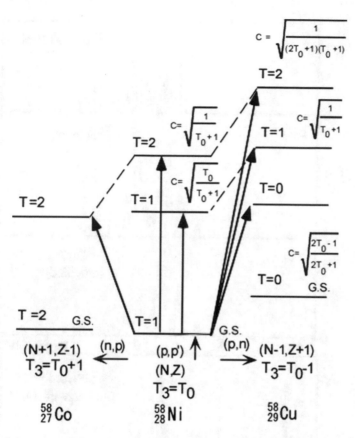

Fig. 2.11 Schematic representation of population of various isospin components of 1^+ levels in (n,p) [GTR], (p,p') [M1 resonance] and (p,n) [GTR] reactions starting from a target nucleus with isospin quantum number $T = T_3 = T_0$. Dashed lines connect analogue levels in the parent and daughter nuclei. Isospin CG coefficients that determine the relative excitation of the different components are shown. For ^{58}Ni $T_0 = 1$

In Fig. 2.12a, the $0°$ spectrum of the $^{58}\text{Ni}(^3\text{He},t)^{58}\text{Cu}$ reaction obtained at a bombarding energy of 450 MeV is shown [37, 38]. The energy resolution was 140 keV that was achieved before the WS beam line was constructed. Because of the strong excitation of $\Delta L = 0$ transitions at $0°$ and this high bombarding energy, mainly 1^+ levels are excited in addition to the isobaric analogue state (IAS) at 0.20 MeV excitation energy in ^{58}Cu. This spectrum should be compared to the 1^+ levels observed in $^{58}\text{Ni}(e,e')$. In Fig. 2.12b-1, the B(M1) strength distribution deduced from inelastic electron scattering measurement [39], in which an energy resolution of 30 keV was achieved, is shown. In order to compare with the spectrum obtained in the $^{58}\text{Ni}(^3\text{He},t)^{58}\text{Cu}$ reaction the B(M1) strength distribution was convoluted with the experimental energy resolution of the $(^3\text{He},t)$ experiment; see Fig. 2.12b-2. Before the final comparison is made, one notes that the relative

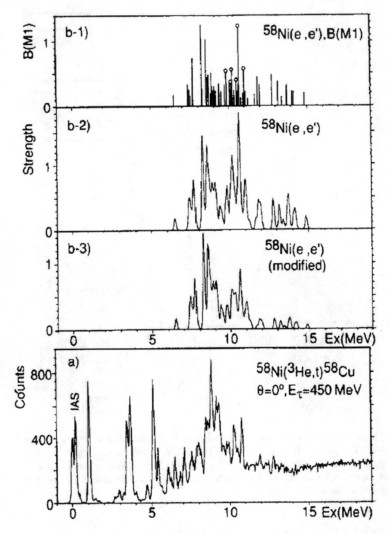

Fig. 2.12 Comparison between the $0°$ ^{58}Ni(^{3}He,t) spectrum (**a**), which is mainly GT strength, and the Ml strength distribution obtained in the ^{58}Ni(e,e') experiment [39]. (**b-1**) The B(M1) distribution deduced in the ^{58}Ni(e,e') experiment. (**b-2**) The reconstructed ^{58}Ni(e,e') spectrum after convoluting with the experimental energy resolution of the (^{3}He,t) experiment. (**b-3**) The same as (**b-2**), but $T = 2$ strength scaled down by a factor of 3. Since the IAS of the ^{58}Ni ground state is observed at 0.20 MeV in ^{58}C, the excitation energy in ^{58}Ni is shifted with 0.20 MeV relative to ^{58}Cu. This figure is from [37, 38]

excitation of the T_0 and $T_0 + 1$ components in ^{58}Cu is $\sigma_{T=1}:\sigma_{T=2} = 3:1$, whereas in ^{58}Ni it is $\sigma_{T=1}:\sigma_{T=2} = 1:1$; see Fig. 2.11. Therefore, one has to identify the $T = 2$ levels in ^{58}Ni and reduce their relative excitation by a factor 3 before the final comparison is made. This was done by comparing to the results of a

^{58}Ni(t,^{3}He)^{58}Co measurement performed at 25 MeV bombarding energy [40, 41] and a recent ^{58}Ni(n,p)^{58}Co measurement at 198 MeV bombarding energy [42], both of which populate only $T = 2$ levels. In the (t,^{3}He) measurement, six 1^+ levels populated by GT transitions were identified and consequently their analogues in ^{58}Ni were identified. These are marked with the small circles in Fig. 2.12b-1. The (n,p) measurement did not have sufficient energy resolution to resolve discrete levels. However, the $T = 2$ GT strength was found to exist continuously between 1 and 5 MeV excitation energy. This corresponds in ^{58}Ni to the excitation-energy region above 11.5 MeV. Therefore, the relative strength of the six 1^+ states and the region above 11.5 MeV were scaled down by factor of 3. The resulting spectrum is shown in Fig. 2.12b-3.

By comparing Figs. 2.12a and b-3 and knowing the $T = 2$ GT distribution from the (t,^{3}He) and (n,p) measurements, Fujita et al. were able to determine the B(GT) strength distributions for all three isospin components. These are shown in Fig. 2.13 after convoluting with the experimental energy resolution of the (^{3}He,t) experiment. It can be seen that the observed B(GT) strength is fragmented for all isospin components. Furthermore, the integrated strengths for the three isospin components are in better agreement with the predictions of the simplified shell-model calculations than with the predictions purely on basis of the CG coefficients.

With the construction of the WS beam line at RCNP, Osaka, it became possible to achieve much better energy resolution through use of the dispersion-matching

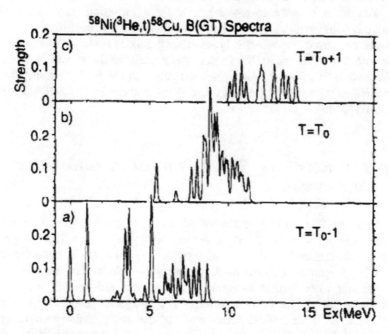

Fig. 2.13 The B(GT) strength distributions for the three isospin components as could be deduced from comparison of various probes as discussed in detail in the text [37, 38]

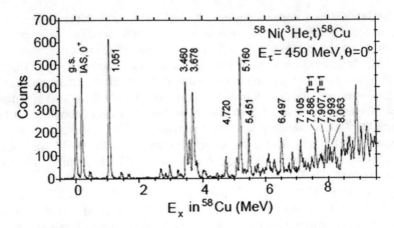

Fig. 2.14 A ^{58}Ni(^{3}He,t)^{58}Cu spectrum taken at $0°$ and 450 MeV incident energy is shown; the energy resolution achieved is about 50 keV. States with $L = 0$ character are indicated by their excitation energies [37, 38]

technique. In Fig. 2.14, a $0°$ spectrum of the ^{58}Ni(^{3}He,t)^{58}Cu reaction taken at a bombarding energy of 450 MeV is shown. A high resolution of 50 keV was achieved. By comparing this spectrum with that of Fig. 2.12a, one can observe the great improvement in resolving the low-lying levels and the fine structure of the GTR. With this achieved high-resolution, the (^{3}He,t) reaction becomes a powerful tool to study spin-isospin modes and, in particular, the GT strength in great detail. It has indeed been used very effectively in the last decade in studies of the spin-isospin modes with the facility at RCNP, Osaka. Many nuclei in the sd and fp shells have been studied in this period allowing the extraction of the B(GT$^-$) distributions in the final nuclei with very high energy resolution of about 35 keV full width at half maximum [43–45].

2.3.3 Determination of GT$^+$ Strength and Its Astrophysical Implications

In a heavy star that initially is formed of hydrogen gas but with a mass larger than ten solar masses (M > M$_\odot$), the collapse due to gravitational pressure increases the temperature in the star interior to $\sim 10^7$–10^8 K initiating thus a fusion of hydrogen through the weak process: $pp \rightarrow de^+\nu$. The thermal pressure keeps the star from further collapsing while all the hydrogen in the star interior is converted to ^{4}He through different reaction paths essentially converting: $4p \rightarrow \, ^4$He$2e^+2\nu + 26.7$ MeV. After about 10^6–10^7 years, hydrogen burning ends in the core of the star. The star starts contracting again under gravitational pressure and the temperature increases to 2×10^8 K resulting in a red giant wherein He-burning occurs in the core while hydrogen burning starts in the surrounding region at temperatures appreciably exceeding 10^7 K. After about 5×10^5 years, the processes of burning heavier elements at higher temperatures occur resulting in an onion-shell structure of the star with a core constituted mainly of Fe and Ni isotopes at a mass of around 1.4 $M_\odot$ surrounded by shells filled consecutively with Si, O, C, He and H. At this stage, gravitational pressure increases but is balanced by the pressure due to a degenerate electron gas which keeps the star from completely collapsing up to the Chandrasekhar limit, which is about 1.4 $M_\odot$. Above this limit, the electron gas pressure cannot keep the star from collapsing. The collapse starts when the temperature reaches T = 10^9 K and a density $\rho = 3 \times 10^7$ g/cm^3. This is accelerated by the neutronization (de-leptonization) of the star core through electron capture processes on nuclei of Fe and Ni isotopes. The reduction of Y_e, the electron-to-baryon ratio, through the electron capture process leads to reduction in the pressure of the degenerate gas and to accelerated collapse followed by a re-bounce from the star core resulting in a supernova explosion.

The electron-capture (EC) process on fp-shell and, in particular, Fe and Ni nuclei plays a dominant role during the last few days of the life (pre-supernova stage) of a heavy star [23]. The rate for EC is governed by the GT$^+$ strength distributions at low excitation energy; which are not accessible to β^+-decay. First estimates of EC rates in stellar environments were made in the early 1980s by FFN using a simple single-particle model [24–27]. In the late 1990s, shell-model codes were developed and new computer hardware became available that allowed to perform large-basis shell-model calculations [46]. This made it possible to calculate relevant GT strength distributions in fp-shell nuclei using a very large number of configurations. The results obtained in these large-scale shell-model calculations [46–50] marked deviations from FFN EC rates and indicated that the EC rates are in general smaller than was calculated by FFN [24–27]. These conclusions have led to revisiting pre-supernova models [51, 52]. The new evaluations resulted in a smaller mass of the iron core of the pre-supernova star and a larger value for the electron-to-baryon ratio Y_e as compared to calculations by FFN.

It became evident that stellar weak reaction rates should be determined with improved reliability compared to FFN. Large-scale shell model calculations provided the possibility to improve the prediction of EC rates by tuning these to reproduce GT$^+$ strength measured in, e.g., (n,p) reaction. Calculations of the EC rates relied on (n,p) data from TRIUMF, which have a rather poor energy resolution [42, 53–56]. The calculations showed agreement with the data after folding with the energy resolution of 1 MeV for the (n,p) reaction; see Fig. 2.15.

At low temperatures, the EC rates depend sensitively on the discrete structure of the low-lying GT strength. Therefore, it is important to resolve the low-lying GT transitions. To test this a case study was made to determine the low-lying 1$^+$ levels of ^{58}Ni in a reaction with which a much better energy resolution can be obtained compared to the (n,p) reaction [57, 58]. For this reason, the $(d,^2He)$ reaction was chosen which in addition to a better energy resolution offers other advantages; see

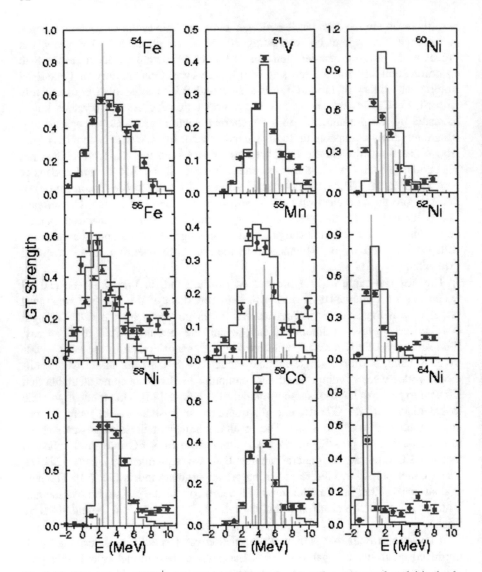

Fig. 2.15 Comparison of GT$^+$ data obtained with the (n,p) reaction on several nuclei in the fp-shell [42, 53–56] with calculated strength distributions making use of large-scale shell-model calculations [46]. The theoretical predictions are shown by discrete lines which are folded with the experimental energy resolution to obtain the plotted histograms; figure obtained from [46]

Fig. 2.16. In this figure, a schematic representation of the charge-exchange $(d,^2\mathrm{He})$ reaction is shown. The deuteron ground state is a $^3S_1$ state ($S = 1$) with isospin $T = 0$. The unbound di-proton system (isospin $T = 1$) is referred to as $^2\mathrm{He}$, if the two protons are in a relative $^1S_0$ state. The $^1S_0$ state dominates if the relative 2-proton kinetic energy $\varepsilon < 1$ MeV. Experimentally, an almost pure $^1S_0$ state can be

Fig. 2.16 Schematic representation of the $(d,^2\mathrm{He})$ charge-exchange reaction on a nucleus A with atomic number Z. The deuteron has $S = 1$, $T = 0$, and the unbound $^2\mathrm{He}$ has $S = 0$, $T = 1$

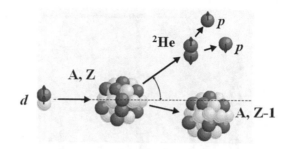

selected by limiting the relative energy of the di-proton system to less than 1 MeV, as in this case the contribution of higher-order partial waves is of the order of a few percent only. Furthermore, the reaction mechanism forces a spin–flip and isospin–flip transition $(\Delta S = \Delta T = 1)$. Therefore, the $(d,^2\mathrm{He})$ reaction is an (n,p)-type probe with exclusive $\Delta S = 1$ character and at small momentum transfers, mainly probes the GT^+ transition strengths. However, a disadvantage of this reaction near $0°$ is the tremendous background originating from d-breakup.

The $^{58}\mathrm{Ni}(d,^2\mathrm{He})$ reaction was studied at KVI using the Big-Bite Spectrometer [59] and the EuroSuperNova (ESN) focal-plane detection system [60–62]. The focal-plane detection system consists of two vertical-drift chambers (VDC), with the capability to determine the position of incidence of a scattered particle in two-dimensions, and a scintillation detector for particle identification and timing purposes. The angle of incidence can be determined from the positions of incidence on the two VDCs, and this can be converted to scattering angle of the particle through ray-tracing techniques. Because of the large background due to d-breakup the VDCs need to have fast readout which is realised with pipeline TDCs and decoding using imaging techniques. Furthermore, the VDCs have to have good double-tracking capability which is essential for detecting the two protons from $^2\mathrm{He}$ down to very small separation distances at the focal plane. Using the VDC information obtained in this way, good phase-space coverage for small relative proton energies is assured making it possible to select the $^1\mathrm{S}_0$ state of the two coincident protons.

In the $^{58}\mathrm{Ni}(d,^2\mathrm{He})$ experiment, a 170 MeV deuteron beam was used to bombard a self-supporting $^{58}\mathrm{Ni}$ target. The experimental details can be found in [57, 58]. Using the dispersion-matching technique, an energy resolution of 130 keV was achieved, which is an order of magnitude better than has been achieved in the (n,p) experiments [42, 53–56]. The experimental method and data-reduction techniques have been described in [63]. In Fig. 2.17, the double-differential cross section up to $E_x = 10$ MeV is shown and compared to data from the $^{58}\mathrm{Ni}(n,p)$ reaction obtained with 1-MeV energy resolution [42]. The superiority of the spectrum obtained with the $(d,^2\mathrm{He})$ is evident.

In order to determine the B(GT) distribution, the double-differential cross sections have to be determined as a function of scattering angle. The cross section due to $\Delta L = 0$ transfer has to be determined through a DWBA analysis in order

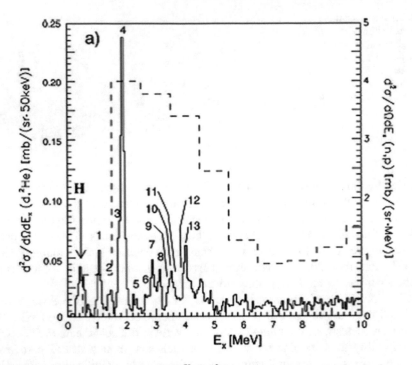

Fig. 2.17 A spectrum obtained with the ^{58}Ni(d,^{2}He) charge-exchange reaction in the angular range $\theta = 0°-1°$. The peak caused by contamination of the target with hydrogen is indicated by 'H'. The dashed histogram represents the spectrum from the (n,p) reaction at $\theta = 0°$ [42]

to subtract any $\Delta L = 2$ contribution which could be mediated by the tensor part of the NN interaction. The cross section is then extrapolated to $q = 0$ transfer (see Eq. given in last paragraph of Subsect. 2.3.1) and then the B(GT) value is determined.

For the region $E_x < 4.1$ MeV, the experimental angular distributions were obtained for all the peaks indicated with numbers in Fig. 2.17 by fitting the spectra with peaks at 6 angular bins in the range between $\theta = 0°$ and $6.5°$ [57, 58]. Peaks with $\Delta L = 0$ strength corresponding to $J^\pi = 1^+$ states can be identified based on their enhanced cross sections near a scattering angle of $0°$ as compared to more backward scattering angles (*i.e.*, $\theta_{c.m.} = 4°$).

As discussed earlier, the B(GT) values have to be normalised to a B(GT) value known, *e.g.*, from β-decay. This was performed for the (d,^{2}He) charge-exchange reaction using the GT transitions to low-lying states in ^{12}B and ^{24}Na [63]. This is illustrated in Fig. 2.18 for ^{24}Na. In this figure, the ^{24}Mg(d,^{2}He) spectrum [63] obtained at $\theta = 0.4°$ and 170 MeV incident energy (Fig. 2.18a) is compared to the ^{24}Mg(p,n) spectrum [64] obtained at $\theta = 0.2°$ and 135 MeV incident energy (Fig. 2.18b). These two reactions lead to mirror nuclei and therefore to mirror transitions with the same B(GT) values. Since the B(GT) values determined from (p,n) have been calibrated making use of several B(GT) values deduced from β-decay in even-

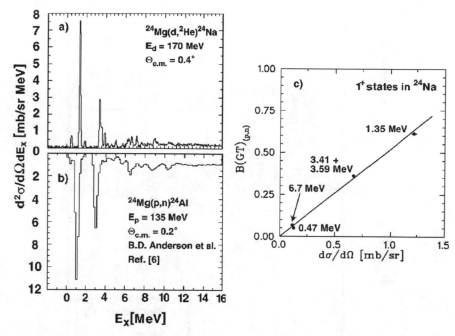

Fig. 2.18 (a) ^{24}Mg$(d,^2$He) spectrum [63] obtained at $\theta = 0.4°$ and 170 MeV incident energy. (b) ^{24}Mg(p,n) spectrum [64] obtained at $\theta = 0.2°$ and 135 MeV incident energy. (c) The B(GT) values determined from the ^{24}Mg(p,n) reaction for several levels in ^{24}Al are plotted versus the differential cross sections extrapolated to $q = 0$ for the mirror levels in ^{24}Na determined in the ^{24}Mg$(d,^2$He) reaction

even sd-shell nuclei, they could be used to calibrate the ones determined from the ^{24}Mg$(d,^2$He) reaction. In Fig. 2.18c, the B(GT) values determined from the ^{24}Mg(p,n) reaction for several levels in ^{24}Al are plotted versus the differential cross sections extrapolated to $q = 0$ for the mirror levels in ^{24}Na determined in the ^{24}Mg$(d,^2$He) reaction. There is a linear proportionality relationship between the two quantities indicating that extracting the B(GT) values for ^{58}Co as described above will be reliable.

In Fig. 2.19, a comparison is made between the B(GT$^+$) values obtained from the ^{58}Ni$(d,^2$He) experiment [57, 58] with those obtained from the (n,p) experiment [42]. The dots in the upper panel represent the B(GT$^+$) values determined for the numbered peaks as obtained from the fitting procedure. The grey histogram represents the B(GT$^+$) values as obtained from MDA taking 1-MeV energy bins. Up to an excitation energy of 4 MeV, the integrated B_{exp}(GT$^+$) strength obtained through the two different methods is consistent. Taking into account systematic errors, the integrated B_{exp}(GT$^+$) strength up to $E_x = 4$ MeV of 2.1 ± 0.4 from ^{58}Ni$(d,^2$He) is in agreement with the integrated B_{exp}(GT$^+$) strength of 2.7 ± 0.3 deduced from the (n,p) reaction. However, the two distributions differ significantly

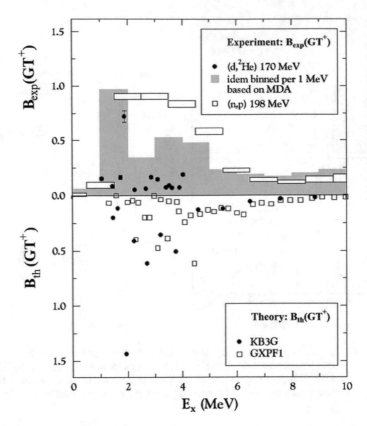

Fig. 2.19 Upper panel: the experimental GT$^+$ strength distribution displayed per peak (dots) and per bin of 1 MeV (grey histogram) and the results from the (*n,p*) reaction [42] displayed per bin of 1 MeV. Lower panel: the results from calculations using the KB3G interaction [47–50, 57, 58] and those using the GXPF1 interaction [65]

in shape. In the lower panel of Fig. 2.19, the results of large-scale shell-model calculations for the GT strength distribution are presented. The open squares show the results of the calculations performed by Honma et al. [65] which have been obtained using the GXPF1 effective interaction. The results displayed as dots have been obtained using the KB3G effective interaction [47–50]. It should be noted that the theoretical results have been calculated using a quenching factor of 0.74 for the GT operator. Both calculations reproduce fairly well the GT strength distribution obtained from the (*d*,^{2}He) reaction. However, the enhancement of the GT strength in the region around 1.9 MeV is only reproduced using the KB3G interaction. The results from the earlier calculations [46] (not shown here) do not reproduce as well the GT strength distribution obtained from the (*d*,^{2}He) reaction.

The experimental data shown in Fig. 2.19 were used to check whether the new results have an impact on the electron-capture rates in the stellar environment. The method to calculate electron-capture (EC) rates follows the formalism derived by

FFN [24–27]. The formula for determining the EC rate is:

$$\lambda_{ec} \approx \sum_i B_i(GT) \int_{\omega_1}^{\infty} \omega p (Q_i + \omega)^2 F(Z, \omega) S_e(\omega, T)\, d\omega$$

Here, $B_i(GT)$ is the Gamow–Teller strength distribution as determined from experiment or theory, ω and p are the energy and momentum of electrons, respectively, $F(Z,\omega)$ is the relativistic Coulomb barrier factor and $S_e(\omega,T)$ is the Fermi-Dirac distribution for an electron gas at temperature T. As representative values for the temperature, density and Y_e the conditions following silicon depletion are taken for the model labelled LMP listed in Table 2.2 of [51, 52], i.e. $T_9 = 4.05$ where T_9 measures temperatures in units of 10^9 K, $\rho = 3.18 \times 10^7$ gcm^{-3}, and $Y_e = 0.48$. This corresponds to the evolution of the core of a 25 M$_\odot$ star using the weak-interaction rates of [47–50].

At finite temperatures, there is a finite probability that nuclei are found in their excited states as well. In such a case, EC occurs not only through the ground state of a nucleus but also through the populated excited states. At the conditions present in a hot heavy star, the contribution of transitions starting from excited nuclear states is thus non-negligible. For the particular conditions mentioned above, the shell-model calculations predict that the rate for EC through excited states of ^{58}Ni is 50% the rate as calculated using only the ground-state contribution [47–50]. This contribution was fixed and added to the EC capture rate through the ground state of ^{58}Ni. The latter contribution was recomputed using the GT-strength distribution measured in the (n,p) and $(d,^2\text{He})$ reactions or computed using the various available theoretical model calculations including the one with KB3G interaction. The results of the rates are displayed in Fig. 2.20 as a function of temperature and for the values of density and Y_e mentioned above. Because of the high sensitivity to the Q-value for EC at low temperatures, the calculated theoretical rates use experimental excitation energies for the first two 1^+ states in ^{58}Co. The details of the GT-strength distribution (excitation energies and B(GT)-values) have a strong impact on the electron-capture rate. This can be seen in Fig. 2.20, where the rates calculated applying the results from the MDA (see Fig. 2.19) are substantially larger than the rates using the high-resolution data. They also differ especially at the low temperatures from the rates based on the (n,p) data [42]. Also, the rates calculated with the large-scale shell-model calculations with the KB3G interaction [57, 58] differ substantially from the earlier ones by Caurier et al. [46] and the ones based on the results of FFN [24–27].

The electron-capture rates in Fig. 2.20 are shown in Fig. 2.21 relative to the ones computed using the $(d,^2\text{He})$ data for the ground-state contribution. It can be seen that the rates based on the large-scale shell-model calculations with the KB3G interaction reproduce the rates based on the $(d,^2\text{He})$ data rather well over the whole temperature range only exceeding them by a constant factor of about 30%. These rates based on the KB3G interaction are also in good agreement with those based on the (n,p) data in the relevant temperature range $(3 < T_9 < 4)$. The rate based on the calculations of Caurier et al. [46] deviate by a factor 3 or

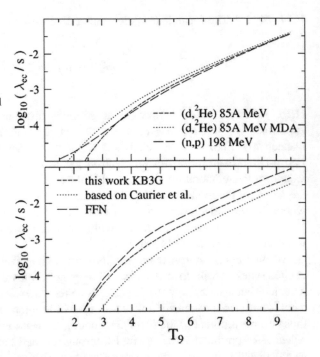

Fig. 2.20 Upper panel: EC rates based on B(GT) distributions from the (n,p) measurement [42] and the $(d,{}^2He)$ measurement [57, 58]. Lower panel: EC rates obtained from the shell-model calculations using the KB3G interaction [47–50, 57, 58], the earlier works of FFN [24–27] and Caurier et al. [46]; see text for details

more for the lowest temperatures. For the relevant temperature range, the FFN rate deviates by about a factor 2. It should be noted that this FFN rate [24–27] does not include any quenching factor. If such a factor is taken into account, it will cause the rates to become smaller. It is clear from Fig. 2.21, that high-resolution experimental information can provide opportunities to decide between different effective interactions used in large-scale shell-model calculations at zero temperature. This gives, of course, more confidence in these calculations at higher temperatures when they reproduce the data at $T = 0$.

It is important to note that there are many nuclei which contribute to the electron-capture rate in the pre-supernova stage. It is therefore quite important to check the shell-model predictions for other nuclei. In particular, the difference between the FFN [24–27] and the large-scale shell-model calculations calls for high-resolution determination of GT strength distributions relevant to the pre-supernova evolution using the $(d,{}^2He)$ reaction to test both calculations.

At KVI, several experiments have been performed to provide such tests and a few of the results have been published: ${}^{51}V(d,{}^2He){}^{51}Ti$ [66], ${}^{50}V(d,{}^2He){}^{50}Ti$ [67] and ${}^{64}Ni(d,{}^2He){}^{64}Co$ [68] . Several other target nuclei of importance for the EC process in the pre-supernova stage have also been investigated with the $(d,{}^2He)$ reaction such as: ${}^{56}Fe$, ${}^{57}Fe$, ${}^{61}Ni$ and ${}^{67}Zn$ [69]. The latter studies have not been published as separate journal articles, but summarised in a talk by Frekers [69]. As an example, we show in Fig. 2.22 a comparison between the spectrum of extracted $B(GT^+)$

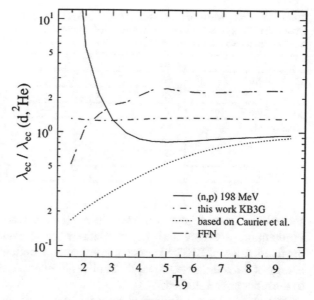

Fig. 2.21 Electron-capture rates relative to the rate obtained from the ^{58}Ni($d,^2$He) data. For the details on the various curves representing experimental and theoretical EC rates, see also caption of Fig. 2.20

Legend in figure:
— (n,p) 198 MeV
· – · this work KB3G
······· based on Caurier et al.
—· – FFN

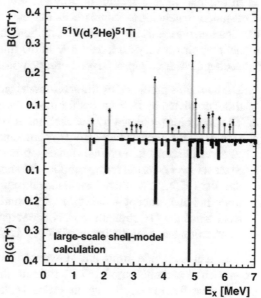

Fig. 2.22 Comparison between the B(GT$^+$) distribution deduced from the ($d,^2$He) reaction (top) and the results from the large-scale shell-model calculation (bottom) [47–50, 66]. The error bars in the top figure are statistical only

^{51}V($d,^2$He)^{51}Ti

large-scale shell-model calculation

strength obtained in the ^{51}V($d,^2$He)^{51}Ti reaction and the results from large-scale shell-model calculations.

The ground state of the initial (i) nucleus ^{51}V has $J_i^\pi = 7/2^-$, which allows GT transitions to $J_f^\pi = 5/2^-$, $7/2^-$, and $9/2^-$ in the final (f) nucleus ^{51}Ti. The shell-model calculation was performed in the complete fp shell employing the KB3G interaction [47–50, 66]. It should be noted that the shell-model calculations

Table 2.3 Comparison of the centroids (given in MeV) of the GT^+ strength distributions obtained from FFN [24–27], from large-scale shell-model calculations (SM) [47–50] and from the $(d,^2He)$ reaction on the various nuclei: ^{51}V [66], ^{50}V [67] and all others [69] (Exp.)

	Nucleus	FFN	SM	Exp.
Even-Even	^{56}Fe	3.8	2.2	1.9
	^{58}Ni	3.8	3.6	3.4
Odd A-Odd Z	^{51}V	3.8	4.7	4.1
Odd A-Odd N	^{57}Fe	5.3	4.1	2.9
	^{61}Ni	3.5	4.6	4.2
	^{67}Zn	4.4	–	3.4
Odd-Odd	^{50}V	9.7	8.5	8.8

show little strength above 6 MeV in agreement with the experimental results. Furthermore, the data and the calculations are in excellent agreement as far as the distribution of $B(GT^+)$ strength is concerned, both in magnitude and excitation energy of the fragmented strength. Similar observations could be made with respect to the other nuclei studied.

In order to have a feeling of how the results of the FFN and the large-scale shell-model calculations compare to the experimentally deduced GT^+ strength distribution, the centroids of these distributions are compared in Table 2.3 for several nuclei that play a very important role in the final steps of the pre-supernova stage preceding the star collapse. Several observations can be made:

1. The centroids predicted by the shell-model calculations are in general lower than those predicted by FFN except for the odd-A nuclei ^{51}V and ^{61}Ni.
2. The centroids predicted by the shell-model calculations are all in better agreement with the ones deduced from experiments except possibly for ^{51}V, where the FFN result seems to reproduce the experimental result slightly better.
3. Since a lower centroid energy implies enhanced EC rates, then one would expect that the EC rates predicted by the shell model will be higher thereby lowering the value of Y_e except in the cases of ^{51}V and ^{61}Ni. This is, however, not the full story since the GT strength in FFN is not quenched whereas in the shell-model distribution a quenching factor of $(0.74)^2$ is included.

One can conclude from the above comparison and the comparison of the experimental and shell-model GT^+ strength distributions of the different nuclei, case by case, that the large-scale shell-model calculations with the KB3G interaction reproduces these experimental distributions very well. Therefore, it is important to take the large-scale shell-model calculations and use them to calculate the EC rates in stellar scenarios at finite temperatures. These can be then used in large nuclear networks of pre-supernova models to determine the course that the star follows until the moment of collapse.

The collapse models describe the final collapse and the explosion phase. At the resulting high temperature, all reactions mediated by the strong and electromagnetic interactions are in equilibrium. The star collapse continues until the central density becomes substantially larger than the nuclear density. Then nuclear pressure slows down the collapse and finally stops it before it rebounds and a shock starts. The equation of state (EoS) plays a decisive role in this stage of the collapse. The important parameters of the EoS are the incompressibility of infinite nuclear matter and the coefficient of the symmetry term of the incompressibility, both of which could be determined from studies of the compression modes: the ISGMR and ISGDR.

2.4 Conclusions and Outlook

The study of the compression modes in stable nuclei furnished information on the incompressibility of nuclear matter. It has been fairly well determined from the ISGMR and ISGDR data on the spherical doubly-closed shell nuclei, ^{90}Zr and ^{208}Pb, by comparison of experimental results with microscopic calculations. The incompressibility of nuclear matter is determined in this way to be $K_\infty = 240 \pm 10$ MeV. The asymmetry term in the nuclear incompressibility has been addressed in measurements on a series of stable Sn and Cd isotopes. First, the values of the nucleus incompressibilities, K_A, were determined using the moment ratios $\sqrt{m_1/m_{-1}}$ for E_{ISGMR}. The data were then fitted with a quadratic function of $(N - Z)/A$ yielding a value $K_\tau = 550 \pm 100$ MeV, i.e. with a much larger uncertainty than for K_∞.

An experimental observation, which has not been discussed above, is the fact that experimentally observed excitation energies of the ISGMR in Sn and Cd fall substantially below those predicted with microscopic calculations using nucleon-nucleon interactions with a nuclear matter incompressibility as deduced from ^{90}Zr and ^{208}Pb, i.e. $K_\infty = 240 \pm 10$ MeV. This observed softness of the Cd and Sn isotopes has not yet been explained and it does require some further theoretical efforts as well as more experimental data on nuclei in different regions of the nuclear chart.

The determination of the asymmetry term in the expansion of the nucleus incompressibility K_τ requires studies of the compression modes in an isotopic chain spanning a wide range of $\delta = (N - Z)/A$ values. As mentioned above, this has been done recently for the stable Sn nuclei [16, 17] and a values of $K_\tau = -550 \pm 100$ was determined. It also was performed on Cd nuclei [70], where a value of $K_\tau = -555 \pm 75$ MeV was obtained. With the advent of radioactive ion beam facilities and the availability of exotic neutron-rich and neutron-deficient nuclei, it becomes possible to cover a wider range in δ-values. This will allow a more accurate determination of K_τ and possibly also to resolve the inconsistency of K_∞ obtained from ^{208}Pb and ^{90}Zr compared to that from Sn and Cd nuclei, because of the slight dependence of these two parameters when extracting them from a limited set of data. Furthermore, in these exotic nuclei new phenomena emerge such as pygmy resonances that have multipole strengths reflecting the collectivity due to the extra neutron-skin or proton-skin relative to the core.

The study of giant resonances in unstable nuclei is quite involved experimentally and has up until recently been restricted to the study of the isovector giant dipole resonance (IVGDR) in a number of nuclei [71, 72]. In pioneering experiments, Monrozeau et al. [73], Vandebrouck et al. [74, 75] and Bagchi et al. [76] measured the isoscalar giant resonances, in particular the ISGMR, ISGDR and ISGQR, in the exotic ^{56}Ni and ^{68}Ni nuclei and determined their properties, i.e. excitation energies, widths and fraction of EWSRs. These experiments were performed at GANIL for deuteron and alpha scattering in inverse kinematics making use of the MAYA active-target detector [77, 78]. In another pioneering experiment, Zamora et al. [79] performed an inelastic α-scattering experiment in inverse kinematics in the experimental storage ring (ESR) at GSI. A stable ^{58}Ni beam was used as a proof of principle of the method. The parameters of the ISGMR determined in this experiment were in perfect agreement with measurements made in normal kinematics on a ^{58}Ni target. Both these techniques, the active-target method and the storage-ring method, promise to be very effective in the study of isoscalar giant resonances in exotic nuclei in the future.

Pre-supernova models depend sensitively on electron-capture (EC) rates on fp-shell nuclei. In turn, GT$^+$ strength distributions in fp-shell nuclei play a decisive role in determining EC rates and thus provide input into modelling of explosion dynamics of massive stars. In principle, these GT$^+$ strength distributions can be determined experimentally through the high-resolution $(d,^2\text{He})$ reaction at intermediate energies. However, these new high-resolution $(d,^2\text{He})$ experiments should be considered as providing essential tests for the shell-model calculations at zero temperature. For finite temperatures as exist in heavy stars, it is important to perform large-scale shell-model calculations. This can yield the EC rates as function of T. These large-scale shell-model calculations lead to smaller EC rates than FFN as a result of the interplay between centroids and quenching of GT$^+$ strength and thus they lead to larger Y_e (electron-to-baryon ratio) and smaller iron core mass.

In addition, the use of radioactive-ion beams to study the GT strength distribution is of importance, especially for nuclei with $(N - Z)/A > 0.1$, where the GT strength moves to even lower excitation energy. At the moment, radioactive ion beams at intermediate energies where it is possible to study GT transitions are available at RIKEN and GSI, and they will become available at NSCL, FAIR, and EURISOL in the future. This will allow to determine GT$^\pm$ strengths in unstable sd and fp shell nuclei. This will provide excellent opportunities to determine electron-capture rates and neutrino-capture rates on key radioactive nuclei. Such a key nucleus is ^{60}Co, since in the FFN model, the most important EC rate is caused by the $^{60}\text{Co}(e^-,\nu)^{60}\text{Fe}$ reaction for a huge range of temperatures and densities during the pre-supernova evolution. This rate is greatly reduced in the shell-model calculations [47–50]. Future high-resolution $(d,^2\text{He})$ data taken in inverse kinematics could provide an accurate determination of the EC rate on ^{60}Co.

References

1. W. Bothe, W. Gentner, Z. Phys. **71**, 236 (1937)
2. A.B. Migdal, J. Phys. (USSR) **8**, 331 (1944)
3. G.C. Baldwin, G.S. Klaiber, Phys. Rev. **71**, 3 (1947)
4. M. Goldhaber, E. Teller, Phys. Rev. **74**, 1046 (1948)
5. H. Steinwedel, J.H.D. Jensen, Phys. Rev. **79**, 1019 (1950)
6. M.N. Harakeh, A. van der Woude, *Giant Resonances: Fundamental High-Frequency Modes of Nuclear Excitations*. (Oxford University Press, New York, 2001), and references therein
7. R. Pitthan, T. Walcher, Phys. Lett. B **36**, 563 (1971)
8. M.N. Harakeh et al., Phys. Rev. Lett. **38**, 676 (1977)
9. D.H. Youngblood et al., Phys. Rev. Lett. **39**, 1188 (1977)
10. M.N. Harakeh et al., Nucl. Phys. A **327**, 373 (1979)
11. J.P. Blaizot, Phys. Rep. **64**, 171 (1980)
12. S. Stringari, Phys. Lett. B **108**, 232 (1982)
13. R.C. Nayak et al., Nucl. Phys. A **516**, 62 (1990), and references therein
14. M. Uchida et al., Phys. Rev. C **69**, 051301 (2004)
15. G. Colò et al., Phys. Rev. C **70**, 024307 (2004)
16. T. Li et al., Phys. Rev. Lett. **99**, 162503 (2007)
17. T. Li et al., Phys. Rev. C **81**, 034309 (2010)
18. S.K. Patra et al., Phys. Rev. C **65**, 044304 (2002)
19. H. Sagawa et al., Phys. Rev. C **76**, 034327 (2007)
20. F. Petrovich, W.G. Love, Nucl. Phys. A **354**, 499 (1981)
21. C. Gaarde, in *Proc. Niels Bohr Centennial Conf., Copenhagen*, eds. by R. Broglia, G. Hageman, B. Herskind (North-Holland, Amsterdam, 1985)
22. J. Rapaport, E. Sugarbaker, Ann. Rev. Nucl. Part. Sci. **44**, 109 (1994), and references therein
23. H.A. Bethe et al., Nucl. Phys. A **324**, 487 (1979)
24. G.M. Fuller, W.A. Fowler, M.J. Newman, Astrophys. J. Suppl. Ser. **42**, 447 (1980)
25. G.M. Fuller, W.A. Fowler, M.J. Newman, Astrophys. J. Suppl. Ser. **48**, 279 (1982)
26. G.M. Fuller, W.A. Fowler, M.J. Newman, Astrophys. J. **252**, 715 (1982)
27. G.M. Fuller, W.A. Fowler, M.J. Newman, Astrophys. J. **293**, 1 (1985)
28. T.N. Taddeucci et al., Nucl. Phys. A **469**, 125 (1987)
29. I. Bergqvist et al., Nucl. Phys. A **469**, 648 (1987)
30. M. Fujiwara et al., Phys. Rev. Lett. **85**, 4442 (2000)
31. T.A. Kirsten, Nucl. Phys. B **77**, 26 (1999)
32. V.N. Gavrin, Nucl. Phys. B **77**, 20 (1999)
33. J.N. Bahcall, Nucl. Phys. B **77**, 64 (1999)
34. M. Bhattacharya et al., Phys. Rev. Lett. **85**, 4446 (2000)
35. M. Fujiwara et al., Nucl. Instrum. Methods Phys. Res. A **422**, 484 (1999)
36. T. Wakasa et al., Nucl. Instrum. Methods Phys. Res. A **482**, 79 (2002)
37. Y. Fujita et al., Phys. Lett. B **365**, 29 (1996)
38. Y. Fujita et al., Eur. Phys. J. A **13**, 411 (2002)
39. W. Mettner et al., Nucl. Phys. A **473**, 160 (1987)
40. F. Ajzenberg-Selove et al., Phys. Rev. C **30**, 1850 (1984)
41. F. Ajzenberg-Selove et al., Phys. Rev. C **31**, 777 (1985)
42. S. El-Kateb et al., Phys. Rev. C **49**, 3128 (1994)
43. Y. Fujita et al., Phys. Rev. C **67**, 064312 (2003)
44. Y. Fujita et al., Phys. Rev. Lett. **92**, 062502 (2004)
45. Y. Fujita, Nucl. Phys. A **805**, 408 (2008), and reference therein
46. E. Caurier et al., Nucl. Phys. A **653**, 439 (1999), and reference therein
47. K. Langanke, G. Martínez-Pinedo, Nucl. Phys. A **673**, 481 (2000)
48. K. Langanke, G. Martínez-Pinedo, At. Data Nucl. Data Tables **79**, 1 (2001)
49. K. Langanke, G. Martínez-Pinedo, Rev. Mod. Phys. **75**, 819 (2003)

50. A. Poves et al., Nucl. Phys. A **694**, 157 (2001)
51. A. Heger et al., Phys. Rev. Lett. **86**, 1678 (2001)
52. A. Heger et al., Astrophys. J. **560**, 307 (2001)
53. W.P. Alford et al., Nucl. Phys. A **514**, 49 (1990)
54. M.C. Vetterli et al., Phys. Rev. C **40**, 559 (1989)
55. W.P. Alford et al., Phys. Rev. C **48**, 2818 (1993)
56. A.L. Williams et al., Phys. Rev. C **51**, 1144 (1995)
57. M. Hagemann et al., Phys. Lett. B **579**, 251 (2004)
58. M. Hagemann et al., Phys. Rev. C **71**, 014606 (2005)
59. A.M. van den Berg, Nucl. Instrum. Methods. Phys. Res. B **99**, 637 (1995)
60. S. Rakers et al., Nucl. Instrum. Methods. Phys. Res. A **481**, 253 (2002)
61. M. Hagemann et al., Nucl. Instrum. Methods. Phys. Res. A **437**, 459 (1999)
62. V.M. Hannen et al., Nucl. Instrum. Methods Phys. Res. A **500**, 68 (2003)
63. S. Rakers et al., Phys. Rev. C **65**, 044323 (2002)
64. B.D. Anderson et al., Phys. Rev. C **43**, 50 (1991)
65. M. Honma et al., Phys. Rev. C **69**, 034335 (2004)
66. C. Bäumer et al., Phys. Rev. C **68**, 031303 (2003)
67. C. Bäumer et al., Phys. Rev. C **71**, 024603 (2005)
68. L. Popescu et al., Phys. Rev. C **75**, 054312 (2007)
69. D. Frekers, Prog. Part. Nucl. Phys. **57**, 217 (2006)
70. D. Patel et al., Phys. Lett. B **718**, 447 (2012)
71. A. Leistenschneider et al., Phys. Rev. Lett. **86**, 5442 (2001)
72. P. Adrich et al., Phys. Rev. Lett. **95**, 132501 (2005)
73. C. Monrozeau et al., Phys. Rev. Lett. **100**, 042501 (2008)
74. M. Vandebrouck et al., Phys. Rev. Lett. **113**, 032504 (2014)
75. M. Vandebrouck et al., Phys. Rev. C **92**, 024316 (2015)
76. S. Bagchi et al., Phys. Lett. B **751**, 371 (2015)
77. C.E. Demonchy et al., Nucl. Instr. Methods Phys. Res. A **583**, 341 (2007)
78. C.E. Demonchy, Nucl. Instr. Methods Phys. Res. A **573**, 145 (2007)
79. J.C. Zamora et al., Phys. Lett. B **763**, 16 (2016)

Chapter 3
Alpha Decay and Beta-Delayed Fission: Tools for Nuclear Physics Studies

P. Van Duppen and A. N. Andreyev

Abstract α decay and β-delayed fission are two important decay modes of heavy exotic nuclei. Experimental α decay and β-delayed fission studies deliver significant nuclear-structure information in regions of the nuclear chart with limited accessibility. This information is important to improve the predictability of contemporary nuclear models used for e.g. nuclear astrophysics calculations. The basic principles and the current understanding of α and β-delayed fission decay are introduced. Examples of recent experiments and their impact on the understanding of heavy nuclei are presented.

3.1 Introduction and Physics Motivation

As emphasized in the previous volumes of The Euroschool on Exotic Beams Lecture Notes (volume I–IV), nuclides with a proton-to-neutron ratio different from stable nuclei are ideal laboratories to study the strong and weak interaction acting in the nuclear medium (see e.g. [1]). Combining experimental and theoretical work improves our understanding and, in particular, increases the predictive power of nuclear models. The latter is essential as, in spite of the tremendous progress in

P. Van Duppen (✉)
Instituut voor Kern- en Stralingsfysica, Department of Physics and Astronomy, KU Leuven, Leuven, Belgium
e-mail: Piet.VanDuppen@kuleuven.be

A. N. Andreyev
Department of Physics, University of York, York, UK
e-mail: Andrei.Andreyev@york.ac.uk

© Springer International Publishing AG, part of Springer Nature 2018
C. Scheidenberger, M. Pfützner (eds.), *The Euroschool on Exotic Beams - Vol. 5*,
Lecture Notes in Physics 948, https://doi.org/10.1007/978-3-319-74878-8_3

radioactive ion beam research, certain regions of the nuclear chart, e.g. heavy nuclei close to the neutron drip line or the region of the superheavy elements, still remain poorly accessible for experimental studies. Understanding the nuclear properties is not only essential for nuclear physics but also for nuclear astrophysics and for applications.

Exotic nuclei can be investigated with several different methods like e.g. decay and in-beam studies, mass and laser spectroscopy, Coulomb excitation and nuclear reactions.

In the present Lecture Notes, we will discuss two radioactive decay modes: α decay and β-delayed fission (βDF). α decay, known since the beginning of the twentieth century, was one of the first 'nuclear' decay modes to be discovered. Fission was discovered in 1938. Interestingly, soon after the respective discoveries, a sufficiently detailed qualitative understanding of the decay mechanism was realized—α decay in 1928 [2–4] and fission literary a few months after its discovery [5, 6].

α decay and βDF happen in heavy nuclei, have quantum-mechanical tunneling through a potential barrier in common and are pivotal for nuclear physics research and other related fields. For example, in nuclear astrophysics studies, α-capture reactions (equivalent to the inverse α-decay process) are important for nucleosynthesis and βDF, together with other fission modes, determine the so-called "fission recycling" in the r-process nucleosynthesis (see e.g. [7, 8]).

These notes report on the current understanding of α and βDF decay and discuss recent examples of experimental studies and their impact on nuclear structure. They are not aimed to give a concise review of the theory of these processes, but rather to supply the essential elements necessary to understand the way nuclear-structure information is deduced from these experiments.

A general introduction on α decay starting from the well-known Geiger-Nuttall (GN) rule and its modern versions followed by a global discussion on semi-classical and microscopic approaches are presented. New insights in the success of the GN rule have surfaced recently and will be discussed along with selected examples where α decay played a key role.

βDF is a sub-class of beta-delayed particle emission processes. It is an exotic process coupling beta decay and fission. A special feature of βDF is that it can provide low-energy fission data (excitation energy of the fissioning nucleus <10 MeV) for very neutron-deficient and neutron-rich nuclei which do not fission spontaneously and which are difficult to access by other methods. Some of these nuclei have recently become reachable for experiments due to new development in production techniques of radioactive beams [9], especially in the lead region. Already these first exploratory experiments led to surprising discoveries which will be highlighted in these lecture notes.

This chapter is structured in the following way. Section 3.2.1 discusses the basics of the α-decay process while Sect. 3.2.2 presents its current understanding from a semi-classical and a microscopic viewpoint. The experimental methods are summarized in Sect. 3.2.3 and recent examples of α-decay studies are given in Sect. 3.2.4. Section 3.3.1 gives an introduction in low-energy fission while the mechanism of βDF is discussed in Sect. 3.3.2. An overview of the production methods of βDF nuclei is presented in Sects. 3.3.3–3.3.5 discuss recent results obtained at radioactive beam facilities. A discussion on the βDF rates and their use to investigate the fission barrier height is presented in Sect. 3.3.6. A summary and conclusion are given in Sect. 3.4.

3.2   Alpha Decay

3.2.1   Basics of the α-Decay Process

In this section, the basics of the α-decay process will be shortly reviewed. More detailed descriptions and discussions are available from a number of excellent nuclear-physics and nuclear-chemistry textbooks (see e.g. [10–12]), review papers [13] and in the previous volumes of these lecture notes series [14, 15].

Every radioactive decay process is characterized by its energy balance and its transition probability. The α decay of the parent nucleus ($^A_Z X_N$) to the daughter nucleus $^{A-4}_{Z-2} Y_{N-2}$ can be represented by:

$$^A_Z X_N \rightarrow ^{A-4}_{Z-2} Y_{N-2} + ^4_2 He_2 + Q_\alpha \qquad (3.1)$$

where A, Z and N are, respectively, the atomic mass number, proton number and neutron number of the α-decaying nucleus. The Q_α value represents the negative of the α-particle binding energy of the parent nucleus and can be obtained from mass differences or differences in binding energy of the parent nucleus and the sum of the daughter nucleus and the α particle:

$$Q_\alpha = (M(^A_Z X_N) - (M(^{A-4}_{Z-2} Y_{N-2}) + M_\alpha))c^2 \qquad (3.2)$$

$$Q_\alpha = (BE(A-4, Z-2) + 28.3\,MeV) - BE(A, Z) \qquad (3.3)$$

where M represents the mass and BE the total binding energy expressed in MeV. As the differences in electron binding energies between parent and daughter are small, the nuclear masses can be replaced by atomic masses which are tabulated in the Atomic Mass Evaluation (AME) tables [16]. When the α-binding energy becomes negative, resulting in a positive Q_α value, the nucleus can undergo spontaneous α decay. As shown in e.g. Fig. 4.2 from [10], along the chart of nuclides this happens for nuclei with mass atomic number $A \geq 150$, though, a small island of alpha-decaying nuclei also exists north-east of doubly-magic 100Sn. However, because of the strong (exponential) dependence of the α-decay probability on the Q_α-value, α decay only becomes the dominant decay mode for some of the more heavy nuclei. Interesting to note is that even the doubly magic ^{208}Pb nucleus is unstable with respect to α decay with a $Q_\alpha = 517.2(1.3)$ keV [17]. Based on the Geiger-Nuttall rule (see further) this would lead to a partial α-decay half life of

$\sim 10^{123}$ years, much beyond the age of the Universe which justifies the notation that ^{208}Pb is a stable nucleus. The longest half-life measured for an α-emitting nucleus has been obtained for ^{209}Bi where, using bolometric techniques, a half life of $1.9(2) \times 10^{19}$ years and an α-decay energy of 3.137 MeV were measured [18].

α decay is a two-body process and the energy released is shared between the α particle and the recoiling nucleus following conservation of energy and momentum. The relation between Q_α and the kinetic energy of the α particle (E_α) is given as:

$$Q_\alpha = E_\alpha \left(\frac{M_d + M_\alpha}{M_d} \right) \tag{3.4}$$

$$\sim E_\alpha \left(\frac{A}{A - 4} \right) \tag{3.5}$$

where M_d is the mass of the daughter nucleus. After α decay, the typical recoil energy is about 100 keV for an $A = 200$ parent nucleus and a Q_α value of 5 MeV. The influence of the recoil effect has to be considered carefully in experiments with e.g. thin α sources as part of the daughter nuclei recoils out of the source sample risking to influence the measurements (see Sect. 3.2.3).

From the half life of the nucleus ($T_{1/2}$) and the α-decay branching ratio (b_α), the partial α-decay half life ($T_{1/2,\alpha}$) or mean lifetime (τ_α), decay constant (λ_α) and decay width (Γ_α) can be deduced:

$$T_{1/2,\alpha} = \left(\frac{T_{1/2}}{b_\alpha} \right) = \tau_\alpha \, ln2 \tag{3.6}$$

$$\tau_\alpha \Gamma_\alpha = \frac{\Gamma_\alpha}{\lambda_\alpha} = \hbar \tag{3.7}$$

A connection between the α-decay energy and the partial α-decay half-life was first established by Geiger and Nuttall who discovered a linear relationship between the logarithm of the α-decay constant and the logarithm of the range of α particles in matter [19, 20]. This led to the well known Geiger-Nuttall (GN) rule for α decay which, in its simplified form, can be written as:

$$LogT_{1/2,\alpha} = \frac{A}{\sqrt{Q_\alpha}} + B \tag{3.8}$$

where A and B are constants, deduced from a fit to the experimental data. The first successful theoretical explanation of this dependence was given by Gamow [2] and independently by Condon and Gurney [3, 4], who explained α decay as the penetration (tunneling) through the Coulomb barrier. This was the first application of quantum mechanics to a nuclear-physics problem. This so-called semi-classical description of the α-decay process assumes that the α particle is 'pre-formed' inside the nucleus with a certain probability and it 'collides' with the Coulomb barrier. Upon every collision, it has a finite probability to tunnel through the barrier

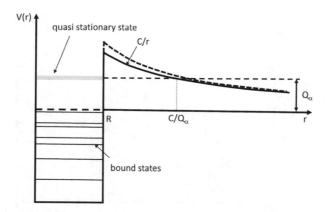

Fig. 3.1 A one-dimensional representation of the α-decay process. The full line shows the nuclear square well and Coulomb potential (see Eqs. (3.10) and (3.11)). R is the nuclear radius and C $= \frac{2(Z-2)e^2}{4\pi\epsilon_0}$ represents the constant due to the Coulomb interaction, with ϵ_0 the vacuum permittivity. The dashed line shows the sum of the Coulomb potential and the centrifugal part of the potential $(C/r + l_\alpha(l_\alpha + 1)\hbar^2/2\mu r^2)$ with l_α representing the transferred angular momentum in the α-decay process and μ the reduced mass of the α particle and daughter nucleus

[10–12]. Figure 3.1 gives a schematic representation. The α particle moves in an attractive, effective potential created by the strong interaction (V_N) and the Coulomb interaction (V_C) and is quasi-bound by the Coulomb barrier (details are given in Sect. 3.2.2). The square well potential (V_N) and the Coulomb interaction as deduced from a spherical charge distribution are approximations; however, they give a reasonable description of the basic features of the α-decay process.

As α decay involves quantum mechanical tunneling through the Coulomb barrier, its partial half life or transition probability depends exponentially on the Q_α value as represented by the GN rule. Figure 3.2 shows the logarithm of the experimental partial α-decay half lives (in s) for the even-even Yb-Ra nuclei with neutron number $N < 126$ as a function of $Q_\alpha^{-1/2}$ (in MeV$^{-1/2}$). For most of the isotopic chains considered, the GN rule gives a reliable description except when crossing the magic $N = 126$ shell and, as will be discussed later, for the case of the polonium ($Z = 84$) isotopic chain. Note that the A and B parameters are dependent on Z and experimental data suggest a linear dependence on Z as shown in Fig. 3.3. This was already introduced as a generalization of the GN rule by Viola and Seaborg [21]. Recent studies, which investigate the Geiger-Nuttall rule from a microscopic perspective, give rise to different parameterizations or dissimilar Z-dependencies of the A and B parameters. However, as will be discussed further, the basic linear $\log(T_{1/2})$ versus $Q_\alpha^{-1/2}$ dependence, deduced from systematic studies remains (see e.g. [22–24]).

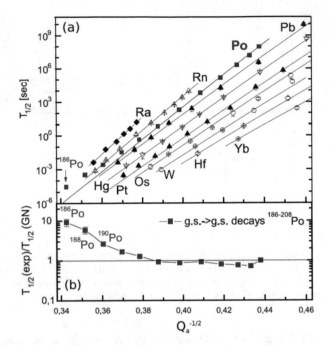

Fig. 3.2 (**a**) Geiger-Nuttall plot for selected isotopic chains. The straight lines show the description of the GN rule from which the $A(Z)$ and $B(Z)$ values are fitted for each isotopic chain. The results are presented in Fig. 3.3. (**b**) The deviation of the experimental α-decay half-lives from those predicted by the GN rule for the light Po isotopes. Adapted from [25]

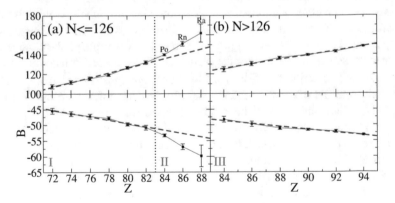

Fig. 3.3 (**a**) The coefficients $A(Z)$ and $B(Z)$ for even-even nuclei with $Z \leq 82$ (region I) and, $Z > 82$ and $N \leq 126$ (region II). (**b**) Same as (**a**) but for nuclei with $Z > 82$ and $N > 126$ (region III). The red dashed lines are linear fits through the data of region I (**a**) and III (**b**). Adapted from [25]

Next to the ground-state to ground-state α decay also decay to excited states can happen, giving rise to what is referred to as fine structure in the α-decay energy spectra. Studying the fine structure in the α decay reveals the energies of excited states, while from the relative intensities for different fine structure decay branches information on the structure of different states can be obtained (see e.g. Sect. 3.2.4). As angular momentum needs to be conserved, the α particle might have to carry angular momentum depending of the spin difference between the parent and the daughter states. As in the ^{4}He nucleus the protons and neutrons are coupled to spin and parity $I^\pi = 0^+$, the angular momentum carried away by the α particle is solely of orbital character which introduces a parity selection rule. If the α particle's orbital angular momentum is represented by l_α, the parity change associated with the α decay should then be $(-1)^{l_\alpha}$. For example, the α decay from the 0^+ ground state of an even-even nucleus towards an excited 2^+ state requires the α particle to take away two units of angular momentum and is allowed, however the decay towards a 1^+ state is forbidden.

α decay involving non-zero orbital angular momentum gives rise to an effective potential barrier consisting of a Coulomb and a centrifugal part that is proportional to $l_\alpha(l_\alpha + 1)\hbar^2/2\mu r^2$. This modified potential, shown in Fig. 3.1 by the dashed line, becomes higher and the barrier thicker, which in turn decreases the α-decay probability. As will be shown in Sect. 3.2.4, the study of the fine structure in e.g. the neutron-deficient lead region has allowed to identify low-lying 0^+ states and supported the appearance of shape coexistence at very low excitation energy.

3.2.2   Theoretical Approaches to the α-Decay Process

3.2.2.1   Semi-Classical Approach

In this section we will discuss the semi-classical description of the α-decay process. Following the approach similar to [26], the parameters of the GN rule will be extracted and, more generally, an expression will be deduced to evaluate the α-formation probability from experimental data.

The α-decay constant can be written as:

$$\lambda_\alpha = p_\alpha \nu T \tag{3.9}$$

where p_α represents the probability to form an α particle inside the nucleus, ν the collision frequency of the α particle's assault against the barrier and T the tunneling probability through the barrier. The one-dimensional potential, as shown in Fig. 3.1, gives all the ingredients necessary to discuss the α-decay process.

The tunneling probability can be calculated using different levels of complexity in the assumed potentials (see for example [10, 11] and for a recent review see [24]). Following [26] and neglecting the centrifugal barrier, thus only considering l_α=0, the potential can be written as

$$V(r) = -V_N + C/R \quad (r < R) \tag{3.10}$$

$$V = \frac{C}{r} \quad (r > R) \tag{3.11}$$

$$C = \frac{2(Z-2)e^2}{4\pi\epsilon_0} \tag{3.12}$$

where Z is the parent's proton number, e is the electron charge (in Coulomb), ϵ_0 the vacuum permittivity and R the radius of the daughter nucleus (see Fig. 3.1). The potential in the nuclear interior ($r < R$) is dominated by the nuclear part, while outside the Coulomb part takes over. Note that at this stage a constant Coulomb potential was assumed inside the nucleus which is an over-simplification. One can then calculate the α-decay width using the Wentzel-Kramers-Brillouin (WKB) approach as:

$$\Gamma_\alpha = \left(\frac{p_\alpha \hbar^2 K}{2\mu R}\right) exp\left(-2\int_R^{C/Q_\alpha} k(r)dr\right) \tag{3.13}$$

where μ represents the reduced mass of the α particle and daughter nucleus, while K and k(r) are the wave numbers in the internal ($r < R$) and barrier regions ($R < r < C/Q_\alpha$), respectively,

$$K = \left(\frac{2\mu}{\hbar^2}\left(Q_\alpha + V_N - \frac{C}{R}\right)\right)^{1/2} \tag{3.14}$$

$$k(r) = \left(\frac{2\mu}{\hbar^2}\left(\frac{C}{r} - Q_\alpha\right)\right)^{1/2} \tag{3.15}$$

leading finally to an expression of the partial α-decay half life,

$$T_{1/2,\alpha} = \frac{2ln2}{p_\alpha}\left(\frac{\mu R}{\hbar K}\right) exp\left(2\int_R^{C/Q_\alpha} k(r)dr\right) \tag{3.16}$$

One recognizes the α-particle formation probability (p_α), a measure for the collision time (the term between the first brackets) and the tunneling probability as represented by the exponential function of the integral. Buck et al. were able to describe experimental α-decay half lives by adjusting the inner radius (R) to reproduce the Q_α value using the Bohr-Sommerfeld quantization condition:

$$R = \frac{\pi}{2} \frac{G+1}{K} \tag{3.17}$$

Here, the 'global' quantum number G equals $2n + l_\alpha$, with n is the harmonic oscillator quantum number and l_α the angular momentum transfer. This description assumes that the α particle is constructed from the valence protons and neutrons in the heavy nucleus (see Sect. 3.2.2.2). The value of G was also adapted to achieve the best agreement between the calculated and experimental half lives. For example, from a simple perspective, the α decay of ^{212}Po to ^{208}Pb involves the clustering of two protons above the $Z = 82$ ($n_\pi = 5$) and two neutrons above the $N = 126$ ($n_\nu = 6$) shell, which, for $n = n_\nu + n_\pi$, results in values of G around 22 [26]. By adjusting V_N to 140 MeV and assuming $p_\alpha = 1$, a good overall agreement between experimental and calculated values was obtained provided G was changed from 22 to 24 for nuclei with $N \leq 126$ and $N > 126$, respectively [26, 27]. The influence of more 'realistic' barriers, including a centrifugal barriers, was also evaluated [28]. The success of these and other cluster-based calculations supports the assumption of α-cluster formation inside nuclei.

Before turning to a more microscopically-based description of the formation amplitude, let us evaluate the tunneling through the barrier by calculating the integral in (3.16). This leads to the following general relation [10, 11],

$$LogT_{1/2,\alpha} = aZ\sqrt{\frac{\mu}{Q_\alpha}} + b\sqrt{\mu Z} + c \tag{3.18}$$

This expression resembles very much the Geiger-Nuttall rule (3.8), however, the linear dependence of the coefficients a, b and c on Z, as extracted from the experimental data [21, 25], is not observed.

Rasmussen introduced a reduced α-decay width deduced from a combination of the experimental α-decay constant and a calculated tunneling probability [30]. This gave rise to:

$$\lambda_\alpha = \delta^2 \frac{T}{h} \tag{3.19}$$

The reduced α-decay width, δ^2, is expressed in keV, and is used in many papers to calculate hindrance factors in the fine structure α-decay studies.

Recently, a general description of α and cluster decay using the R-matrix approach was proposed [31, 32]. From R-matrix theory, one obtains the following expression for the decay half life,

$$T_{1/2,\alpha} \simeq \frac{ln2}{v}\left|\frac{H}{RF_\alpha(R)}\right|^2 \tag{3.20}$$

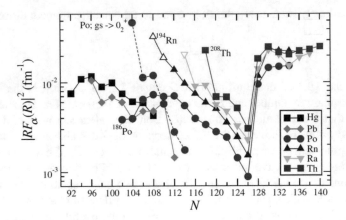

Fig. 3.4 α-particle formation probability for the decays of even-even isotopes in the mercury to thorium region as a function of neutron number of the parent nucleus. The values were obtained using Eq. (3.20) combined with experimental Q_α and $T_{1/2,\alpha}$ values. The symbols connected by a dashed line show the $|RF_\alpha(R)|^2$ values for fine structure α decays of polonium isotopes to the 0_2^+ states in the corresponding lead daughter nuclides. Adapted from [29]

with H the Coulomb-Hankel function [33], v the outgoing velocity of the α particle and $|RF_\alpha(R)|^2$ the α-formation probability at radius R. Note that the velocity is equivalent to $\hbar K/\mu$ in (3.16). Qi et al. realized that the Coulomb-Hankel function, that describes the tunneling process, can be approximated by an analytical expression and they obtained the following global expression for the α-decay half life,

$$LogT_{1/2,\alpha} = a2Z\sqrt{\frac{\mathcal{A}}{Q_\alpha}} + b\sqrt{2Z\mathcal{A}(A_d^{1/3} + 4^{1/3})} + c \qquad (3.21)$$

with A_d the atomic mass number of the daughter nucleus and $\mathcal{A} = 4A_d/(A_d + 4)$ proportional to the reduced mass. Values for the constants a, b and c are reported [31, 32], but important to note is that a and b are of similar magnitude, that a is positive, and b and c are negative with a value of $c = -21.95$. Again the GN rule dependence on Q_α is reproduced; however, a similar $\sqrt{Z}$ dependence of the second term as in (3.18) is observed. Expression (3.20) allows to extract the experimental formation probability using the experimental Q_α values and $T_{1/2,\alpha}$ values. This gives a more precise and unambiguous assessment of the clustering process compared to the reduced α-decay width from [30]. Figure 3.4 shows the formation probability for the even-even nuclei from mercury to thorium. A behavior similar to the reduced α-width plots (see e.g. Fig. 3 from [34]) is observed. In [25] it was discussed that the GN rule is broken as soon as the $\log|RF_\alpha(R)|^2$ dependence on neutron number is not linear. Therefore a separate GN fit for $N > 126$ and $N < 126$ is performed (see Fig. 3.3).

It was furthermore underlined that the previously noticed success of the GN rule to describe the available data was somehow due to the limited data range within each isotopic series. A notorious exception are the polonium isotopes where a deviation from the GN rule up to a factor of ten has been reported for the most neutron deficient isotope ^{186}Po (see Fig. 3.2). Because of the large data range— from ^{218}Po ($N = 134$) to ^{186}Po ($N = 102$)—deviations from the linear behavior in the formation probability could be observed. In addition, as will be discussed later, the fine structure in the α decay and nuclear-structure effects reveal an even larger discrepancy.

Finally, the linear Z-dependence of the $A(Z)$ and $B(Z)$ parameters in the GN rule deserves some further attention. It can be easily shown that this leads to a critical Q_α value where the α-decay half lives are equal irrespective of Z: all lines from the GN rule converge to the same point as can be seen by extrapolating the linear fit curves in Fig. 3.2. Using the coefficients reported in [25], for the isotopes with $N < 126$, this corresponds to $Q_\alpha = 20.2\,\mathrm{MeV}$ and $\log T_{1/2,\alpha} = -21.5$. This is an un-physical effect and reinforces the limited validity range of the GN rule or other adaptations. Interesting to note, however, is that the critical $\log T_{1/2,\alpha}$ value is close to the logarithm of typical semi-classical collision time of -21.1 (see [11]). By invoking similar conditions using the Universal Decay Law approach [31, 32] gives a value of -21.95. A closer look to the expressions from [26] (see Eq. (3.16)) and of the Universal Decay Law [31, 32] (see Eq. (3.21)) reveals that this condition corresponds to $Q_\alpha \simeq (2Z_d)/(A_d^{1/3} + 4^{1/3})$ for the former and R $= C/Q_\alpha$ for the latter. These are Q_α values corresponding to the top of the one-dimensional Coulomb barrier.

3.2.2.2 Microscopic Description

Determining the formation probability of an α particle inside a heavy nucleus from basic principles is challenging. Several microscopic calculations using shell model and cluster formation approaches, based on the early work by Mang [35] and Fliessbach et al. [36], have been undertaken. In these calculations the α-particle formation probability at the surface of the daughter nucleus is calculated starting from the single-particle shell-model wave functions. Shell-model calculations applied for nuclei around the doubly magic ^{208}Pb nucleus reveal that the calculated α decay width underestimates the experimental value at least by one order of magnitude [36]. Further improvements complementing shell-model with cluster-model approaches in the framework of R-matrix calculations [37] to determine the cluster formation probability, lead to consistency between the calculated and the experimental α-decay width of ^{212}Po [33, 38]. Following the description of [33, 38, 39], the key ingredient is a careful evaluation of the formation amplitude that represents an overlap integral of the parent α-decaying state with the daughter nucleus coupled to an α particle [25, 39]. Several methods have been developed and applied to describe the α-decay process from a microscopic view point, including deformed nuclei [40], but a detailed review of such efforts is outside the scope of the present lecture notes and we refer the reader to [24, 33, 41]. One general conclusion from these works is that they support the large degree of clusterisation happening inside the nucleus which explains the success of the semi-classical approaches where cluster formation is explicitly assumed.

A lot of the work has been concentrated on the α decay of ^{212}Po to the doubly magic nucleus ^{208}Pb [33, 38] which is considered as a text-book example of the fastest α decay observed. Recently, however, the super-allowed character of the α decay of ^{104}Te has been investigated theoretically [42]. This work was triggered by the experimental data obtained in the ^{109}Xe-^{105}Te-^{101}Sn α-decay chain [43, 44] (see Sect. 3.2.4) and the importance of these studies was also emphasized in [14]. As the protons and neutrons that form the α particle are occupying similar single-particle orbitals for the $N = Z$ nuclei (around ^{100}Sn) but different ones for ^{212}Po, an increase in proton-neutron correlation and cluster formation probability is expected. The calculations reported in [42] support this interpretation.

In closing this part of the notes, we mention that several attempts have been undertaken to develop a common description of α decay, cluster decay (whereby the parent nucleus emits a heavier nucleus like e.g. ^{14}C) and fission. The reader is referred to [45, 46] and references therein.

3.2.3 Experimental Approaches and Observables

The α-decaying medium heavy and heavy isotopes are typically produced using heavy-ion fusion evaporation reactions at beam energies around and above the Coulomb barrier, or high-energy proton induced spallation reactions. In order to purify the reaction products from the primary and possibly other secondary beams, two complementary approaches are used: the 'in-flight' separation technique and the 'Isotope Separator On Line (ISOL)' technique. Heavy-ion beams are used for the former as 'in-flight' separation makes use of the reaction kinematics, while high-energy protons impinging on thick actinide targets (^{238}U or ^{232}Th) are used for the latter. α-decay studies have been performed at several different in-flight recoil separators like SHIP and TASCA at GSI (Germany), RITU and MARA at JYFL (Finland), FMA and AGFA at Argonne, BGS at Berkeley (USA), GARIS/GARIS2 at RIKEN (Japan), VASSILISSA/SHELS and DGFS at Dubna (Russia) (for a review on these facilities, see [47]). The two main ISOL-based facilities where α-decay studies are performed are ISOLDE [48] and TRIUMF [49]. For completeness, we note interesting results on the production and decay properties of suburanium isotopes following projectile fragmentation of ^{238}U at 1 GeV/u beams [50, 51]. Details on the ISOL and fragmentation approach have been described in previous chapters of these lecture notes series [9, 52].

In Sect. 3.3.5, an experimental set-up used at ISOLDE-CERN for α and beta-delayed fission studies will be presented. In this section, dedicated to α decay, we will demonstrate the use of the method of α-α correlations which is extensively applied in modern experiments to study α decay and e.g. to synthesize new isotopes and elements. As an example, we discuss the α-decay study performed at the SHIP velocity filter at GSI to identify the new isotope ^{194}Rn [55]. The SHIP velocity filter [53, 56, 57] is shown schematically in Fig. 3.5. The nuclei of interest were produced in the heavy-ion fusion-evaporation reaction ^{52}Cr $+$ ^{144}Sm $\rightarrow$ ^{194}Rn $+$ 2n (evaporation of two neutrons) at several beam energies between 231 and 252 MeV and a typical intensity of 500–700 pnA. Eight enriched ^{144}Sm targets of 400 μg/cm^2 thickness were mounted on a rotating target wheel. The target rotation allows to spread the incoming beam over a larger target area to reduce the target temperature. This is especially necessary when using high beam intensities of heavy ions to reach e.g. the super heavy elements (see further). SHIP acts as a velocity filter using electrical and magnetic fields in a perpendicular configuration like a Wien filter. Due to the difference in kinematical properties of the primary beam, passing through the thin target, and of the nuclei of interest, recoiling from the target, they can be

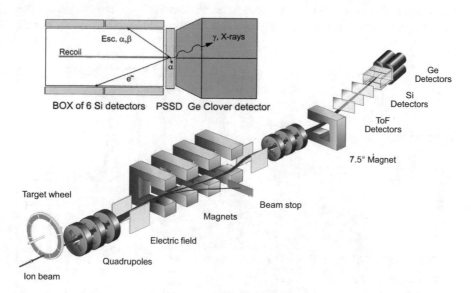

Fig. 3.5 Schematic layout of the SHIP velocity filter at GSI. The heavy ion beam enters SHIP from the left, hits the targets mounted on a rotating wheel. The recoil products, leaving the target with a rather broad angular distribution, are focused by the first triplet of quadrupole lenses. The sequence of perpendicular electric and magnetic fields, serving as a Wien filter, separates the recoils from the projectiles. Because of the large velocity difference between recoil products and the primary heavy ion beam, the latter is deviated from the former and stopped in the beam stop. The recoils are further focused by the second triplet of quadrupole lenses and finally implanted in the PSSD at the focal plane of SHIP. The inset shows the silicon box and germanium detection system surrounding the PSSD, and typical decay channels (see text for details). Adapted from [53, 54]

spatially separated by the filter. The primary beam is dumped in the beam dump, while the recoiling nuclei are implanted in a 300 μm thick 35 × 80 mm² 16-strip position-sensitive silicon detector (PSSD) at the focal plane of SHIP. This detector measures the energy, time and position of recoils and their subsequent decays, e.g. α, fission and conversion electrons. Upstream of the PSSD, six silicon detectors of similar shape were mounted in a box geometry, see Fig. 3.5. The box detector can be used to reconstruct the energy of α particles and conversion electrons that escape from the PSSD in the backward hemisphere [54]. This increases the total detection efficiency for decays, up to ∼80%. Due to the position resolution of ∼0.5 mm along the strips, the whole PSSD can be considered as an array of 35 mm/0.5 mm ×16 strips ∼1100 effective detector pixels, which allows to reduce the rate of random correlations. As the identification is based on correlating the recoil implantation with subsequent decays (α, fission or electrons), this high degree of pixelation allows to extend the correlation times and thus the accessible range of half-lives. Nowadays, most advanced focal plane implantation detectors are double-sided strip detectors (DSSD), which have up to 120 strips × 60 strips = 7200 effective pixels. Combined with digital electronics read-out this results in a further performance increase.

The time and position correlations between the recoil implantation and the subsequent decays makes the identification of the isotopes unambiguous. For example, as shown in Fig. 3.6, by searching for fourfold recoil-α_1-α_2-α_3 events,

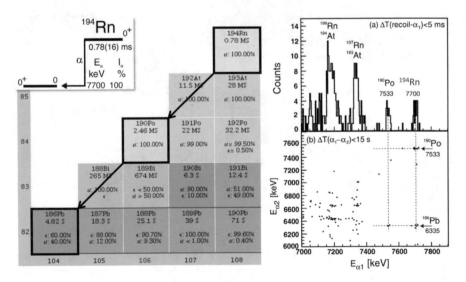

Fig. 3.6 Principles of identification of the new isotope ^{194}Rn in the reaction ^{52}Cr + ^{144}Sm → ^{194}Rn + 2n. (Left panel) The decay scheme of ^{194}Rn with the decay information and the triple α-decay sequence ^{194}Rn → ^{190}Po → ^{186}Pb→ ^{182}Hg, used for identification. (Right panel—top) The α_1 decay spectrum obtained with a constraint on the time between the implantation of the recoiling nuclei and the subsequent first α decay. (Right panel—bottom) A two-dimensional correlation $E_{\alpha 1}$ versus $E_{\alpha 2}$ plot with the correlation time between α_1 and α_2 less than 15 s. Adapted from [55].

observed within the same position of the PSSD, allowed to identify the new isotope ^{194}Rn. The identification of unknown α decays of ^{194}Rn was performed via establishing their time and position correlations with the *known* α decays of the daughter ^{190}Po ($T_{1/2} = 2.5$ ms) and granddaughter ^{186}Pb ($T_{1/2} = 4.8$ s).

The α-decay spectrum observed with a time condition between the arrival time of the recoil and the observation of the first α particle (ΔT(recoil-α_1) < 5 ms) reveals α decays from known nuclei (e.g. ^{190}Po, ^{196,197}Rn,...) produced in the studied reaction or in the reactions on heavier samarium isotopes present as impurities in the target. The α peak at 7.7 MeV could be identified as due to α decay of the new isotope ^{194}Rn by using the two dimensional α-α correlation plot. It was constructed by imposing as extra condition that the first α decay (α_1) is followed by a second α decay (α_2) within 15 s in the same position as the α_1 decay. In this way a correlation between the α decay of ^{194}Rn and ^{190}Po/^{186}Pb and between ^{190}Po and ^{186}Pb could be established. The correlation times were chosen as a compromise between statistics (a few times the half life of the longest lived isotope) and random correlated events.

Essentially the same principles are used in the search for new elements in the region of the Super Heavy Elements (see Sect. 3.2.4.3). For example, element $Z = 113$ was observed with a cross section as low as 20 fb using cold fusion reaction with ^{208}Pb and ^{209}Bi targets, and $Z = 118$ was identified in fusion reactions using beams of ^{48}Ca on actinide target with cross sections around 0.6 pb [53, 57, 58].

3.2.4   Examples of Nuclear-Structure Information Extracted from α-Decay Studies

Recent experiments in the very neutron-deficient nuclides around $Z = 82$ and in the heavy and superheavy element region have considerably expanded the number of α-decaying isotopes. In this section, a number of selected examples where α decay plays an important role will be discussed.

3.2.4.1   Super-Allowed α Decay Around ^{100}Sn

The α decay of ^{212}Po is a textbook example of 'allowed' α decay as it involves two protons and neutrons just outside the respective shell closures at $Z = 82$ and $N = 126$. From a reduced width [30] or α-particle formation probability [29] point of view (see Fig. 3.4), it is indeed the fastest α decay observed. However, from the fact that the protons and neutrons occupy different shell-model orbitals, it was conjectured that α decay of nuclei along the $N = Z$ line could be "super-allowed". This is because in the latter case the protons and neutrons occupy the same shell-model orbitals, which could lead to enhanced clustering into an α particle. Thus α-decay studies of nuclei close to ^{100}Sn nucleus reveal interesting information on the nuclear-structure of $N = Z \sim 50$ nuclei and on the α-decay process itself [14, 59].

The peculiar issue with these studies is that, because of the stabilizing effect of the proton and neutron shell closures, the Q_α values of the daughter nuclei increase and thus their half life decrease. This is similar to the region 'north-east' of ^{208}Pb as e.g. illustrated by the sequence: ^{224}Th (E_α = 7.17 MeV, $T_{1/2}$ = 1.05 s) → ^{220}Ra (7.45 MeV, 18 ms) → ^{216}Rn (8.05 MeV, 45 µs) → ^{212}Po (8.79 MeV, 0.3 µs) → ^{208}Pb(stable). The same happens around ^{100}Sn where the double α-decay chain ^{110}Xe (3.73 MeV, 105 ms) → ^{106}Te (4.16 MeV, 70 µs) → ^{102}Sn was reported from a study at the GSI on-line separator [59]. Experiments at the Holifield Radioactive Ion Beam Facility Recoil Mass Spectrometer extended these studies towards the ^{109}Xe (4.06 MeV, 13 ms) → ^{105}Te (4.88 MeV, 0.62 µs) → ^{101}Sn decay chain [43]. This experiment required special attention to the signal treatment because of the very short-half life of ^{105}Te. This half life is much shorter compared to the typical time processing of silicon detector signals using analog electronics. The latter is of the order of µs due to the shaping applied in analog amplifiers. Thus using spectroscopy techniques, the α-decay signal from ^{109}Xe would pile up with the one from ^{105}Te. Digital electronics, where the signal traces from the preamplifiers are sampled every 25 ns, allowed to separate these two decays as is shown in Fig. 3.7. In a follow up experiment, fine structure in the α decay of ^{105}Te was observed and it was concluded that the α line originally assigned to the ground-state to ground-state transition decays in fact towards an excited state at 171.7 keV in ^{101}Sn [44]. This state was already reported from an in-beam γ-ray experiment at the Argonne Fragment Mass Analyser separator [60] and was interpreted as a $7/2^+$ to $5/2^+$ transition, proposing the ground-state spin and parity of ^{101}Sn to be I^π = $5/2^+$. However from the α-decay probability, as determined in [44], it was concluded that the order of two states should be reversed, suggesting an I^π = $7/2^+$ ground state spin and parity for ^{101}Sn. The order of these two states in ^{101}Sn, one neutron outside the N = 50 shell closure, has a profound impact on the nuclear-structure understanding of the ^{100}Sn region as it determines the effective neutron single-particle energy and can discriminate between shell-model calculations using different effective interactions [44, 60]. This controversy could possibly be solved using other experimental tools like e.g. laser-spectroscopy studies as described in [61, 62].

The reduced α-decay width (see Sect. 3.2.2 and [30]) of ^{105}Te was compared with the 'equivalent' decay of ^{213}Po (2 protons plus 3 neutrons above the doubly-magic cores of ^{100}Sn and ^{208}Pb, respectively) and an enhancement factor of $\sim$3 was observed hinting to the super-allowed nature of the former decay. Recently this has been interpreted in the framework of a full microscopic shell-model based calculation to analyze the formation probabilities and subsequent α decay of ^{212}Po and ^{104}Te [42]. The calculated α-particle formation probability in ^{104}Te is 4.85 times larger compared to ^{212}Po and thus, based on these results, the α decay towards ^{100}Sn can be considered as 'super allowed'.

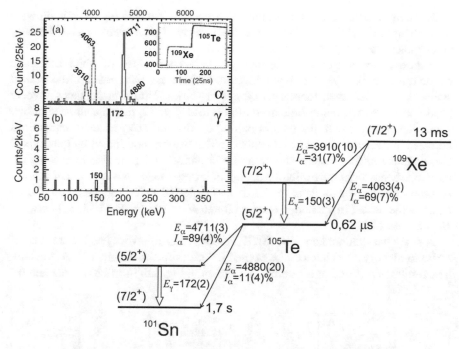

Fig. 3.7 (Left panel, top) The α-decay energy spectrum obtained after implantation of ^{109}Xe in a DSSD and filtered for double-pulse recording is shown together with an example of a preamplifier trace from the DSSD detector. Recording the full trace allowed to separate the parent from the daughter decay and to determine the α energy precisely. The weak ground-state to ground-state transition at 4880 keV and a much stronger fine structure decay at 4711 keV are marked in the spectrum. The panel below provides the coincident γ-ray spectrum showing evidence for populating the excited level at 171 keV in ^{101}Sn. (Right panel) A partial α-decay scheme for the decay of ^{109}Xe $\rightarrow$ ^{105}Te $\rightarrow$ ^{101}Sn. The fine structure decay of ^{105}Te is strongly suggesting that the structure of the connected states is similar. In contrast, the much smaller intensity of the ground-state to ground-state decay suggests different structures for the parent and daughter states. Adapted from [44]

3.2.4.2   Shape Coexistence in the Neutron-Deficient Lead Region

Shape coexistence in atomic nuclei is a topic introduced in 1956 by Morinaga [63] when the first excited state at 6.049 MeV in ^{16}O, which has spin and parity $I^\pi = 0^+$, was interpreted as a deformed state. The observation of a strongly deformed state in a doubly-magic nucleus came as a big surprise. Its structure was deduced as due to proton/neutron multi-particle multi-hole excitations across the closed proton and neutron shell gaps at $N = Z = 8$. Shape coexistence is manifested by the appearance of so-called 'intruder' states with different deformation situated at relatively low-excitation energy in the atomic nucleus. Since the work of [63], shape coexistence has been identified in most regions of the nuclear chart and recent reviews on the subject can be found in [64, 65]. α decay has played an important role in the study of shape coexistence in the neutron-deficient lead region as it allowed, through the study of the fine structure in the α decay of even-even nuclei, to identify and characterize low-lying 0^+ states.

For even-even nuclei these are low-lying 0^+ states on top of which collective bands can be build—see e.g. Fig. 1 from [66] where the systematics of the excited states in the even-even mercury isotopes is shown. While these states involve proton multi-particle multi-hole excitation across $Z = 82$, they are situated at low excitation energy due to the gain in pairing energy and extra residual interactions between the valence protons and neutrons. The latter is dominated by quadrupole correlations which induce quadrupole deformation and give rise to the typical parabolic behavior of the excitation energy of intruder states as a function of valence neutron numbers. In the lead nuclei, it was expected that these shape coexisting states come lowest in excitation energy around $N = 104$ (^{186}Pb) midshell between $N = 82$ and 126.

A study was undertaken at the SHIP velocity filter to investigate the α decay of ^{190}Po to identify and characterize excited 0^+ states in the daughter ^{186}Pb nucleus. The results of this and other studies are shown in Fig. 3.8. In ^{186}Pb two excited 0^+

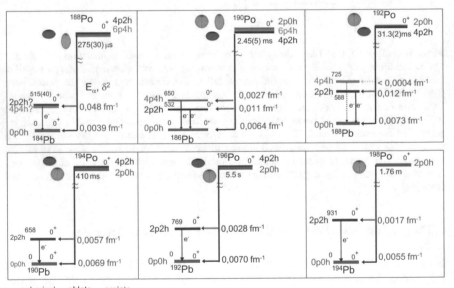

-- spherical, -- oblate, -- prolate

Fig. 3.8 The results from the α-decay fine structure study of the neutron-deficient even-even $^{188-198}$Po isotopes are shown. These studies revealed the existence of low-lying 0^+ states in the corresponding daughter lead isotopes. In ^{186}Pb it could be shown that the first three states have 0^+ spin and parity. The excitation energy of the 0^+ states is given in keV and the half life of the polonium isotopes is shown. Next to the arrows indicating the α-decay branch, the α particle formation probability ($|\mathrm{RF}_\alpha(R)|^2$) is given as calculated following [31, 32] (see (3.20)). The color code indicates the nuclear-model dependent interpretation of the deformation of the 0^+ states (see also Fig. 3.9)

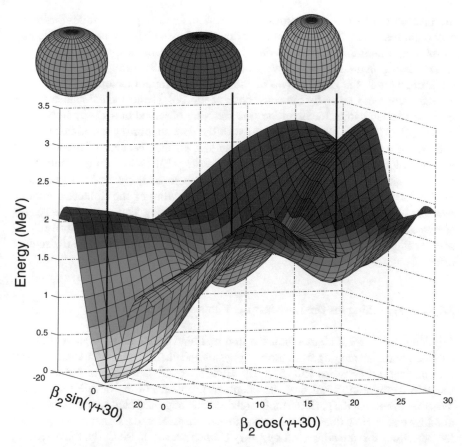

Fig. 3.9 The total potential energy surface of ^{186}Pb showing three minima in the beta/gamma deformation plane. The minima of the spherical (0p-0h) (blue), oblate (2p-2h) (red) and prolate (4p-4h) (green) shapes are indicated. The 2p-2h and 4p-4h denote the number of protons excited across the $Z = 82$ shell closure, 2 and 4 respectively. Adapted from [67]

states were identified below the first excited 2^+ state making this nucleus unique in the nuclear chart as this is the only nucleus which has three lowest states as 0^+ [67]. By using different types of calculations, like e.g. mean-field, beyond mean-field as well as symmetry based calculations [64], which provide a robust picture of shape coexistence, the three 0^+ states were interpreted as spherical, oblate and prolate deformed. Figure 3.9 shows the calculated total potential energy surface of ^{186}Pb and demonstrates the three minima, that give rise to these three different shapes. The α-formation probability ($|RF_\alpha(R)|^2$) for the individual α decay towards the three 0^+ states could be deduced (see Eq. (3.2.2)) and is indicated in Fig. 3.8. From this we learn that the $|RF_\alpha(R)|^2$ feeding the excited 0_2^+ state increases over one order of magnitude when going from ^{198}Po to ^{188}Po (see also Fig. 3.4). Differences in charge radii as deduced from laser spectroscopy studies show that the neutron-deficient lead isotopes stay spherical in their ground states while for polonium the onset of deformation is evidenced for the lightest isotopes [68, 69]. Combining this information with data from in-beam gamma spectroscopy studies in lead and polonium [64, 70] we can infer a gradually stronger mixing of the intruder states in the ground state of the lightest polonium isotopes. This mixing is pictorially represented by the color code of Fig. 3.8. Dedicated calculations have been performed with a focus on the fine structure observed in α decay in the lead region [71–75]. They show agreement with the data supporting the importance of α decay as a probe for nuclear-structure studies. Further, because of the increased mixing in the polonium ground state [70], the $|RF_\alpha(R)|^2$ value of the ground-state to ground-state α decay of the lightest polonium isotopes is reduced as can be seen in Fig. 3.4. Finally, the smaller α formation probability of the lightest polonium isotopes to populate the ground state of the corresponding lead isotopes causes the strong deviation of the lightest polonium isotopes from the Geiger-Nuttall rule as seen in Fig. 3.8 which can now be understood as being due to a nuclear-structure effect.

3.2.4.3 Searching for the Heaviest Elements

Studying the heaviest elements at the end of Mendeleev's table impacts different research domains ranging from atomic physics and chemistry, over nuclear physics and nuclear astrophysics [53, 57, 76]. How do relativistic effects influence the electron configuration and thus the chemical properties of the heaviest elements? Where is the end of the periodic table and does the long-predicted island of increased stability exist? How does the nucleosynthesis r-process end? These are only a few key questions that are addressed by heavy element research. Recently, the discovery of four new elements have been acknowledged and names and symbols have been given by the International Union of Pure and Applied Chemistry (IUPAC): Nihonium (Nh) for element 113, Moscovium (Mc) for element 115, Tennessine (Ts) for element 117 and, Oganesson (Og) for element 118. In a recent Nobel Symposium NS160, entitled "Chemistry and Physics of Heavy and Superheavy Elements", the field was reviewed and the reader is referred to the individual contributions for a detailed overview of the current state of the art [77].

Because of its high sensitivity to detect decay events of single, individual atoms, α decay plays a pivotal role in the discovery and study of the heavy and superheavy elements (see Sect. 3.2.3). Cold fusion reactions combining projectiles around and above doubly-magic ^{48}Ca with targets of doubly-magic ^{208}Pb (or ^{209}Bi), are used to produce the new elements up to $Z = 113$. These reactions result in low-excitation energy of the compound nucleus and thus in the emission of at most one or two neutrons. The identification of the reaction products could be established using the recoil-α-α time and position correlations method and α-decay chains that link to known regions of the nuclear chart [78]. This is similar to the example of ^{194}Rn shown in Fig. 3.6. For the discovery of the more heavy elements fusion reactions using ^{48}Ca beams on long-lived actinide targets (that are more neutron-

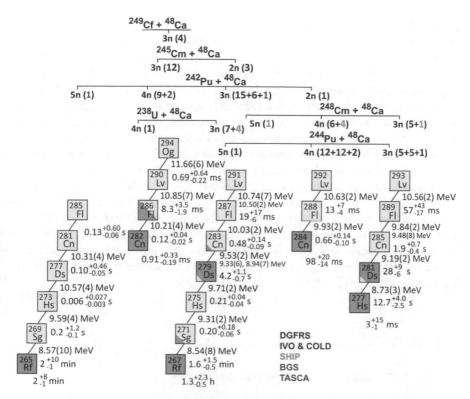

Fig. 3.10 Different α-decay chains supporting the discovery of the new $Z = 114, 116, 118$ elements (Fl, Lv and Og, respectively). These were obtained using ^{48}Ca reactions on different actinide targets. The number of decay chains observed in the different experiments is indicated. The color code indicates the facility where the events were identified and the reader is referred to the original paper for further information. The average α-decay energy and half lives are given for the α emitters observed (yellow square). The green squares indicate spontaneously-fissioning nuclei. Figure from [58]

rich as compared to ^{208}Pb) were used [58]. These reactions lead to higher excitation energies in the compound nucleus and thus to the emission of typically three to four neutrons (hot fusion). Still this results in more neutron-rich isotopes, than those produced in cold-fusion reactions, from which the α-decay chains end in so-far unknown spontaneously fissioning nuclei (see Fig. 3.10). Therefore, often, other means as e.g. cross bombardment with different beam/target combinations (see Fig. 3.10) and measurements of excitation functions have to be used to firmly establish the Z and A of the newly discovered element and isotope.

The fact that α-decay energies can be determined with good accuracy (typical up to a few tens of keV uncertainty) allows to deduce accurate masses and binding energies of these heavy isotopes provided the mass of one of the isotopes in the α-decay chain is measured with good accuracy. Penning trap mass spectroscopy is an excellent tool to measure absolute values of atomic masses [79] and it has been applied to the heavy mass region, in particular for $^{252-254}$No ($Z = 102$) [80, 81]. These measurements revealed new masses and binding energies for a wide range of isotopes giving information on the deformed shell closures around $N = 152$ and $N = 162$ [81]. Other mass spectroscopic techniques like multi-reflection time of flight mass spectrometer (MR-ToF-MS) have recently been commissioned and are now used in the super heavy element region [82]. It should be noted, however, that the use of α-decay energies to link isotopes with accurately known mass with other isotopes only works if the ground-state to ground-state α-decay energy is known. While the latter is obvious for even-even isotopes as their α decay is dominated by the 0^+–0^+ α decay, it is less straightforward for odd mass and odd-odd isotopes. Because of the fine structure in the α decay of most nuclei and of the potential presence of long living isomers, it might not be possible to extract the ground-state to ground-state α-decay energy from the α spectra. Therefore detailed α-γ spectroscopy studies are necessary [83]. In turn, through these fine-structure α-decay studies excited states can be identified and their characteristics determined. The latter data serve as important input to improve theoretical models and to make predictions more reliable [84].

3.3 Beta-Delayed Fission

3.3.1 Introduction to Low-Energy Fission

Nuclear fission, discovered in 1938 [5], provides one of the most dramatic examples of nuclear decay, whereby the nucleus splits preferentially into two smaller fragments releasing a large amount of energy [85, 86]. Figure 3.11a shows the simplified concept of the fission process within the so-called macroscopic 'Liquid Drop Model' (LDM), as used in 1939 to explain fission [6]. The nuclear potential energy surface (PES) for a ^{238}U nucleus is shown as a function of two important parameters in fission: elongation (e.g. quadrupole deformation) of the initial nucleus and the mass asymmetry of resulting fission fragments. Within the LDM, the nucleus elongates along the line of zero mass asymmetry during the fission process. This is shown as a red line in Fig. 3.11a. Thus, initially the nucleus increases its potential energy, until at some moment the maximum of the potential energy surface is reached. This point is called the saddle point (the top of the fission barrier). Afterwards, at even further elongation, the nucleus reaches the scission point and splits in two equal fission fragments.

While the classical LDM was able to qualitatively explain why fission is one of the important decay modes of heavy nuclei, it failed to describe e.g. the experimentally-observed asymmetric mass split in two un-equal fragments, which was realised already in 1938. Following the recognition of the importance of the microscopic shell corrections, which arise due to specific/non-uniform occupancies of levels by protons and neutrons, the description of the fission process was

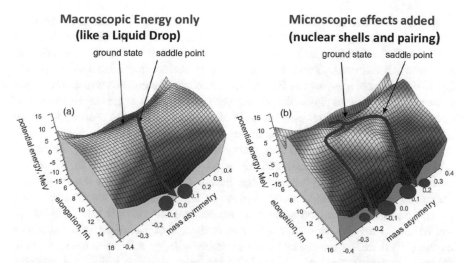

Fig. 3.11 (a) Macroscopic, V_{macro}(LDM), and (b) total, $V_{total} = V_{macro}$(LDM)+V_{micro}(Shells) potential energy surface for the ^{238}U nucleus as a function of elongation and fission-fragment mass asymmetry. The most probable fission paths (or 'fission valleys'), which follow the lowest energy of the nucleus, are shown by the red lines with arrows. While in the LDM approach only symmetric fission can happen along the single 'symmetric' valley, the introduction of microscopic shell effects produces the asymmetric fission valleys. The mass asymmetry parameter is defined as the ratio $\frac{A1-A2}{A1+A2}$, where $A1$ and $A2$ are masses of the two fission fragments (thus, the mass asymmetry =0 for symmetric mass split). The energy difference between the ground state and the saddle point defines the quantity, dubbed as 'fission barrier height'. Figure modified from [87]

discussed within the so-called macroscopic-microscopic framework [88–90]. In this model, the total potential energy becomes the sum of macroscopic (LDM) and microscopic shell effects (Shells) energy: $V_{total} = V_{macro}$(LDM)+V_{micro}(Shells). This naturally led to the appearance of the asymmetric fission valleys on the potential energy surface (see, Fig. 3.11b), thus to the asymmetric fission-fragment mass distribution (FFMD).

Therefore, fission is a unique tool for probing the nuclear potential-energy landscape and shell effects at extreme values of deformation, and as a function of mass asymmetry, spin, and excitation energy. In particular, fission enables the study of nuclear-structure effects in the heaviest nuclei [91], has direct consequences on their creation in nuclear explosions [92] and in the astrophysical r-process [93–95]. The latter is terminated by fission, thus fission has a direct impact on the abundance of medium-mass elements in the Universe through so-called "fission recycling". The fragments, resulting from fission of very heavy nuclei, become the new seed nuclides for the r-process. Apart of its importance for fundamental studies, fission has many practical applications, such as the generation of energy and the production of radioisotopes for science and medicine. Fission is also a very powerful mechanism to produce nuclei far from the stability line [96].

It is important to note the strong dependence of microscopic effects on the temperature (or, excitation energy) of the nucleus. Indeed, the shell effects are 'washed out' at sufficiently high excitation energies, which leads to the disappearance of asymmetric fission valleys on the potential energy surface, reverting it to the smooth LDM-like surface, as shown in Fig. 3.11a. Therefore, the nucleus will again fission symmetrically. This explains the strong need for fission studies as a function of excitation energy, and also as a function of proton and neutron number to reveal specific nuclear-structure effects on fission.

As a function of the excitation energy of the fissionning nucleus, the fission process is often broadly classified either as high-energy fission, in which the excitation energy strongly exceeds the fission barrier height, or as low-energy fission. In contrast to high-energy fission, in which the microscopic effects are washed out, the interplay between macroscopic and microscopic effects in fission can be sensitively explored at low excitation energy. In particular, in spontaneous fission (SF) from the ground state, the excitation energy is $E^* = 0$ MeV, while in SF from isomeric states or in thermal neutron-induced fission it does not exceed a few MeV [85]. However, being the ultimate tool for low-energy fission studies, SF is limited to heavy actinides and trans-actinides [97]. Recently, by using Coulomb excitation (often dubbed as "Coulex") of relativistic radioactive beams, fission studies became possible in new regions of the Nuclidic Chart, earlier unexplored by low-energy fission [98, 99], see Sect. 3.3.5.4. In this case, the excitation energy is centered around $E^* \sim 11$ MeV. In terms of the excitation energy, the beta-delayed fission (βDF), which is the topic of this section, is intermediate between SF and Coulex-induced fission. As will also be shown below, βDF allows to study low-energy fission of some of the most exotic nuclides, which are not yet accessible to fission studies by other techniques, and which do not fission spontaneously from their ground state.

3.3.2   Mechanism of Beta-Delayed Fission, Conditions to Occur, Observables

The exotic process of β-delayed fission can be considered as a sub-class of a more general group of beta-delayed decay processes, which includes β-delayed proton (denoted βp), β-delayed neutron (βn), β-delayed gamma ($\beta\gamma$) and β-delayed α particle ($\beta\alpha$) emission, see detailed reviews [100, 101]. All these decay modes proceed via two steps, first the β (β^+/EC or β^-) decay happens, which can be accompanied, under certain conditions, by the subsequent emission of one of the particles (or more particles, in case of multiparticle β-decay processes, e.g., β2n for β-delayed two neutron emission). Similarly, β-delayed fission, discovered in 1965–1966 [102–104] is a two-step nuclear-decay process that couples β decay and fission. In these lecture notes we will review the most salient features of βDF and discuss some of the most recent methods and results, while we refer the reader to a comprehensive review [105] for a detailed discussion of this topic.

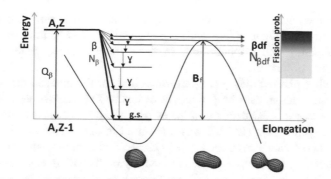

Fig. 3.12 Schematic representation of the βDF process on the neutron-deficient side of the nuclear chart. The Q_β value of the parent (A, Z) nucleus is indicated, while the curved line shows the potential energy of the daughter $(A, Z - 1)$ nucleus with respect to nuclear elongation. The fission barrier B_f is also shown. The color code on the right-hand side represents the probability for excited states, with excitation energies close to B_f, to undergo fission; the darker colors correspond to higher probabilities. The plot is from [110]

In βDF, a parent nucleus first undergoes β decay, populating excited state(s) in the daughter nuclide. In the case of neutron-deficient nuclei, electron capture (EC) or β^+ decay is considered (referred further as β^+/EC), while β^- decay happens on the neutron-rich side of the Nuclidic Chart. Figure 3.12 shows a simplified diagram of βDF for the case of neutron-deficient nuclei, which are mostly considered in this study. If the excitation energy of these states, E^*, is comparable to or greater than the fission barrier height, B_f, of the daughter nucleus ($E^* \sim B_f$), then fission may happen in competition with other decay modes, e.g. γ decay and/or particle emission (neutron, proton or α), depending on which side of the β-stability valley the parent nucleus is situated. Therefore, the special feature of βDF is that fission proceeds from excited state(s) of the daughter nuclide, but the experimentally-observed time behavior of the βDF events is determined by the half-life of the parent nucleus (as with any β-delayed particle decays). As in most cases the β-decay half-lives are longer than tens of ms, it makes βDF more easily accessible for experimental studies. The experimental observables from βDF include: half-life, probability ($P_{\beta DF}$, see Sect. 3.3.6), kinetic energies of resulting fission fragments, and energies and multiplicities of e.g. γ rays and neutrons, accompanying the fission process. It is important to stress that the unique experimental signature for βDF is the coincidences between the x rays (usually, K x rays) of the *daughter* nucleus (which arise after the EC decay of the parent) with the fission fragments. This method was used by the Berkeley group in some of their experiments to directly confirm the βDF process in e.g. ^{232,234}Am [106–108] and in ^{228}Np [109].

Two main conditions must be satisfied for βDF to occur in measurable quantities. First of all, the parent nucleus must possess a non-zero β-branching ratio ($b_\beta > 0$). Secondly, the Q_β-value of the parent must be comparable to or be higher than the fission barrier of the daughter Q_β(Parent) $\sim B_f$(Daughter). These conditions are fulfilled for three regions of nuclei (a) very neutron-deficient isotopes in the lead region (see Sects. 3.3.5.1–3.3.5.3), (b) very neutron-deficient nuclides in the neptunium-to-mendelevium (Np–Md) region, and (c) very neutron-rich isotopes between francium and protactinium. In these three regions, βDF was indeed experimentally observed.

Historically, βDF studies were performed first for the neutron-deficient nuclei in the uranium region, as a by-product of extensive searches for new elements in the 1960s. Since the EC decay dominates over the β^+ decay in these cases, the term EC-delayed fission (ECDF) was often used in the literature, while the term βDF was predominantly reserved for β^--delayed fission in the neutron-rich nuclei. However, since in the neutron-deficient lead region the β^+ decay can effectively compete with the EC decay, we will use throughout this review the term βDF for both the neutron-rich and neutron-deficient nuclei.

A distinctive feature of βDF is its *low-energy* character. Previously, low-energy fission studies were limited to nuclei along the valley of stability from around thorium (Th) to rutherfordium (Rf) and above, mostly using SF and fission induced by thermal neutrons. In βDF, the *maximum* excitation energy of the daughter nucleus is limited by the Q_{EC} ($Q_{\beta-}$ in case of neutron-rich nuclei) of the parent nuclide. The typical Q_{EC} ($Q_{\beta-}$) values are in the range of 3–6 MeV and 9–12 MeV for the known βDF nuclei in the trans-uranium and lead regions, respectively.

The importance of βDF is also highlighted by its ability to provide low-energy fission data for very exotic nuclei which do not decay by SF and which are difficult to access by other techniques. For example, the recently-studied βDF nuclei in the lead region possess very different neutron to proton ratios, e.g. $N/Z = 1.23$–1.25 for ^{178,180}Hg (see Sect. 3.3.5.1) in contrast to a typical ratio of $N/Z = 1.55$–1.59 in the uranium region, where numerous SF and βDF cases were long-known. This allows to investigate potential differences in the βDF process and its observables in the two regions, which differ in many nuclear-structure properties, see discussion in Sects. 3.3.5.1–3.3.5.3.

Finally, the potential role of βDF for the r-process termination by fission (along with neutron-induced and spontaneous fission) is the subject of on-going discussion [93–95], also in view of its possible implications for the determination of the age of the Galaxy by the actinide geochronometers [111, 112].

We refer the reader to Table 1 from [105, 110] where all 26 βDF nuclei, known by 2013, are summarized. In [105] a review of production/identification methods and of the βDF probability values is provided, while [110] discusses the partial βDF half-lives. Since then, three new βDF nuclides—^{230}Am, ^{236}Bk and ^{240}Es were discovered, see Sect. 3.3.4. The data, shown in these tables, demonstrate that so far all known βDF emitters are odd-odd nuclei. To explain this point and to highlight common phenomena relevant to βDF, we analyse in Fig. 3.13 an example of βDF of neutron-deficient At isotopes. The calculated Q_{EC}(At) and B_f(Po) values in the region of our interest for the Finite Range Droplet Model/Finite Range Liquid Drop Model (FRDM/FRLDM) [113, 114] and for the Thomas-Fermi (TF) model [115] are compared to the extrapolated or experimental Q_{EC}(At) values from Atomic Mass Evaluation (AME2012) [17]. However, the latter values should be considered with caution since in most of the lightest odd-odd astatine isotopes, e.g. in ^{192,194}At

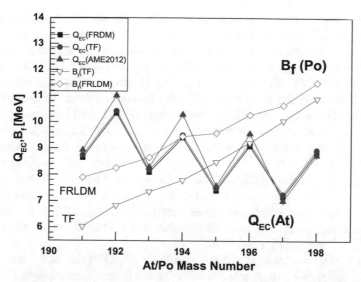

Fig. 3.13 Calculated Q_{EC}(At) (closed circle and square symbols) and B_f(Po) (open symbols) values from the FRDM/FRLDM [113, 114] and from the TF model [115]. The Q_{EC} values from AME2012 [17] are shown by the closed triangles. The plot is from [116]

[117, 118], there are more than one long-lived nuclear state with unknown relative excitation energy and β-branching ratios. Furthermore, it is not always known which of them is the ground state and for which the experimental mass determination was quoted. This is quite a general issue in the odd-odd nuclei, which might influence the derivation of fission fragments mass distributions and probability values in βDF studies.

A few important features are evident in Fig. 3.13. First of all, the good agreement for the Q_{EC}(At) values between the two mass models is most probably because the Q_{EC} values are deduced as a difference of the calculated parent and daughter masses. Therefore, even if the two models give masses, systematically shifted by some value, this shift will largely cancel out in their difference.

Secondly, due to the odd-even staggering effect in masses, the Q_{EC} values of the odd-odd (thus, even-A) parent astatine isotopes are on average $\sim$1.5–2 MeV larger than for their odd-A neighbors. This is one of the main reasons why so far all nuclei where βDF was observed experimentally are odd-odd. Another reason is that after the β decay of an odd-odd isotope, an even-even daughter is produced, which is expected to fission easier than an odd-A neighbor, produced after the β decay of an odd-A precursor. The very strong (several orders of magnitude) hindrance for SF of the odd-A and odd-odd nuclei in comparison with the even-even nuclides is a well-established experimental fact, see e.g. [97]. As the excitation energy of the fissioning daughter nucleus in βDF is relatively low, similar fission hindrance factors could be also expected for βDF, however this is still an open question.

Another important feature shown in Fig. 3.13 is that while both models predict a fast decrease of the calculated fission barriers as a function of lowering neutron number, the rate of decrease is different. As a result, the respective calculated $Q_{EC}(At) - B_f(Po)$ values are quite different, e.g. -0.08 MeV (FRDM/FRLDM) and $+1.7$ MeV (TF) for ^{194}At. Thus, one could expect that the chance of β-decay feeding to states in the vicinity or above the fission barrier in the daughter ^{194}Po should be higher in the TF model in comparison to FRDM/FRLDM model, however the specific β-decay strength function needs to be considered, see Sect. 3.3.6. For ^{198}At, both models predict large negative $Q_{EC}(At) - B_f(Po)$ values, thus only sub-barrier fission of ^{198}Po can happen, which will most probably result in negligible/unmeasurable βDF probability. On the other hand, both models predict large positive $Q_{EC}(At) - B_f(Po)$ values for ^{192}At, thus higher probabilities are expected in this case, which was indeed confirmed in the SHIP experiments [119]. A similar effect occurs for several βDF cases in the lead region, as discussed for ^{178}Tl [120], ^{186,188}Bi [121] and ^{196}At.

Several βDF studies explored the sensitivity of βDF probability values to the $Q_{EC} - B_f$ differences, to infer information on the fission barrier height [122–125], thus to check the validity of different fission models far from the β-stability line, this will be discussed in Sect. 3.3.6.

3.3.3 Production of βDF Nuclei and Determination of Their Properties

Historically, four methods were exploited to produce nuclei which exhibit the βDF decay, they are described in more details in review [105].

– Charged-particle induced reactions, typically fusion-evaporation or transfer reactions at beam energies in the vicinity of the Coulomb barrier. E.g. the first βDF cases in ^{228}Np and ^{232,234}Am were identified in the fusion-evaporation reactions ^{10,11}B $+$ ^{230}Th $\rightarrow$ ^{232,234}Am and ^{22}Ne $+$ ^{209}Bi $\rightarrow$ ^{228}Np [102, 103].
– Reactions induced by γ rays and neutrons, see e.g. [126, 127].
– Radiochemical separation from naturally-occurring long-lived, usually neutron-rich, precursors, see e.g. [126, 128].
– Spallation reactions with 1-GeV protons on a thick uranium target, followed by mass separation with an electromagnetic mass-separator [9]. The recent novel experiments at the ISOLDE mass separator at CERN (Geneva) [48] used the same method to identify βDF of ^{178,180}Tl [120, 129, 130], ^{194,196}At [119] and ^{200,202}Fr [131], they will be discussed in details below.

In the earlier βDF experiments in 1960–1980s, which mostly exploited the first two production methods described above, rather simple 'mechanical' techniques, e.g. the rotating drum or a thick catcher foil [132] were used to collect *all* reaction products for subsequent observation of their decays. The main drawbacks of such experiments were discussed in [105] and are briefly summarized here. First of all, many other reaction channels can be open, including those involving possible impurities in the target, which could produce other nuclei, that possess a fission branch (e.g. SF). Therefore, in the absence of direct determination of A and/or Z of the fissionning parent nucleus, the assignment of the observed activity to βDF to a specific isotope could often be ambiguous. To partially overcome this issue, a tedious series of cross-bombardments to produce the same βDF candidate in different projectile-target combinations had to be performed, also as a function of the projectile energy, to obtain the so-called excitation functions, see e.g. [133]. The latter can be used to differentiate between different reaction channels.

The chemical separation of the element of interest from the collected sample (e.g. from the catcher foil) in some experiments allowed both to get a purer sample and to achieve the Z-value determination of the fissioning parent nucleus, see e.g. [108]. The use of He-jet technique to transport the fissionning activity from the target region with high radiation levels to a detection system situated some distance (meters) away in a shielded environment is another common method to reduce the background to measure rare decay events [108, 134].

On the detection side, often the so-called mica foil detectors were used in the earlier experiments to register fission events [132]. Such detectors allow to measure the fission rate (thus, production cross sections/excitation functions) and also half-lives of the fissionning activities. However, mica foils fully lack information on the kinetic energy of coincident fission fragments, thus e.g. FFMD's measurements were impossible, as they require the detection of coincident events.

The use of silicon detectors in the earlier experiments was not wide-spread, and at most only single silicon detectors were used. This prohibited the measurements of coincidence fission fragments. From the mid-90s, more complex arrangements of silicon detectors, e.g. the merry-go-around detection system at Berkeley [135, 136], were introduced in fission research, which provided coincidence measurements of fission fragments.

From the mid-90s, the use of kinematic separators (see Sect. 3.2.3) to study the decay of heavy elements and, in particular, of βDF was introduced, see an example of βDF study of ^{194}At in Fig. 3.14 and Sect. 3.3.4. The use of ISOL-type separators for βDF studies in the lead region, e.g. ISOLDE at CERN, was introduced since the last decade and brought substantial progress to βDF studies in the lead region, see Sect. 3.3.5.

3.3.4   Recent βDF Results at Recoil Separators

Here we will discuss the βDF study of ^{194}At performed at SHIP as one example of βDF studies at recoil separators. The principles of radioactive ion beam production and identification at SHIP were briefly explained in Sect. 3.2.3. Figure 3.14 shows the identification of βDF of ^{194}At [119]. Despite its success, this study also demonstrated several drawbacks of the present βDF experiments at recoil

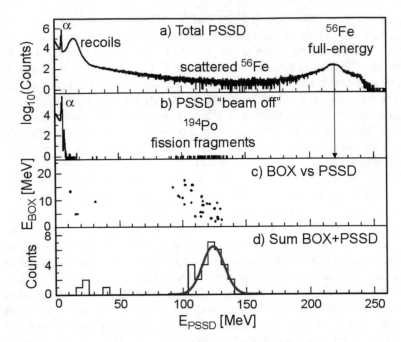

Fig. 3.14 Identification of βDF of ^{194}At at SHIP in the reaction ^{141}Pr(^{56}Fe, 3n)^{194}At. (**a**) The total energy spectrum in the PSSD. The primary beam is provided with a pulsed structure of 5 ms 'beam on'/15 ms 'beam off', the spectrum shows all data, with no condition on the beam. Note the α decays around 5 MeV, the recoiling nuclei around 20 MeV. The spectrum at higher energies is dominated by the ^{56}Fe beam that leaks through SHIP or is scattered on its way to the PSSD. (**b**) same as (**a**) but obtained solely during the 15 ms 'beam off' period. Note the very strong suppression of the primary ^{56}Fe beam and the recoils. Only the α spectrum as well as events due to beta-delayed fission of ^{194}At remain. (**c**) Two dimensional E_{BOX} versus E_{PSSD} of coincident events detected in the PSSD and in the BOX detector. The group of events in the middle of the plot corresponds to detection of coincident fission fragments, whereby one of the fragments is registered in PSSD, while the second (escaping) fragment—in the BOX detector. (**d**) The sum energy $E_{PSSD} + E_{BOX}$, providing the TKE determination for the fission of daughter nuclide ^{194}Po. A Gaussian fit is shown by the red solid line. Adapted from [119]

separators. Indeed, 66 fission events were clearly identified in the experiment at SHIP, see Fig. 3.14b. However, the interpretation of these decays was hindered due to the presence of two states in ^{194}At with similar half-lives of 310(8) ms and 253(10) ms [118] with yet unknown β-decay branchings. Therefore, the assignment of the observed fission events to a specific (or to both) isomeric states in ^{194}At is not yet clear, see detailed discussion in [119]. Another drawback of such experiments is that the parent nuclei are implanted in the silicon detector at a depth of several micrometers. This prevents an accurate measurement of the individual energies of the coincident fission fragments, as the fission fragment escaping the PSSD detector in the backward hemisphere leaves some of its energy in the PSSD. This energy will be summed up with the energy of the second fission fragment stopped in the PSSD.

Therefore, the measured energies of coincident fission fragments in the PSSD and BOX will be distorted, depending on implantation depth and fission fragments angle relative to the detector surface. There is no clear procedure how to account for these effects. As a result, no FFMD's derivation is yet possible in such experiments, as it requires the knowledge of unperturbed kinetic energies of both fission fragments. This is because, in the first approximation, the ratio of fission fragments masses is inversely proportional to the ratio of their kinetic energies. Only information on total fission kinetic energy (TKE) with a rather large uncertainty can be extracted via a tedious calibration procedure, see Fig. 3.14d.

Using the same method, 23 βDF events of two isomeric states in ^{192}At were identified at SHIP [119], with similar difficulties encountered with respect of their assignment to a specific isomeric state, as in case of ^{194}At.

The difficulty of isomer separation can, however, be overcome by using the technique of laser-assisted isomer separation, as discussed in the next section.

To conclude this section, we mention that recently, βDF of ^{236}Bk and ^{240}Es was identified at the gas-filled separator RITU at JYFL [137], while βDF of ^{230}Am was observed at the GARIS gas-filled separator at RIKEN [138, 139]. For ^{236}Bk and ^{240}Es, quite large βDF probabilities, of the order of $\sim$10% were derived, while a lower limit of $P_{\beta DF} >$30% was derived for ^{230}Am. The latter value is the highest so far among all measured βDF probabilities. Both experiments used a technique, very similar to the SHIP study, but employed more advanced multi-strip DSSSD's.

3.3.5 A New Approach to Study βDF at the ISOLDE Mass Separator at CERN

About a decade ago, a new technique to study βDF nuclei in the lead region was developed at the mass separator ISOLDE [48], by using low-energy 30–60 keV radioactive beams. This method allows to extend low-energy fission studies to very exotic neutron-deficient and neutron-rich nuclei, which are difficult to access by other techniques.

As an example, Fig. 3.15 (see also Fig. 3.16) provides a brief overview of the production method of the isotope ^{180}Tl in the pilot βDF study at ISOLDE [129]. A novel and unique feature of this βDF experiment was the combination of Z-selective ionization of a specific element (Tl, in this case) with RILIS [9, 140, 141] and subsequent mass separation at $A = 180$ with ISOLDE. After selective ionization, acceleration up to 30 keV, and mass separation, a pure beam of ^{180}Tl with an intensity of $\sim$150 ions/s was analyzed by the Windmill (WM) detection system [129]. Here, the radioactive beam was deposited on a thin carbon foil, surrounded by two silicon detectors (Si1 and Si2), along with HPGe detectors for coincident particle—γ-ray measurements. The use of two silicon detectors in a compact geometry allowed both singles α/fission decays and double-fold fission-fragment coincidences to be efficiently measured. The same method was later used for βDF studies of ^{178}Tl [120], ^{194,196}At [116, 131] and of ^{200,202}Fr [131]. A detailed discussion of the results will be given in Sect. 3.3.5.3.

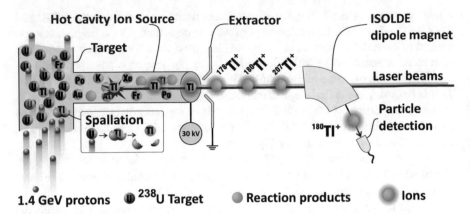

Fig. 3.15 Schematic view of the ISOLDE and Resonance Ionization Laser Ion Source (RILIS) operation as applied in the βDF study of ^{180}Tl [129]. The 1.4-GeV 2 μA proton beam impinges on a thick 50 g/cm^2 ^{238}U target, producing a variety of reaction products via spallation and fission reactions. The neutral reaction products diffuse towards the hot cavity, where the thallium atoms are selectively ionized to the 1^+ charge state by two overlapping synchronized laser beams, precisely tuned to provide thallium ionization in a two-color excitation and ionization scheme. The ionized thallium ions are extracted by the high-voltage potential of 30 kV, followed by the $A = 180$ mass separation with the ISOLDE dipole magnet. The mass-separated ^{180}Tl ions are finally implanted in the carbon foils of the Windmill system (see Fig. 3.16), for subsequent measurements of their decays with the silicon and germanium detectors, as described in the main text. Plot modified from [142]

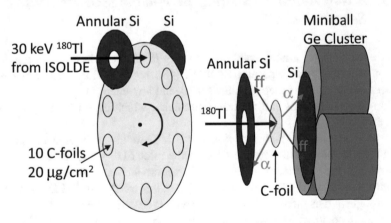

Fig. 3.16 Left side: The Windmill setup used in the experiments at ISOLDE to study βDF of ^{178,180}Tl, ^{194,196}At and ^{200,202}Fr; right side: a zoom of the detector arrangement. The case of ^{180}Tl is shown as an example, with mass-separated ^{180}Tl ions being implanted through a hole in an annular silicon detector (300 μm thickness) into a thin carbon foil of 20 μg/cm^2 thickness. A second Si detector (300 μm thickness) is placed 3 mm behind the foil. A set of Ge detectors is used for γ- and K X-ray measurements in coincidence with particle decays. Plot is taken from [129]

The uniqueness of this technique is the unambiguous A and Z identification of the precursor, via the combination of the Z-selection by the RILIS and the mass-separation by ISOLDE. Other advantages include a point-like radioactive source, the implantation in a very thin foil whereby both fission fragments can be efficiently measured with little deterioration of their energies, and the proximity of germanium detectors for γ-ray spectroscopy. Simultaneous measurement of fission and α decays in the same detectors removes some of the systematic uncertainties for branching ratio determination.

3.3.5.1 New Island of Asymmetric Fission in the Neutron-Deficient Mercury Region: The Case of ^{180}Tl

In this section, as an example of what can be learned from βDF studies, the results of the βDF experiments at ISOLDE will be discussed, in which βDF of ^{178,180}Tl [66, 120, 129], ^{194,196}At [116, 119] and ^{202}Fr [131] was studied, resulting in the low-energy fission data for daughter (after β decay) isotopes ^{178,180}Hg, ^{194,196}Po and ^{202}Rn, respectively.

Historically, the first βDF study at ISOLDE was performed for the isotope ^{180}Tl, whose production method was described in Sect. 3.3.5. The $Q_{EC}(^{180}\text{Tl}) = 10.44$ MeV and the calculated fission barrier is $B_f(^{180}\text{Hg}) = 9.81$ MeV, thus $Q_{EC}(^{180}\text{Tl}) - B_f(^{180}\text{Hg}) = 0.63$ MeV, which allows for some above-barrier fission to happen. Despite this, a rather low βDF probability of $P_{\beta DF}(^{180}\text{Tl}) = 3.2(2) \times 10^{-5}$ was deduced [105].

In a ~ 50-h long experiment, 1111 singles and 356 coincidence fission events were observed and attributed to the βDF of ^{180}Tl, see Fig. 3.17. The mass distribution for fission fragments of ^{180}Hg is clearly asymmetric with the most abundantly-produced fragments being ^{100}Ru and ^{80}Kr and their neighbors. No commonly-expected symmetric split in two semi-magic ^{90}Zr ($Z = 40$, $N = 50$) nuclei was found, and the observation of "new type of asymmetric fission in proton-rich nuclei", which differs from asymmetric fission in the trans-uranium region, was claimed [129, 143].

3.3.5.2 Mass-Asymmetry in ^{180}Tl and in ^{238}U: What Is the Difference?

This discovery caused an intense interest from the theory community, whereby very different approaches, such as the macroscopic-microscopic model [129, 143, 144], two modern versions of the scission-point model [145–148] and two fully self-consistent models, HFB-D1S and HFB-SkM* [149, 150], were used to shed light on the observed phenomenon.

In particular, the five-dimensional (5D) macroscopic-microscopic model [143, 151] was first to be applied to explain the observed asymmetric mass split of the fission fragments of ^{180}Hg. Figure 3.18 from [143] shows two-dimensional potential-energy surfaces (PES) for ^{180}Hg and ^{236}U and highlights the crucial

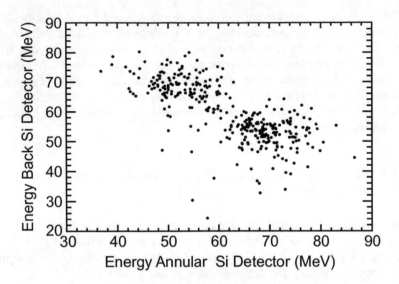

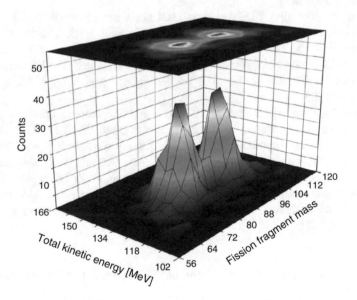

Fig. 3.17 Top panel: a coincidence energy spectrum for βDF of ^{180}Tl measured by two silicon detectors of the Windmill setup. The two-peaked structure originates because the two fission fragments have different energies, a direct result of the asymmetric mass distribution; Bottom panel: The derived fission-fragment distribution of the daughter isotope ^{180}Hg as a function of the fragment mass and the total kinetic energy. The conversion from energy to mass spectra relies on the conservation of mass and energy, and assumes that the masses of fission fragments add up to the mass of the parent nucleus, see details in [129]. Plots are taken from [129, 130]

Contrasting Fission Potential-Energy Surfaces Hg↔U

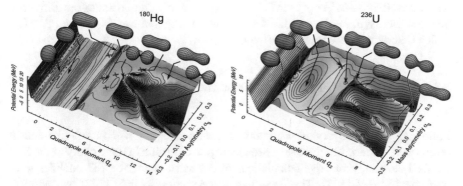

Fig. 3.18 Calculated PES surfaces for ^{180}Hg and ^{236}U, as a function of the dimensionless quadrupole moment and the mass asymmetry, as defined in Fig. 3.11. The shapes of the nuclei at several key deformations are drawn, connected to the points on the surface by arrows. The saddle points are indicated with red crosses. The plots are modified from [143]

differences in the nature of asymmetric fission for proton-rich nuclei in the lead region compared to the classical region of asymmetric fission in the actinide region around ^{236}U.

The PES for ^{236}U shows features common to many actinide nuclei with $226 \le A \le 256$, such as a deformed ground state, a relatively low two- or three-humped fission barrier, and most prominently, well-separated symmetric ($\alpha_g = 0$) and asymmetric ($\alpha_g \sim 0.2$) valleys. The latter valley is usually attributed as being due to the strong shell effects (spherical and/or deformed) of fission fragments in the vicinity of the double-magic ^{132}Sn. Fission starts from the ground state on the left-hand side of the figure, passes through the nearly symmetric first saddle point to the symmetric fission isomer minimum around $q_2 \sim 2$. Then the mass asymmetry begins to increase as the nucleus passes over the mass-asymmetric second saddle point, through a shallow third minimum, and finally over a third asymmetric saddle at the head of the asymmetric valley to a shape near the asymmetric scission point. The higher symmetric saddle point reduces the probability of entering the symmetric valley by requiring barrier penetration for systems with near-threshold energies.

In contrast, the PES for ^{180}Hg is very different, with only a single pronounced symmetric valley corresponding to separated semi-magic ^{90}Zr nuclei, and no deep asymmetric valley extending to scission. The dominant symmetric valley is, however, inaccessible due to the high barrier along the symmetric path from the ground state. The symmetric valley remains separated from a shallow asymmetric valley by a high ridge in the potential. It is important to note that, within this model description, by the time the separating ridge disappears, a quite well-defined mass-asymmetric di-nuclear system has already developed with two nascent fission fragments still connected by a narrow neck. However, such a system does not have the possibility for mass equilibration towards an energetically more favorable mass-symmetric split, due to the very small neck size at this moment.

A similar result can also be seen in Fig. 3.19, which shows the PES calculations within the HFB-SKM* approach [150]. This plot nicely demonstrates the difference between the asymmetric fission of ^{180}Hg and nearly symmetric fission of ^{198}Hg, which is known from earlier experiments.

3.3.5.3   Multimodal Fission in the Transitional Neutron-Deficient Region Above $Z = 82$

Following the discovery of asymmetric fission of ^{178,180}Hg, further theoretical efforts were undertaken to cover a broader region of fission in the vicinity of the $Z = 82$ shell closure. Extensive calculations of the mass yields for 987 nuclides were performed in Ref. [152]. The Brownian shape-motion method [153] was applied, which involves the "random walks" to determine the most probable fission path on the previously calculated five-dimensional potential-energy surfaces [151]. One of the aims of this study was to establish theoretically whether ^{178,180}Hg represent separate cases of asymmetric fission in this region, or whether they belong to a broad contiguous region of asymmetric fission, and if so, its extent. Figure 3.20 shows the map of expected asymmetric and symmetric fission, whereby a broad island of asymmetric fission in the neutron-deficient lead region is predicted. In agreement with the experimental data, this new region of asymmetric fission also includes ^{178,180}Hg, though they are predicted to lie on its left-most border. Furthermore, this island is separated from the classical region of asymmetric fission in the actinides by an extended area of symmetric fission (see the large blue region in figure).

Therefore, to explore further this region, and especially the predicted asymmetric-to-symmetric transition of FFMDs between ^{178,180}Hg and e.g. ^{204,208}Rn, ^{210}Ra (which are known to fission symmetrically), the βDF experiments at ISOLDE aimed at studies of transitional βDF isotopes of ^{194,196}At and ^{202}Fr [116, 131]. The low-energy fission data for daughter (after β decay) isotopes ^{194,196}Po and ^{202}Rn, respectively, were collected. A summary of the results is shown in Fig. 3.21.

Experimentally, the mixture of two fission modes can be directly manifested by the two observables: the appearance of three peaks in the fission fragments energy distribution and by the skewness/broadness of the TKE distribution for fission fragments.

To illustrate these effects, for a reference, the top panel in the leftmost column of Fig. 3.21 shows the two-dimensional Si1-Si2 energy plot of coincident FFs of the daughter isotope ^{180}Hg (the same plot as in Fig. 3.17). The dominant asymmetric fission of ^{180}Hg is clearly demonstrated by a double-humped structure seen in this plot, with practically no events in between the peaks. The respective single-peaked and quite narrow Gaussian-like TKE distribution, depicted in the middle panel of the same column, indicates that a single fission mode dominates in ^{180}Hg. Finally, the deduced clearly asymmetric FFMD is depicted in black in the bottom panel, whereby the most probable fission fragments were found in the vicinity of ^{80}Kr and ^{100}Ru.

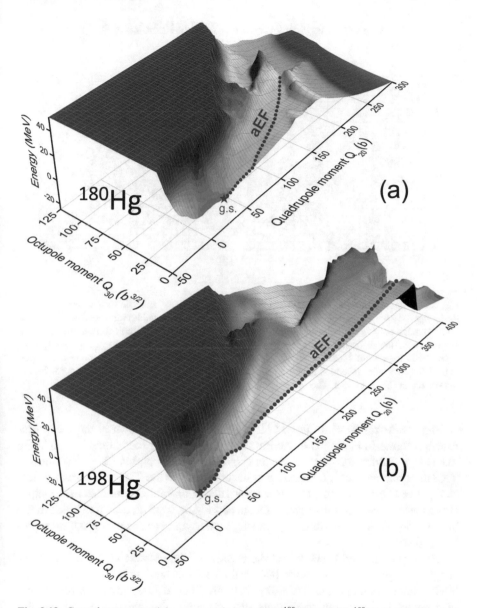

Fig. 3.19 Ground-state potential-energy surfaces for (**a**) ^{180}Hg and (**b**) ^{198}Hg in the (Q_{20}, Q_{30}) plane calculated in HFB-SkM*. The Q_{20} value represents the elongation along the axial symmetry, while Q_{30} parameter defines the deviation from the axial symmetry. The static fission pathway aEF corresponding to asymmetric elongated fragments is marked. The difference between the two nuclei is mainly seen in the magnitude of the final fission fragments mass asymmetry, corresponding to very different Q_{30} values. The figure is taken from [150]

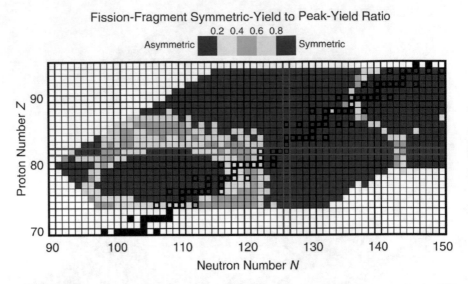

Fig. 3.20 Calculated ratios of the intensities of symmetric to asymmetric fission modes for 987 fissioning isotopes. Black squares (open in colored regions, filled outside) indicate β-stable nuclei. Two extended regions of predominantly asymmetric fission (small symmetry-to-asymmetry ratios) are drawn in the red color, the one in the left bottom corner is the region of a new type of asymmetric fission and includes ^{178,180}Hg, while the previously known asymmetric fission region in the heavy actinides is seen in the top right corner. The region of predominantly symmetric fission in between is shown in the blue. The figure is taken from [152]

The results for βDF studies of ^{194,196}At and ^{202}Fr are shown in the second-to-fourth columns in Fig. 3.21. In contrast to ^{180}Hg, a single broad hump is seen in the 2D energy distribution of ^{194,196}Po and ^{202}Rn (top row, columns 2–4). In addition, TKE distributions are significantly broader compared to the ^{180}Hg reference (middle row), as can be concluded from the standard deviation values, extracted from single-Gaussian fits, see [131] for details. Deduced FFMDs spectra, drawn in black in the bottom row, exhibit a mixture of symmetry with asymmetry, resulting in their triple-humped shape.

The triple-humped FFMDs and the breadth of the extracted TKE distributions suggest the presence of at least two distinct fission modes in these nuclei, each having different mass and TKE distributions. This feature was therefore further investigated by discriminating between fission events with high or low TKE, similar to the method used in Refs. [154, 155] to illustrate the bimodal fission in the transfermium region. In Fig. 3.21, FFMDs of fission events with respectively higher or lower TKE in comparison to a certain threshold energy E_{thres} are shown by respectively the dashed blue and full green lines. The value E_{thres} was arbitrarily taken as the mean TKE value and is indicated by a dashed red line on the TKE distributions and the 2D energy plots. Remarkably, the ^{194,196}Po cases exhibit a narrow symmetric distribution for fragments with higher TKE, while a broader, possibly asymmetric structure is observed for lower TKE. In contrast, this feature

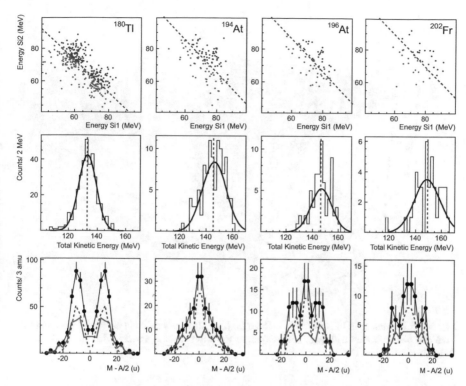

Fig. 3.21 Summary plot of the ISOLDE experiments to study βDF of ^{180}Tl, ^{194,196}At and ^{202}Fr. 2D energy distribution of coincident fission fragments in two silicon detectors (top), total kinetic energy (middle) and mass distributions (bottom) of the investigated nuclei are shown. The green and blue curves represent data below and above the average TKE values for each case shown by the red dashed lines in the first and second rows of the plot. Details are given in the main text. Figure is taken from [116, 131]

is absent in ^{180}Hg in which only one asymmetric fission mode was identified. In the case of ^{202}Rn, statistics prohibit drawing definitive conclusions.

These results establish a multimodal fission for these three isotopes, lying in the transitional region between the asymmetry of ^{178,180}Hg and symmetry of e.g. ^{204,208}Rn and ^{210}Ra. Self-consistent PES calculations performed within the HBF-D1S framework [150] provide a clear insight in the underlying reasons for the occurrence of the multimodal fission in this region. As an example, Fig. 3.22 shows the PES for ^{196}Po, where two distinct competing paths—an asymmetric and symmetric—are marked. Beyond $Q_{20} = 250$ b, the PES flattens in such a way that a mildly asymmetric fission pathway competes with the symmetric pathway, which allows multimode fission to happen.

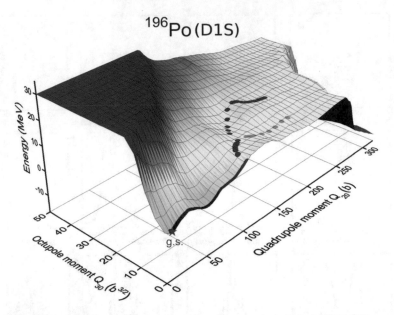

Fig. 3.22 Ground-state PES for ^{196}Po in the (Q_{20}, Q_{30}) plane calculated in the HFB-D1S approach. Two competing fission pathways corresponding to different mass asymmetries are marked. The figure is taken from [150]

Such a flat, relatively structure-less PES is expected to represent quite a general behavior of PES's in this region of nuclei, and it is very different from the typical PES's in the actinides, where a dominant asymmetric valley is usually present, as discussed in respect of Fig. 3.18. It is also very different from the PES's for e.g. ^{180}Hg, which might indicate that the fission below and above the shell closure at $Z = 82$ also has some inherent differences. Clearly, the outcome of any FFMD calculations on such flat PES's will strongly depend on specific details of a subtle and complex interplay between several degrees of freedom, including a yet not fully understood dependence on the excitation energy. In such a way, these data provide a crucial text of the modern fission approaches.

3.3.5.4 Complementary Approaches to Study Low-Energy Fission in the Lead Region

Unfortunately, due to relatively low fission rates of only up to some tens fissions per hour in the present βDF experiments at ISOLDE, no further details could be extracted, unless much longer experiments are performed. In this respect, earlier and recent experiments exploiting the Coulomb excitation of relativistic radioactive beam followed by fission at the SOFIA setup at GSI [156], have all potential to establish a complementary way for low-energy fission studies in this region. One

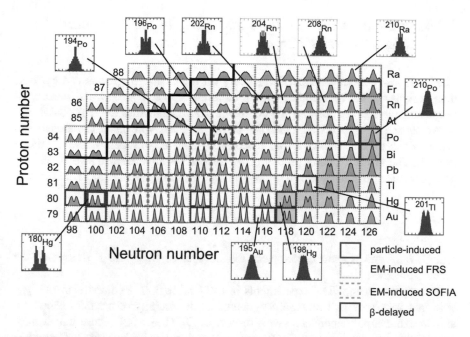

Fig. 3.23 Calculated FFMDs (gray), with fission-fragment masses on the horizontal and their relative yields on the vertical axis, for even-N neutron-deficient isotopes between gold and radium at excitation energies slightly above the theoretical fission-barrier heights $B_{f,th}$ from Ref. [114]. The calculated yields are compared with selected experimental FFMDs from particle-induced (blue symbols, [157, 158]), βDF (red, [120, 129, 131]) and Coulex-induced fission from FRS@GSI (green, [98, 99]) and SOFIA (dashed light green, [156]). The isotopes ^{180,190}Hg [159], ^{182}Hg [160] and ^{179,189}Au [161], recently measured by fusion-fission reactions are also marked in blue. The border of the lightest known isotopes is shown by the thick solid line, β-stable nuclei are shown on a gray background. Figure is modified from [131]

of the main advantages of SOFIA is its access to all types of nuclides—odd-odd, even-even and odd-A, while only the odd-odd cases can be studied via βDF, see Sect. 3.3.2. The feasibility of this approach was already confirmed by the first SOFIA campaign, which reached some of the neutron-deficient Hg isotopes.

This complementarity is demonstrated by Fig. 3.23, which shows a subset of the calculated data from Fig. 3.20, but in the mass-yield representation, with selected examples (solely due to the space limitation) of the measured FFMDs via βDF, Coulex and fusion-fission approaches. A good agreement between measured and calculated FFMDs can be noted for many nuclides, shown in the plot, e.g. for ^{180}Hg, ^{201}Tl, ^{210}Po, ^{204,208}Rn, ^{210}Ra. On the other hand, one also notices a clear discrepancy for e.g. ^{195}Au and ^{198}Hg, for which a strongly asymmetric mass division is predicted, while experimentally a symmetric mass split was observed, see also [131] for further details.

3.3.6 β-Delayed Fission Rates and Their Use to Determine Fission Barrier Heights

The previous sections concentrated mainly on the fission fragments energy/mass distributions and their theoretical interpretation. In this section, we will discuss what interesting physics conclusions can be drawn from the measured βDF probabilities.

At present, it is believed that βDF could, together with neutron-induced and spontaneous fission, influence the fission-recycling in r-process nucleosynthesis [93, 94]. Therefore, a reliable prediction of the relative importance of βDF in nuclear decay, often expressed by the βDF probability $P_{\beta DF}$, is needed. $P_{\beta DF}$ is defined as

$$P_{\beta DF} = \frac{N_{\beta DF}}{N_\beta}, \tag{3.22}$$

where $N_{\beta DF}$ and N_β are respectively the number of βDF and β decays of the precursor nucleus.

Before the recent βDF experiments in the lead region, a comparison of $P_{\beta DF}$ data in a relatively narrow region of nuclei in the vicinity of uranium showed a simple exponential dependence with respect to Q_β [136, 162]. These nuclei have comparable and relative low $Q_\beta(Parent)$ and fission-barrier heights $B_f(Daughter)$ values, $Q_\beta \sim 3$–6 MeV and $B_f \sim 4$–6 MeV, respectively. In addition, these nuclei have a typical N/Z ratio around ~ 1.4–1.5, which is close to that of traditional spontaneous fission of heavy actinides.

The recent βDF studies at ISOLDE allow to further explore such systematic features by including the newly obtained data in the neutron-deficient lead region, in which the βDF nuclides have significantly different N/Z ratios (~ 1.2–1.3), B_f (~ 7–10 MeV) and Q_β values (~ 9–11 MeV) as compared to those in the uranium region.

However, from an experimental point of view, the dominant α-branching ratio ($\geq 90\%$) in most βDF precursors in the neutron-deficient lead region [17] makes precise determination of N_β in Eq. (3.22) difficult. Therefore, the partial βDF half-life $T_{1/2,\beta DF}$, as discussed in details in [110], can be useful.

By analogy with other decay modes, $T_{1/2,\beta DF}$ is defined by

$$T_{1/2,\beta DF} = T_{1/2} \frac{N_{dec,tot}}{N_{\beta DF}}, \tag{3.23}$$

where $T_{1/2}$ represents the total half-life and $N_{dec,tot}$ the number of decayed precursor nuclei. The relation between $T_{1/2,\beta DF}$ and $P_{\beta DF}$ can be derived from Eqs. (3.22) and (3.23) as

$$T_{1/2,\beta DF} = \frac{T_{1/2}}{b_\beta P_{\beta DF}}, \tag{3.24}$$

with b_β denoting the β-branching ratio. If the α-decay channel dominates, as is often the case in the neutron-deficient lead region, one can safely approximate $N_{\text{dec,tot}}$ in Eq. (3.23) by the amount of α decays N_α. This removes part of the systematic uncertainties as the fission fragments and the alpha decay are detected in the same silicon detector with identical detection efficiency.

Following [132, 163], the expression for $P_{\beta DF}$, as a function of excitation energy E, is given by

$$P_{\beta DF} = \frac{\int_0^{Q_\beta} S_\beta(E)F(Q_\beta - E)\frac{\Gamma_f(E)}{\Gamma_{\text{tot}}(E)}dE}{\int_0^{Q_\beta} S_\beta(E)F(Q_\beta - E)dE}, \qquad (3.25)$$

whereby the β-strength function of the parent nucleus is denoted by S_β and the Fermi function by F. The fission and total decay widths of the daughter, after β decay, are respectively given by Γ_f and Γ_{tot}. Equation (3.24) can be combined with Eq. (3.25) to deduce the decay constant of βDF, defined as $\lambda_{\beta DF} = \ln(2)/T_{1/2,\beta DF}$, as

$$\lambda_{\beta DF} = \int_0^{Q_\beta} S_\beta(E)F(Q_\beta - E)\frac{\Gamma_f(E)}{\Gamma_{\text{tot}}(E)}dE. \qquad (3.26)$$

This expression shows a few important points. First of all, the strong dependence on the fission barrier B_f of the daughter nucleus (via the Γ_f) allows the determination of its fission barrier. This method was indeed used in several studies, see e.g. [122, 124, 125, 164]. It is important to note that, by exploring four variants of β strength function, it was suggested that a specific choice of S_β only weakly influences the calculation of $P_{\beta DF}$ [125]. The Fermi function F can be fairly well described by a function linearly proportional to $(Q_{EC} - E)^2$ [110] for EC decay.

A dedicated study [110] was performed to verify Eq. (3.26) by using experimental βDF partial half-lives and theoretical values for $(Q_\beta - B_f)$. Tabulated fission barriers from four different fission models were used (see Table 1 in [110]), of which three are based on a macroscopic-microscopic and one on a mean-field approach. The latter model employs the Extended Thomas-Fermi plus Strutinsky Integral (ETFSI) method [165], but tabulated barriers for the most neutron-deficient isotopes of our interest are absent in literature. The microscopic-macroscopic approaches all rely on shell corrections from [113] and describe the macroscopic structure of the nucleus by either a TF model [115], LDM [166] or the FRLDM [114]. The Q_β values were taken from the 2012 atomic mass evaluation tables [17] and are derived from the difference between the atomic masses of parent M_P and daughter M_D nuclei.

Figure 3.24 shows $\log_{10}(T_{1/2,\beta DF})$ against $(Q_\beta - B_f)$ for the fission barriers from the four different models under consideration. Using the same evaluation criteria as proposed in [105] for $P_{\beta DF}$ measurements, 13 reliable $T_{1/2,\beta DF}$ values, were selected. These data points, represented by the closed symbols, are fitted by a linear function. An equal weight to all fit points is given because the experimental uncertainties on

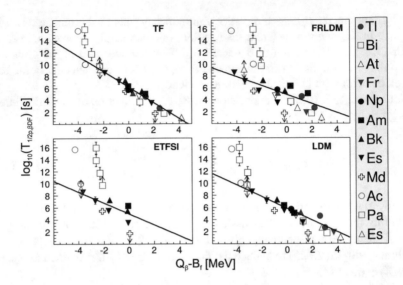

Fig. 3.24 Plots of $T_{1/2,\beta DF}$ versus $(Q_\beta - B_f)$ for four different models, see the main text. The closed symbols, representing "reliable" values for $T_{1/2,\beta DF}$ are used for a linear fit with equal weights to the data points. Other data are indicated by the open symbols. The color code represents the different regions of the nuclear chart for which βDF has been experimentally observed: the neutron-deficient lead region (red), the neutron-deficient (black) and neutron-rich (blue) uranium region

$\log_{10}(T_{1/2,\beta DF})$ are in most cases much smaller than the deviation of the data points with the fitted line. The remaining data points from Table 1 from [110] are shown by open symbols and were excluded from the fit. The color code discriminates between the neutron-deficient lead region (red), neutron-deficient (black) and neutron-rich (blue) uranium region.

Figure 3.24 illustrates a linear dependence of $\log_{10}(T_{1/2,\beta DF})$ on $(Q_\beta - B_f)$ for TF and LDM barriers for over 7 orders of magnitude of $T_{1/2,\beta DF}$. The dependence is somewhat less pronounced for the FRLDM model. A similar linear trend is observed for the ETFSI model, but the lack of tabulated fission barriers in the neutron-deficient region, especially in the lead region, prohibits drawing definite conclusions.

In contrast to a rather good agreement for most neutron-deficient nuclei, all models show a larger systematical deviation from this linear trend for the neutron-rich βDF precursors ^{228}Ac and ^{234,236}Pa. In [105], concerns were raised on the accuracy of the $P_{\beta DF}$ values measured in this region, which could explain this deviation. Note also that the precursors in this region of the nuclear chart undergo β^- decay in contrast to the EC-delayed fission on the neutron-deficient side for which Eq. (3.26) was deduced. In particular, the Fermi function for β^- decay is approximately proportional to $(Q_\beta - E)^5$ [132, 167], in contrast to the quadratic dependence on $(Q_\beta - E)$ for EC decay, and some modifications of the Γ_f/Γ_{tot} ratio might need to be considered, see [110].

3.4   Conclusion and Outlook

Despite α decay was discovered more than a century ago and fission 80 years ago, a full microscopic understanding of these important decay processes is still missing.

α decay is now a standard experimental tool in radioactive beam research programs. As an example, we discussed its use to unravel nuclear-structure effects in the $N \sim Z$ nuclei in the vicinity of ^{100}Sn and the neutron-deficient lead region. α decay is furthermore indispensable for the study and identification of new elements and new superheavy isotopes. While not explicitly addressed in these notes, it plays an important role in the atomic physics and chemistry research program at the end of Mendeleev's table.

βDF represents a unique probe to study low-energy fission and provides new information that improves our understanding of fission. With the availability of intense neutron-deficient beams in the lead region and using the mechanism of βDF, the low-energy fission process could be characterized in a new area of the nuclear chart previously unexplored by low-energy fission. A new region of asymmetric fission was identified and this finding had an impact on the theoretical description of the fission process.

New or upgraded facilities that will become operational in the next decade will have a profound impact on this research field. Facilities like e.g. the Dubna superheavy element factory, the GARIS2 separators at RIKEN, the new heavy ion injector at GSI and the Super Separator Spectrometer (S3) at GANIL, will use higher primary beam intensities and longer beam times. This creates potential to discover new elements and isotopes but it will, at the same time, allow for more detailed and higher precision studies in the heavy and superheavy element region. Also the plans to develop the next generation ISOL facility, called the European Isotope Separator On-Line (EURISOL) facility, will have major impact on α and βDF studies in the heavy element region both on the neutron-deficient and neutron-rich side. For example, provided the necessary degree of isomer selectivity can be reached, βDF studies on isomeric beams might become possible. Moreover, new observables, like prompt neutron and γ-ray energies and multiplicities, will become available as well as more accurate fission fragment yield distributions, total kinetic energies and βDF probabilities for very neutron-rich nuclei. This information will guide refinements and improvements of the nuclear models used to explain e.g. the r-process nucleosynthesis.

Finally, it should be mentioned that α decay and fission are important in practical applications like medical radioisotope and energy production. An in-depth understanding of both processes might also lead to more efficient procedures and new usage in the applied sector.

Acknowledgements PVD acknowledges the support from KU Leuven Research Office, the Research Foundation - Flanders (FWO Vlaanderen) and the IAP network program P8/12 (Belgian Science Policy office). AA acknowledges the support from Science and Technology Facilities Council of the UK.

References

1. M. Huyse, *The Why and How of Radioactive-Beam Research*. Lecture Notes in Physics, vol. 651 (Springer, Berlin, 2004), pp. 1–32
2. G. Gamow, Z. Phys. A: Hadrons Nucl. **51**, 204–212 (1928). https://doi.org/10.1007/BF01343196
3. R. Gurney, E. Condon, Nature **122**, 439 (1928)
4. R.W. Gurney, E.U. Condon, Phys. Rev. **33**, 127–140 (1929)
5. O. Hahn, F. Strassmann, Naturwissenschaften **27**, 11–15 (1939)
6. N. Bohr, J.A. Wheeler, Phys. Rev. **56**, 426–450 (1939)
7. C. Angulo, *Experimental Tools for Nuclear Astrophysics*. Lecture Notes in Physics, vol. 761 (Springer, Heidelberg, 2009), pp. 253–282
8. K. Langanke, F.-K. Thielemann, M. Wiescher, *Nuclear Astrophysics and Nuclei Far from Stability*. Lecture Notes in Physics, vol. 651 (Springer, Berlin, 2004), pp. 383–467
9. P. Van Duppen, *Isotope Separation On Line and Post Acceleration*. Lecture Notes in Physics on Physics, vol. 700 (Springer, Berlin, 2006), pp. 37–77
10. K. Heyde, *Basic Ideas and Concepts in Nuclear Physics: An Introductory Approach*. Series in Fundamental and Applied Nuclear Physics, 3th edn. (Institute of Physics Publishing, Bristol, 2004)
11. S.S.M. Wong, *Introductory Nuclear Physics* (Wiley, New York, 1998)
12. W.D. Loveland, D.J. Morrissey, G.T. Seaborg, *Modern Nuclear Chemistry* (Wiley, New York, 2006)
13. E. Roeckl, *Alpha Decay*. Nuclear Decay Modes (IOP Publishing, Bristol, 1996), pp. 237–274
14. E. Roeckl, *Decay Studies of N/Z Nuclei*. Lecture Notes in Physics, vol. 651 (Springer, Berlin, 2004), pp. 223–261
15. D.S. Delion, *Theory of Particle and Cluster Emission*. Lecture Notes in Physics, vol. 819 (Springer, Berlin, 2010), pp. 3–293
16. G. Audi, M. Wang, A. Wapstra, F. Kondev, M. MacCormick, X. Xu, B. Pfeiffer, Chin. Phys. C **36**, 1287 (2012)
17. M. Wang, G. Audi, A.H. Wapstra, F.G. Kondev, M. MacCormick, X. Xu, B. Pfeiffer, Chin. Phys. C **38**, 1603 (2012)
18. P. de Marcillac, N. Coron, G. Dambier, J. Leblanc, J.-P. Moalic, Nature **422**(6934), 876–878 (2003)
19. H. Geiger, J. Nuttall, Philos. Mag. **22**(130), 613–621 (1911)
20. H. Geiger, Z. Phys. **8**(1), 45–57 (1922)
21. V. Viola, G. Seaborg, J. Inorg. Nucl. Chem. **28**(3), 741–761 (1966)
22. B.A. Brown, Phys. Rev. C **46**, 811–814 (1992)
23. H. Koura, J. Nucl. Sci. Technol. **49**(8), 816–823 (2012)
24. D. Ni, Z. Ren, Ann. Phys. **358**, 108–128 (2015). School of Physics at Nanjing University
25. C. Qi, A. Andreyev, M. Huyse, R. Liotta, P.V. Duppen, R. Wyss, Phys. Lett. B **734**, 203–206 (2014)
26. B. Buck, A.C. Merchant, S.M. Perez, Phys. Rev. Lett. **65**, 2975–2977 (1990)
27. B. Buck, A. Merchant, S. Perez, At. Data Nucl. Data Tables **54**(1), 53–73 (1993)
28. B. Buck, A.C. Merchant, S.M. Perez, Phys. Rev. C **45**, 2247–2253 (1992)
29. A.N. Andreyev, M. Huyse, P. Van Duppen, C. Qi, R.J. Liotta, S. Antalic, D. Ackermann, S. Franchoo, F.P. Heßberger, S. Hofmann, I. Kojouharov, B. Kindler, P. Kuusiniemi, S.R. Lesher, B. Lommel, R. Mann, K. Nishio, R.D. Page, B. Streicher, Š. Šáro, B. Sulignano, D. Wiseman, R.A. Wyss, Phys. Rev. Lett. **110**, 242502 (2013)
30. J.O. Rasmussen, Phys. Rev. **113**, 1593–1598 (1959)
31. C. Qi, F.R. Xu, R.J. Liotta, R. Wyss, Phys. Rev. Lett. **103**, 072501 (2009)
32. C. Qi, F.R. Xu, R.J. Liotta, R. Wyss, M.Y. Zhang, C. Asawatangtrakuldee, D. Hu, Phys. Rev. C **80**, 044326 (2009)
33. R. Lovas, R. Liotta, A. Insolia, K. Varga, D. Delion, Phys. Rep. **294**(5), 265–362 (1998)

34. J. Wauters, P. Dendooven, M. Huyse, G. Reusen, P. Van Duppen, P. Lievens, Phys. Rev. C **47**, 1447–1454 (1993)
35. H.J. Mang, Phys. Rev. **119**, 1069–1075 (1960)
36. T. Fliessbach, H.J. Mang, J.O. Rasmussen, Phys. Rev. C **13**, 1318–1323 (1976)
37. R.G. Thomas, Prog. Theor. Phys. **12**(3), 253 (1954)
38. K. Varga, R.G. Lovas, R.J. Liotta, Phys. Rev. Lett. **69**, 37–40 (1992)
39. C. Qi, A.N. Andreyev, M. Huyse, R.J. Liotta, P. Van Duppen, R.A. Wyss, Phys. Rev. C **81**, 064319 (2010)
40. D.S. Delion, A. Insolia, R.J. Liotta, Phys. Rev. C **46**, 1346–1354 (1992)
41. D.S. Delion, S. Peltonen, J. Suhonen, Phys. Rev. C **73**, 014315 (2006)
42. M. Patial, R.J. Liotta, R. Wyss, Phys. Rev. C **93**, 054326 (2016)
43. S.N. Liddick, R. Grzywacz, C. Mazzocchi, R.D. Page, K.P. Rykaczewski, J.C. Batchelder, C.R. Bingham, I.G. Darby, G. Drafta, C. Goodin, C.J. Gross, J.H. Hamilton, A.A. Hecht, J.K. Hwang, S. Ilyushkin, D.T. Joss, A. Korgul, W. Królas, K. Lagergren, K. Li, M.N. Tantawy, J. Thomson, J.A. Winger, Phys. Rev. Lett. **97**, 082501 (2006)
44. I.G. Darby, R.K. Grzywacz, J.C. Batchelder, C.R. Bingham, L. Cartegni, C.J. Gross, M. Hjorth-Jensen, D.T. Joss, S.N. Liddick, W. Nazarewicz, S. Padgett, R.D. Page, T. Papenbrock, M.M. Rajabali, J. Rotureau, K.P. Rykaczewski, Phys. Rev. Lett. **105**, 162502 (2010)
45. D.N. Poenaru, M. Ivascu, A. Sandulescu, J. Phys. G: Nucl. Phys. **5**(10), L169 (1979)
46. D.N. Poenaru, W. Greiner, *Fission Approach to Alpha-Decay in Superheavy Nuclei* (World Scientific, Singapore, 2012), pp. 321–328
47. C.E. Düllmann, Nucl. Instrum. Methods Phys. Res., Sect. B **266**(19), 4123–4130 (2008). Proceedings of the XVth International Conference on Electromagnetic Isotope Separators and Techniques Related to Their Applications.
48. M. Borge, K. Blaum, J. Phys. G: Nucl. Part. Phys. **45**, 010301 (2018)
49. J. Dilling, R. Krücken, L. Merminga, Hyperfine Interact. **225**, 1–282 (2014)
50. Z. Liu, J. Kurcewicz, P. Woods, C. Mazzocchi, F. Attallah, E. Badura, C. Davids, T. Davinson, J. Döring, H. Geissel, M. Górska, R. Grzywacz, M. Hellström, Z. Janas, M. Karny, A. Korgul, I. Mukha, M. Pfützner, C. Plettner, A. Robinson, E. Roeckl, K. Rykaczewski, K. Schmidt, D. Seweryniak, H. Weick, Nucl. Inst. Methods Phys. Res. A **543**(2), 591–601 (2005)
51. J. Kurcewicz, Z. Liu, M. Pfützner, P. Woods, C. Mazzocchi, K.-H. Schmidt, A. Kelić, F. Attallah, E. Badura, C. Davids, T. Davinson, J. Döring, H. Geissel, M. Górska, R. Grzywacz, M. Hellström, Z. Janas, M. Karny, A. Korgul, I. Mukha, C. Plettner, A. Robinson, E. Roeckl, K. Rykaczewski, K. Schmidt, D. Seweryniak, K. Sümmerer, H. Weick, Nuc. Phys. A **767**, 1–12 (2006)
52. D.J. Morrissey, B.M. Sherrill, *In-Flight Separation of Projectile Fragments*. Lecture Notes in Physics, vol. 651 (Springer, Berlin, 2004), pp. 113–135
53. S. Hofmann, J. Phys. G: Nucl. Part. Phys. **42**(11), 114001 (2015)
54. A.N. Andreyev, D. Ackermann, F.P. Heßberger, S. Hofmann, M. Huyse, G. Münzenberg, R.D. Page, K. Van de Vel, P. Van Duppen, Nucl. Instrum. Methods A **533**(3), 409–421 (2004)
55. A.N. Andreyev, S. Antalic, M. Huyse, P. Van Duppen, D. Ackermann, L. Bianco, D.M. Cullen, I.G. Darby, S. Franchoo, S. Heinz, F.P. Heßberger, S. Hofmann, I. Kojouharov, B. Kindler, A.P. Leppänen, B. Lommel, R. Mann, G. Münzenberg, J. Pakarinen, R.D. Page, J.J. Ressler, S. Saro, B. Streicher, B. Sulignano, J. Thomson, R. Wyss, Phys. Rev. C **74**, 064303 (2006)
56. G. Münzenberg, W. Faust, S. Hofmann, P. Armbruster, K. Güttner, H. Ewald, Nucl. Instrum. Methods **161**(1), 65–82 (1979)
57. S. Hofmann, *Superheavy Elements*. Lecture Notes in Physics, vol. 761 (Springer, Berlin, 2009), pp. 203–252
58. Y.T. Oganessian, V.K. Utyonkov, Rep. Prog. Phys. **78**(3), 036301 (2015)
59. Z. Janas, C. Mazzocchi, L. Batist, A. Blazhev, M. Górska, M. Kavatsyuk, O. Kavatsyuk, R. Kirchner, A. Korgul, M. La Commara, K. Miernik, I. Mukha, A. Płochocki, E. Roeckl, K. Schmidt, Eur. Phys. J. A **23**(2), 197–200 (2005)

60. D. Seweryniak, M.P. Carpenter, S. Gros, A.A. Hecht, N. Hoteling, R.V.F. Janssens, T.L. Khoo, T. Lauritsen, C.J. Lister, G. Lotay, D. Peterson, A.P. Robinson, W.B. Walters, X. Wang, P.J. Woods, S. Zhu, Phys. Rev. Lett. **99**, 022504 (2007)
61. R. Neugart, G. Neyens, *Nuclear Moments*. Lecture Notes in Physics, vol. 700 (Springer, Berlin, 2006), pp. 135–189
62. R. Ferrer, A. Barzakh, B. Bastin, R. Beerwerth, M. Block, P. Creemers, H. Grawe, R. de Groote, P. Delahaye, X. Fléchard, S. Franchoo, S. Fritzsche, L.P. Gaffney, L. Ghys, W. Gins, C. Granados, R. Heinke, L. Hijazi, M. Huyse, T. Kron, Y. Kudryavtsev, M. Laatiaoui, N. Lecesne, M. Loiselet, F. Lutton, I.D. Moore, Y. Martínez, E. Mogilevskiy, P. Naubereit, J. Piot, S. Raeder, S. Rothe, H. Savajols, S. Sels, V. Sonnenschein, J.-C. Thomas, E. Traykov, C. Van Beveren, P. Van den Bergh, P. Van Duppen, K. Wendt, A. Zadvornaya, Nat. Commun. **8**, 14520 (2017)
63. H. Morinaga, Phys. Rev. **101**, 254–258 (1956)
64. K. Heyde, J.L. Wood, Rev. Mod. Phys. **83**, 1467–1521 (2011)
65. J.L. Wood, K. Heyde, J. Phys. G: Nucl. Part. Phys. **43**(2), 020402 (2016)
66. J. Elseviers, A.N. Andreyev, S. Antalic, A. Barzakh, N. Bree, T.E. Cocolios, V.F. Comas, J. Diriken, D. Fedorov, V.N. Fedosseyev, S. Franchoo, J.A. Heredia, M. Huyse, O. Ivanov, U. Köster, B.A. Marsh, R.D. Page, N. Patronis, M. Seliverstov, I. Tsekhanovich, P. Van den Bergh, J. Van De Walle, P. Van Duppen, M. Venhart, S. Vermote, M. Veselský, C. Wagemans, Phys. Rev. C **84**, 034307 (2011)
67. A.N. Andreyev, M. Huyse, P. Van Duppen, L. Weissman, D. Ackermann, J. Gerl, F.P. Heßberger, S. Hofmann, A. Kleinbohl, G. Munzenberg, S. Reshitko, C. Schlegel, H. Schaffner, P. Cagarda, M. Matos, S. Saro, A. Keenan, C. Moore, C.D. O'Leary, R.D. Page, M. Taylor, H. Kettunen, M. Leino, A. Lavrentiev, R. Wyss, K. Heyde, Nature **405**(6785), 430–433 (2000)
68. H. De Witte, A.N. Andreyev, N. Barré, M. Bender, T.E. Cocolios, S. Dean, D. Fedorov, V.N. Fedoseyev, L.M. Fraile, S. Franchoo, V. Hellemans, P.H. Heenen, K. Heyde, G. Huber, M. Huyse, H. Jeppessen, U. Köster, P. Kunz, S.R. Lesher, B.A. Marsh, I. Mukha, B. Roussière, J. Sauvage, M. Seliverstov, I. Stefanescu, E. Tengborn, K. Van de Vel, J. Van de Walle, P. Van Duppen, Y. Volkov, Phys. Rev. Lett. **98**, 112502 (2007)
69. T.E. Cocolios, W. Dexters, M.D. Seliverstov, A.N. Andreyev, S. Antalic, A.E. Barzakh, B. Bastin, J. Büscher, I.G. Darby, D.V. Fedorov, V.N. Fedosseyev, K.T. Flanagan, S. Franchoo, S. Fritzsche, G. Huber, M. Huyse, M. Keupers, U. Köster, Y. Kudryavtsev, E. Mané, B.A. Marsh, P.L. Molkanov, R.D. Page, A.M. Sjoedin, I. Stefan, J. Van de Walle, P. Van Duppen, M. Venhart, S.G. Zemlyanoy, M. Bender, P.-H. Heenen, Phys. Rev. Lett. **106**, 052503 (2011)
70. R. Julin, T. Grahn, J. Pakarinen, P. Rahkila, J. Phys. G: Nucl. Part. Phys. **43**(2), 024004 (2016)
71. D. Delion, A. Florescu, M. Huyse, J. Wauters, P. Van Duppen, A. Insolia, R. Liotta, ISOLDE Collaboration et al., Phys. Rev. Lett. **74**(20), 3939 (1995)
72. C. Xu, Z. Ren, Phys. Rev. C **75**, 044301 (2007)
73. D.S. Delion, A. Florescu, M. Huyse, J. Wauters, P. Van Duppen, A. Insolia, R.J. Liotta, Phys. Rev. Lett. **74**, 3939–3942 (1995)
74. D. Karlgren, R.J. Liotta, R. Wyss, M. Huyse, K.V.D. Vel, P.V. Duppen, Phys. Rev. C **73**, 064304 (2006)
75. D.S. Delion, R.J. Liotta, C. Qi, R. Wyss, Phys. Rev. C **90**, 061303 (2014)
76. Y. Oganessian, K. Rykaczewski, Phys. Today **68**(8), 32–38 (2015)
77. D. Rudolph, L.-I. Elding, C. Fahlander, S. Åberg, EPJ Web Conf. **131**, 00001 (2016)
78. S. Hofmann, EPJ Web Conf. **131**, 06001 (2016)
79. G. Bollen, *Traps for Rare Isotopes*. Lecture Notes in Physics, vol. 651 (Springer, Berlin, 2004), pp. 169–210
80. M. Block, D. Ackermann, K. Blaum, C. Droese, M. Dworschak, S. Eliseev, T. Fleckenstein, E. Haettner, F. Herfurth, F.P. Heßberger, S. Hofmann, J. Ketelaer, J. Ketter, H.-J. Kluge, G. Marx, M. Mazzocco, Y.N. Novikov, W.R. Plaß, A. Popeko, S. Rahaman, D. Rodríguez, C. Scheidenberger, L. Schweikhard, P.G. Thirolf, G.K. Vorobyev, C. Weber, Nature **463**(7282), 785–788 (2010)

81. M. Block, EPJ Web Conf. **131**, 05003 (2016)
82. P. Schury, M. Wada, Y. Ito, D. Kaji, F. Arai, M. MacCormick, I. Murray, H. Haba, S. Jeong, S. Kimura, H. Koura, H. Miyatake, K. Morimoto, K. Morita, A. Ozawa, M. Rosenbusch, M. Reponen, P.-A. Söderström, A. Takamine, T. Tanaka, H. Wollnik, Phys. Rev. C **95**, 011305 (2017)
83. F.-P. Heßberger, EPJ Web Conf. **131**, 02005 (2016)
84. P.-H. Heenen, B. Bally, M. Bender, W. Ryssens, EPJ Web Conf. **131**, 02001 (2016)
85. F. Gönnenwein, in *Contribution to the École Joliot Curie*. Neutrons and Nuclei (2014). https:// ejc2014.sciencesconf.org/
86. C. Wagemans (ed.), The Nuclear Fission Process (CRC Press, Boca Raton, 1991)
87. A.V. Karpov, A. Kelić, K.-H. Schmidt, J. Phys. G: Nucl. Part. Phys. **35**(3), 035104 (2008)
88. V.M. Strutinsky, Nucl. Phys. A **95**(2), 420–442 (1967)
89. V.M. Strutinsky, Nucl. Phys. A **122**(1), 1–33 (1968)
90. M. Brack, J. Damgaard, A.S. Jensen, H.C. Pauli, V.M. Strutinsky, C.Y. Wong, Rev. Mod. Phys. **44**, 320–405 (1972)
91. P. Armbruster, Rep. Prog. Phys. **62**, 465 525 (1998)
92. C.O. Wene, S.A.E. Johansson, Phys. Scr. **10A**, 156–162 (1974)
93. I. Panov, E. Kolbe, B. Pfeiffer, T. Rauscher, K.-L. Kratz, F.-K. Thielemann, Nucl. Phys. A **747**(2–4), 633–654 (2005)
94. I. Petermann, K. Langanke, G. Martínez-Pinedo, I. Panov, P.G. Reinhard, F.K. Thielemann, Eur. Phys. J. A **48**, 1–11 (2012)
95. S. Goriely, J.-L. Sida, J.-F. Lemaître, S. Panebianco, N. Dubray, S. Hilaire, A. Bauswein, H.-T. Janka, Phys. Rev. Lett. **111**, 242502 (2013)
96. M.J.G. Borge, Nucl. Instrum. Methods Phys. Res., Sect. B **376**, 408–412 (2016). Proceedings of the XVIIth International Conference on Electromagnetic Isotope Separators and Related Topics (EMIS2015), Grand Rapids, MI, USA, 11–15 May 2015
97. F.P. Heßberger, Eur. Phys. J. A **53**(4), 75 (2017)
98. K.-H. Schmidt, S. Steinhäuser, C. Böckstiegel, A. Grewe, A. Heinz, A. Junghans, J. Benlliure, H.G. Clerc, M. de Jong, J. Müller, M. Pfützner, B. Voss, Nucl. Phys. A **665**(3–4), 221–267 (2000)
99. K.-H. Schmidt, J. Benlliure, A.R. Junghans, Nucl. Phys. A **693**, 169–189 (2001)
100. B. Blank, M.J.G. Borge, Prog. Part. Nucl. Phys. **60**(2), 403–483 (2008)
101. M. Pfützner, M. Karny, L.V. Grigorenko, K. Riisager, Rev. Mod. Phys. **84**, 567–619 (2012)
102. V.I. Kuznetsov, N.K. Skobelev, G.N. Flerov, Yad. Fiz. **4**, 279 (1966) [Sov. J. Nucl. Phys. **4**, 202 (1967)]
103. V.I. Kuznetsov, N.K. Skobelev, G.N. Flerov, Yad. Fiz. **5**, 271 (1967) [Sov. J. Nucl. Phys. **5**, 191 (1967)]
104. N.K. Skobelev, Yad. Fiz. **15**, 444 (1972) [Sov. J. Nucl. Phys. **15**, 249 (1972)]
105. A.N. Andreyev, M. Huyse, P. Van Duppen, Rev. Mod. Phys. **85**, 1541–1559 (2013)
106. H.L. Hall, K.E. Gregorich, R.A. Henderson, C.M. Gannett, R.B. Chadwick, J.D. Leyba, K.R. Czerwinski, B. Kadkhodayan, S.A. Kreek, D.M. Lee, M.J. Nurmia, D.C. Hoffman, Phys. Rev. Lett. **63**, 2548–2550 (1989)
107. H.L. Hall, K.E. Gregorich, R.A. Henderson, C.M. Gannett, R.B. Chadwick, J.D. Leyba, K.R. Czerwinski, B. Kadkhodayan, S.A. Kreek, D.M. Lee, M.J. Nurmia, D.C. Hoffman, C.E.A. Palmer, P.A. Baisden, Phys. Rev. C **41**, 618–630 (1990)
108. H.L. Hall, K.E. Gregorich, R.A. Henderson, C.M. Gannett, R.B. Chadwick, J.D. Leyba, K.R. Czerwinski, B. Kadkhodayan, S.A. Kreek, N.J. Hannink, D.M. Lee, M.J. Nurmia, D.C. Hoffman, C.E.A. Palmer, P.A. Baisden, Phys. Rev. C **42**, 1480–1488 (1990)
109. S.A. Kreek, H.L. Hall, K.E. Gregorich, R.A. Henderson, J.D. Leyba, K.R. Czerwinski, B. Kadkhodayan, M.P. Neu, C.D. Kacher, T.M. Hamilton, M.R. Lane, E.R. Sylwester, A. Türler, D.M. Lee, M.J. Nurmia, D.C. Hoffman, Phys. Rev. C **50**, 2288–2296 (1994)
110. L. Ghys, A.N. Andreyev, S. Antalic, M. Huyse, P. Van Duppen, Phys. Rev. C **91**, 044314 (2015)
111. F.K. Thielemann, J. Metzinger, H. Klapdor, Z. Phys. A **309**, 301–317 (1983)

112. B.S. Meyer, W.M. Howard, G.J. Mathews, K. Takahashi, P. Möller, G.A. Leander, Phys. Rev. C **39**, 1876–1882 (1989)
113. P. Möller, J.R. Nix, W.D. Myers, W.J. Swiatecki, At. Data Nucl. Data Tables **59**(2), 185–381 (1995)
114. P. Möller, A.J. Sierk, T. Ichikawa, A. Iwamoto, M. Mumpower, Phys. Rev. C **91**, 024310 (2015)
115. W.D. Myers, W.J. Świątecki, Phys. Rev. C **60**, 014606 (1999)
116. V.L. Truesdale, A.N. Andreyev, L. Ghys, M. Huyse, P. Van Duppen, S. Sels, B. Andel, S. Antalic, A. Barzakh, L. Capponi, T.E. Cocolios, X. Derkx, H. De Witte, J. Elseviers, D.V. Fedorov, V.N. Fedosseev, F.P. Heßberger, Z. Kalaninová, U. Köster, J.F.W. Lane, V. Liberati, K.M. Lynch, B.A. Marsh, S. Mitsuoka, Y. Nagame, K. Nishio, S. Ota, D. Pauwels, L. Popescu, D. Radulov, E. Rapisarda, S. Rothe, K. Sandhu, M.D. Seliverstov, A.M. Sjödin, C. Van Beveren, P. Van den Bergh, Y. Wakabayashi, Phys. Rev. C **94**, 034308 (2016)
117. A.N. Andreyev, S. Antalic, D. Ackermann, S. Franchoo, F.P. Heßberger, S. Hofmann, M. Huyse, I. Kojouharov, B. Kindler, P. Kuusiniemi, S.R. Lesher, B. Lommel, R. Mann, G. Münzenberg, K. Nishio, R.D. Page, J.J. Ressler, B. Streicher, S. Saro, B. Sulignano, P. Van Duppen, D.R. Wiseman, Phys. Rev. C **73**, 024317 (2006)
118. A.N. Andreyev, S. Antalic, D. Ackermann, L. Bianco, S. Franchoo, S. Heinz, F.P. Heßberger, S. Hofmann, M. Huyse, I. Kojouharov, B. Kindler, B. Lommel, R. Mann, K. Nishio, R.D. Page, J.J. Ressler, P. Sapple, B. Streicher, S. Saro, B. Sulignano, J. Thomson, P. Van Duppen, M. Venhart, Phys. Rev. C **79**, 064320 (2009)
119. A.N. Andreyev, S. Antalic, D. Ackermann, L. Bianco, S. Franchoo, S. Heinz, F.P. Heßberger, S. Hofmann, M. Huyse, Z. Kalaninová, I. Kojouharov, B. Kindler, B. Lommel, R. Mann, K. Nishio, R.D. Page, J.J. Ressler, B. Streicher, S. Saro, B. Sulignano, P. Van Duppen, Phys. Rev. C **87**, 014317 (2013)
120. V. Liberati, A.N. Andreyev, S. Antalic, A. Barzakh, T.E. Cocolios, J. Elseviers, D. Fedorov, V.N. Fedoseeev, M. Huyse, D.T. Joss, Z. Kalaninová, U. Köster, J.F.W. Lane, B. Marsh, D. Mengoni, P. Molkanov, K. Nishio, R.D. Page, N. Patronis, D. Pauwels, D. Radulov, M. Seliverstov, M. Sjödin, I. Tsekhanovich, P. Van den Bergh, P. Van Duppen, M. Venhart, M. Veselský, Phys. Rev. C **88**, 044322 (2013)
121. J.F.W. Lane, A.N. Andreyev, S. Antalic, D. Ackermann, J. Gerl, F.P. Heßberger, S. Hofmann, M. Huyse, H. Kettunen, A. Kleinböhl, B. Kindler, I. Kojouharov, M. Leino, B. Lommel, G. Münzenberg, K. Nishio, R.D. Page, Š. Šáro, H. Schaffner, M.J. Taylor, P. Van Duppen, Phys. Rev. C **87**, 014318 (2013)
122. H.V. Klapdor, C.O. Wene, I.N. Isosimov, Y.W. Naumow, Z. Phys. A **292**, 249–255 (1979)
123. D. Habs, H. Klewe-Nebenius, V. Metag, B. Neumann, H.J. Specht, Z. Phys. A **285**, 53–57 (1978)
124. A. Staudt, M. Hirsch, K. Muto, H.V. Klapdor-Kleingrothaus, Phys. Rev. Lett. **65**, 1543 (1990)
125. M. Veselský, A.N. Andreyev, S. Antalic, M. Huyse, P. Möller, K. Nishio, A.J. Sierk, P. Van Duppen, M. Venhart, Phys. Rev. C **86**, 024308 (2012)
126. Y.P. Gangrsky, G.M. Marinesky, M.B. Miller, V.N. Samsuk, I.F. Kharisov, Yad. Fiz. **27**, 894 (1978) [Sov. J. Nucl. Phys. **27**, 475 (1978)]
127. A. Baas-May, J.V. Kratz, N. Trautmann, Z. Phys. A **322**, 457 (1985)
128. X. Yanbing, Z. Shengdong, D. Huajie, Y. Shuanggui, Y. Weifan, N. Yanning, L. Xiting, L. Yingjun, X. Yonghou, Phys. Rev. C **74**, 047303 (2006)
129. A.N. Andreyev, J. Elseviers, M. Huyse, P. Van Duppen, S. Antalic, A. Barzakh, N. Bree, T.E. Cocolios, V.F. Comas, J. Diriken, D. Fedorov, V. Fedosseev, S. Franchoo, J.A. Heredia, O. Ivanov, U. Köster, B.A. Marsh, K. Nishio, R.D. Page, N. Patronis, M. Seliverstov, I. Tsekhanovich, P. Van den Bergh, J. Van De Walle, M. Venhart, S. Vermote, M. Veselský, C. Wagemans, T. Ichikawa, A. Iwamoto, P. Möller, A.J. Sierk, Phys. Rev. Lett. **105**, 252502 (2010)

130. J. Elseviers, A.N. Andreyev, M. Huyse, P. Van Duppen, S. Antalic, A. Barzakh, N. Bree, T.E. Cocolios, V.F. Comas, J. Diriken, D. Fedorov, V.N. Fedosseev, S. Franchoo, L. Ghys, J.A. Heredia, O. Ivanov, U. Köster, B.A. Marsh, K. Nishio, R.D. Page, N. Patronis, M.D. Seliverstov, I. Tsekhanovich, P. Van den Bergh, J. Van De Walle, M. Venhart, S. Vermote, M. Veselský, C. Wagemans, Phys. Rev. C **88**, 044321 (2013)
131. L. Ghys, A.N. Andreyev, M. Huyse, P. Van Duppen, S. Sels, B. Andel, S. Antalic, A. Barzakh, L. Capponi, T.E. Cocolios, X. Derkx, H. De Witte, J. Elseviers, D.V. Fedorov, V.N. Fedosseev, F.P. Heßberger, Z. Kalaninová, U. Köster, J.F.W. Lane, V. Liberati, K.M. Lynch, B.A. Marsh, S. Mitsuoka, P. Möller, Y. Nagame, K. Nishio, S. Ota, D. Pauwels, R.D. Page, L. Popescu, D. Radulov, M.M. Rajabali, J. Randrup, E. Rapisarda, S. Rothe, K. Sandhu, M.D. Seliverstov, A.M. Sjödin, V.L. Truesdale, C. Van Beveren, P. Van den Bergh, Y. Wakabayashi, M. Warda, Phys. Rev. C **90**, 041301 (2014)
132. V.I. Kuznetsov, N.K. Skobelev, Phys. Part. Nucl. **30**, 666 (1999)
133. Y.A. Lazarev, Y.T. Oganessian, I.V. Shirokovsky, S.P. Tretyakova, V.K. Utyonkov, G.V. Buklanov, EPL (Europhys. Lett.) **4**(8), 893 (1987)
134. S.A. Kreek, H.L. Hall, K.E. Gregorich, R.A. Henderson, J.D. Leyba, K.R. Czerwinski, B. Kadkhodayan, M.P. Neu, C.D. Kacher, T.M. Hamilton, M.R. Lane, E.R. Sylwester, A. Türler, D.M. Lee, M.J. Nurmia, D.C. Hoffman, Phys. Rev. C **49**, 1859–1866 (1994)
135. D.C. Hoffman, D.M. Lee, K.E. Gregorich, M.J. Nurmia, R.B. Chadwick, K.B. Chen, K.R. Czerwinski, C.M. Gannett, H.L. Hall, R.A. Henderson, B. Kadkhodayan, S.A. Kreek, J.D. Leyba, Phys. Rev. C **41**, 631–639 (1990)
136. D.A. Shaughnessy, J.L. Adams, K.E. Gregorich, M.R. Lane, C.A. Laue, D.M. Lee, C.A. McGrath, J.B. Patin, D.A. Strellis, E.R. Sylwester, P.A. Wilk, D.C. Hoffman, Phys. Rev. C **61**, 044609 (2000)
137. J. Konki, J. Khuyagbaatar, J. Uusitalo, P. Greenlees, K. Auranen, H. Badran, M. Block, R. Briselet, D. Cox, M. Dasgupta, A.D. Nitto, C. Düllmann, T. Grahn, K. Hauschild, A. Herzán, R.-D. Herzberg, F. Heßberger, D. Hinde, R. Julin, S. Juutinen, E. Jäger, B. Kindler, J. Krier, M. Leino, B. Lommel, A. Lopez-Martens, D. Luong, M. Mallaburn, K. Nishio, J. Pakarinen, P. Papadakis, J. Partanen, P. Peura, P. Rahkila, K. Rezynkina, P. Ruotsalainen, M. Sandzelius, J. Sarén, C. Scholey, J. Sorri, S. Stolze, B. Sulignano, C. Theisen, A. Ward, A. Yakushev, V. Yakusheva, Phys. Lett. B **764**, 265–270 (2017)
138. D. Kaji, K. Morimoto, H. Haba, E. Ideguchi, H. Koura, K. Morita, J. Phys. Soc. Jpn. **85**(1), 015002 (2016)
139. G.L. Wilson, M. Takeyama, A.N. Andreyev, B. Andel, S. Antalic, W.N. Catford, L. Ghys, H. Haba, F.P. Heßberger, M. Huang, D. Kaji, Z. Kalaninova, K. Morimoto, K. Morita, M. Murakami, K. Nishio, R. Orlandi, A.G. Smith, K. Tanaka, Y. Wakabayashi, S. Yamaki, Phys. Rev. C **96**, 044315 (2017)
140. V.N. Fedosseev, L.E. Berg, D.V. Fedorov, Fink, D., O.J. Launila, R. Losito, B.A. Marsh, R.E. Rossel, S. Rothe, M.D. Seliverstov, A.M. Sjodin, K.D.A. Wendt, Rev. Sci. Instrum. **83**(2), 02A903 (2012)
141. B.A. Marsh, V.N. Fedosseev, D.A. Fink, T. Day Goodacre, R.E. Rossel, S. Rothe, D.V. Fedorov, N. Imai, M.D. Seliverstov, P. Molkanov, Hyperfine Interact. **227**(1), 101–111 (2014)
142. S. Rothe, A.N. Andreyev, S. Antalic, A. Borschevsky, L. Capponi, T. Cocolios, H. De Witte, E. Eliav, D. Fedorov, V. Fedosseev, D. Fink, S. Fritzsche, L. Ghys, M. Huyse, N. Imai, U. Kaldor, Y. Kudryavtsev, U. Koster, J. Lane, J. Lassen, Nat. Commun. **4**(1835), 1 (2013)
143. T. Ichikawa, A. Iwamoto, P. Möller, A.J. Sierk, Phys. Rev. C **86**, 024610 (2012)
144. P. Möller, J. Randrup, A.J. Sierk, Phys. Rev. C **85**, 024306 (2012)
145. S. Panebianco, J.-L. Sida, H. Goutte, J.-F. Lemaître, N. Dubray, S. Hilaire, Phys. Rev. C **86**, 064601 (2012)
146. A.V. Andreev, G.G. Adamian, N.V. Antonenko, Phys. Rev. C **86**, 044315 (2012)
147. A.V. Andreev, G.G. Adamian, N.V. Antonenko, A.N. Andreyev, Phys. Rev. C **88**, 047604 (2013)
148. A.V. Andreev, G.G. Adamian, N.V. Antonenko, Phys. Rev. C **93**, 034620 (2016)
149. M. Warda, A. Staszczak, W. Nazarewicz, Phys. Rev. C **86**, 024601 (2012)

150. J.D. McDonnell, W. Nazarewicz, J.A. Sheikh, A. Staszczak, M. Warda, Phys. Rev. C **90**, 021302 (2014)
151. P. Möller, D.G. Madland, A.J. Sierk, A. Iwamoto, Nature **409**(6822), 785–790 (2001)
152. P. Möller, J. Randrup, Phys. Rev. C **91**, 044316 (2015)
153. J. Randrup, P. Möller, Phys. Rev. Lett. **106**, 132503 (2011)
154. E.K. Hulet, J.F. Wild, R.J. Dougan, R.W. Lougheed, J.H. Landrum, A.D. Dougan, M. Schadel, R.L. Hahn, P.A. Baisden, C.M. Henderson, R.J. Dupzyk, K. Sümmerer, G.R. Bethune, Phys. Rev. Lett. **56**, 313–316 (1986)
155. E.K. Hulet, J.F. Wild, R.J. Dougan, R.W. Lougheed, J.H. Landrum, A.D. Dougan, P.A. Baisden, C.M. Henderson, R.J. Dupzyk, R.L. Hahn, M. Schädel, K. Sümmerer, G.R. Bethune, Phys. Rev. C **40**, 770–784 (1989)
156. J.-F. Martin, J. Taieb, A. Chatillon, G. Bélier, G. Boutoux, A. Ebran, T. Gorbinet, L. Grente, B. Laurent, E. Pellereau, H. Alvarez-Pol, L. Audouin, T. Aumann, Y. Ayyad, J. Benlliure, E. Casarejos, D. Cortina Gil, M. Caamaño, F. Farget, B. Fernández Domínguez, A. Heinz, B. Jurado, A. Kelić-Heil, N. Kurz, C. Nociforo, C. Paradela, S. Pietri, D. Ramos, J.-L. Rodríguez-Sànchez, C. Rodríguez-Tajes, D. Rossi, K.-H. Schmidt, H. Simon, L. Tassan-Got, J. Vargas, B. Voss, H. Weick, Eur. Phys. J. A **51**(12), 1–8 (2015)
157. M. Itkis, N. Kondrat'ev, S. Mul'gin, V. Okolovich, A. Rusanov, G. Smirenkin, Sov. J. Nucl. Phys. **52**, 601 (1990)
158. M. Itkis, N. Kondrat'ev, S. Mul'gin, V. Okolovich, A. Rusanov, G. Smirenkin, Sov. J. Nucl. Phys. **53**, 757 (1991)
159. K. Nishio, A.N. Andreyev, R. Chapman, X. Derkx, C.E. Düllmann, L. Ghys, F. Heßberger, K. Hirose, H. Ikezoe, J. Khuyagbaatar, B. Kindler, B. Lommel, H. Makii, I. Nishinaka, T. Ohtsuki, S. Pain, R. Sagaidak, I. Tsekhanovich, M. Venhart, Y. Wakabayashi, S. Yan, Phys. Lett. B **748**, 89–94 (2015)
160. E. Prasad, D.J. Hinde, K. Ramachandran, E. Williams, M. Dasgupta, I.P. Carter, K.J. Cook, D.Y. Jeung, D.H. Luong, S. McNeil, C.S. Palshetkar, D.C. Rafferty, C. Simenel, A. Wakhle, J. Khuyagbaatar, C.E. Düllmann, B. Lommel, B. Kindler, Phys. Rev. C **91**, 064605 (2015)
161. R. Tripathi, S. Sodaye, K. Sudarshan, B.K. Nayak, A. Jhingan, P.K. Pujari, K. Mahata, S. Santra, A. Saxena, E.T. Mirgule, R.G. Thomas, Phys. Rev. C **92**, 024610 (2015)
162. Shaughnessy, D.A., Gregorich, K.E., Adams, J.L., Lane, M.R., Laue, C.A., Lee, D.M., McGrath, C.A., Ninov, V., Patin, J.B., Strellis, D.A., Sylwester, E.R., Wilk, P.A., Hoffman, D.C., Phys. Rev. C **65**, 024612 (2002)
163. Gangrsky, Y.P., Miller, M.B., Mikhailov, L.V., Kharisov, I.F., Yad. Fiz. **31**, 306 (1980); [Sov. J. Nucl. Phys. **31**, 162 (1980)]
164. A.N. Andreyev, D.D. Bogdanov, S. Saro, G.M. Ter-Akopian, M. Veselsky, A.V. Yeremin, Phys. Lett. B **312**(1–2), 49–52 (1993)
165. A. Mamdouh, J. Pearson, M. Rayet, F. Tondeur, Nucl. Phys. A **679**(3–4), 337–358 (2001)
166. A.V. Karpov, A.S. Denikin, M.A. Naumenko, A.P. Alekseev, V.A. Rachkov, V.V. Samarin, V.V. Saiko, V.I. Zagrebaev, Nucl. Instr. Meth. A **859**, 112 (2017); NRV: Low Energy Nuclear Knowledge Base. nrv.jinr.ru/nrv
167. H. Hall, D.C. Hoffman, Annu. Rev. Nucl. Part. Sci. **42**(1), 147–175 (1992)

Chapter 4
Introduction to Hypernuclear Experiments, and Hypernuclear Spectroscopy with Heavy Ion Beams

Take R. Saito

Abstract The information on a hypernucleus, a bound nuclear system with hyperon(s), contributes essentially to understand the baryon-baryon interaction under the flavoured-SU(3) symmetry. After the discovery of the first event of hypernuclear production and decay in 1952, hypernuclei have been studied extensively with cosmic-rays, secondary meson beams and primary electron beams from accelerators facilities mainly at CERN, BNL, KEK, LNF-INFN, JLab, MAMI C and J-PARC. With these experiments, many interesting properties of hypernuclei have been revealed. Hypernuclei can also be produced by using heavy ion induced reactions. Experiments at LBL and JINR were performed with heavy ion beams bombarded on fixed targets to produce hypernuclei as projectile-like fragments, while the STAR and ALICE collaborations produced and identified hypertriton and anti-hypertriton at mid-rapidity by colliding ultra-relativistic heavy ion beams. Hypernuclear experiments with heavy ion beams bombarding a fixed target have been extended by the HypHI collaboration at GSI, and hypernuclear final states produced by the ^{6}Li + ^{12}C reaction at 2 A GeV have been successfully studied in the HypHI experiment. The method will be also extended by employing a forward spectrometer with an excellent momentum resolution such as FRS at FAIR Phase 0 (GSI) and super-FRS at FAIR Phase 1. In this article, a brief overview of former hypernuclear experiments with cosmic-rays, secondary meson beams and primary electron beams will be presented and discussed. An overview of hypernuclear experiments with heavy ion beams will also be given. Then, details of the HypHI experiment will be discussed as well as future plans at FAIR Phase 0 (GSI) and Phase 1 will be briefly summarized.

T. R. Saito (✉)
GSI Helmholtz Center for Heavy Ion Research, Darmstadt, Germany
e-mail: t.saito@gsi.de

© Springer International Publishing AG, part of Springer Nature 2018   117
C. Scheidenberger, M. Pfützner (eds.), *The Euroschool on Exotic Beams - Vol. 5*,
Lecture Notes in Physics 948, https://doi.org/10.1007/978-3-319-74878-8_4

4.1   Introduction to Hypernuclear Experiments

In this section, a brief overview of the former and on-going hypernuclear experiments will be given. Hypernuclear experiments were started with nuclear emulsions and then continued with counter techniques with meson- and electron-beams at CERN, BNL, KEK, LNF-INFN, JLab, MAMI C and J-PARC. Motivations of hypernuclear experiments with heavy ion beams, that are the main subject of this article, will also be given in this section.

4.1.1   Physics Motivation in Brief

The investigation of fundamental interactions is one of the keys to understand our Nature. All the ordinary materials consist of atomic nuclei, and they are composed by nucleons (neutrons and protons, N). Nucleons are baryons that contain three quarks. Precise knowledge of baryon-baryon interactions with the lightest three quarks, up-, down- and strange-quarks, is an essential issue to understand Nature. Such an interaction can be described by a symmetry group, so called flavored-SU(3). In flavored-SU(3), there are 64 interaction vertexes. Among them, only four vertexes are only with nucleons, and the rest is with hyperon(s) (Y) which is a baryon with at least one strange-quark. The interaction involving only nucleons has been well studied by reaction experiments involving nucleons and nuclei. However, none of the interactions with hyperon(s), YN and YY, has been well known since it is not practical to perform a reaction experiment with a hyperon as a beam or a target because of a short life time of hyperons, $\sim 10^{-10}$ s.

Information on the baryon-baryon interaction with a hyperon can also contribute to the understanding of the nature in the astronomical scale. Many model calculations predict stellar objects which contain a significant fraction of hyperons. In some predictions hyperons are even the dominant component in the center of dense objects like the core of neutron stars. In order to explore the properties and the evolution of such stars, detailed information on the baryon-baryon interaction with hyperons(s) is required. Recently, the neutron star J1614-2230 has been reported to have its mass of approximately two times of the solar mass [1], and this observation excludes the equation of state with hyperons in its core. However, the exclusion also depends on properties of the YN-interactions at large density of nuclear matter, which is still to be investigated.

So far, the only possibility to study YN- and YY-interactions is to employ a hypernucleus as a micro-laboratory. A hypernucleus consists of not only nucleons but also hyperon(s). The conventional method of hypernuclear spectroscopy, however, is so far tightly limited in both production and possible observable of the produced hypernuclei. Especially, information of YN- and YY-interactions in neutron rich nuclear matter can hardly be obtained in the conventional way, although it is highly demanded to understand the extreme astrophysical objects.

4.1.2   Discovery of Hypernuclei and the Nuclear Emulsion Era

The first hypernuclear event was discovered by Danysz and Pniewski in 1952 with the observation of the emission and the subsequent, highly energetic disintegration of a multiply-charged fragment from a cosmic ray interaction in the nuclear emulsion [2]. Figure 4.1 shows a photomicrograph of this event. A cosmic ray (track p) hit the nuclear emulsion, and it interacted with the nucleus in the emulsion at point A. From this point another particle emerged as track f and propagated to point B where it stopped. Then, it decayed into three charged particles (tracks 1, 2 and 3). The path length in the nuclear emulsion of this fragment, indicated as track f, of 90 μm indicated a survival time of a few times 10^{-12} s, many orders of magnitude longer than those expected of an ordinary de-excitation process of such a highly excited nucleus, which indicated a Λ-hyperon is bound and decays inside the nucleus. After this discovery, a variety of light hypernuclei was observed with the nuclear emulsions and they are summarized in [3]. The binding energy B_Λ (or separation energy) of a Λ hyperon in a hypernucleus is defined by the following relation

$$(M_{core} + M_\Lambda)c^2 - B_\Lambda = M_{hyp}c^2 \qquad (4.1)$$

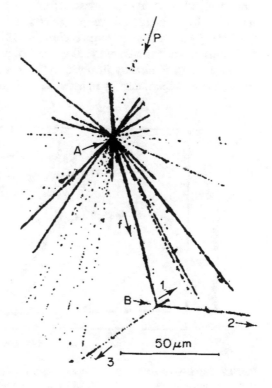

Fig. 4.1 The first hypernuclear events observed in the nuclear emulsion [2]. A cosmic ray (track p) was induced in the nuclear emulsion, and it interacted with the nucleus in the emulsion at point A. From this point a hypernucleus emerged as track f and propagated to point B where it stopped. Then, it decayed into three charged particles (tracks 1, 2 and 3). The picture is taken from [3]

P

A

f

B →

1

2 →

3

50 μm

where M_{core}, M_Λ and M_{hyp} are respectively the masses, for example in unit of MeV/c^2, of the nuclear core in its ground state, of the Λ hyperon and of the hypernucleus, the last being determined from the complete kinematic reconstruction of its decay process in the nuclear emulsions. These determinations of hypernuclear mass, and hence binding energy, have been confined to measurements made in nuclear emulsions of 71 π^--mesonic decays ($\Lambda \rightarrow p + \pi^-$ in hypernuclei). The observed B_Λ values are also summarized in [3], and they show that the Λ-separation energies are significantly smaller than the nucleon-separation energies, being approximately 1/3 of the nucleon separation energy of ordinary nuclei.

4.1.3 Hypernuclear Experiments with Secondary Meson Beams and Counter Techniques

Later in 1973, the production and observation of the hypernucleus $^{12}_\Lambda$C by using low momentum secondary K^- beams at CERN was reported as the first hypernuclear counter experiment [4]. Figure 4.2 shows the observed π^- momentum distribution, which was induced with the secondary K^- beams stopped in a ^{12}C target. Two bumps above 250 MeV/c, that are also shown in the insert with the background subtracted, correspond to π^- mesons from the ground and excited states of the produced $^{12}_\Lambda$C hypernucleus. It was an epoch making experiment for the hypernuclear spectroscopy with beams from accelerators and the counter techniques. With the K^- beams, a production cross section value of hypernuclei is reasonably large since a strange-quark is already involved in K^- mesons as beam particles, and it is a production of hypernuclei by transferring a strange-quark from the K^- beam in the

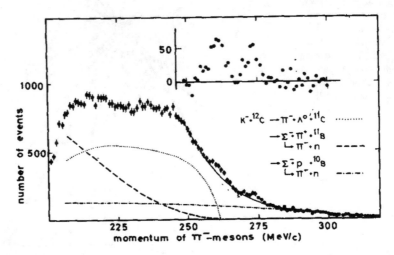

Fig. 4.2 Observed spectrum of π^--mesons with stopped-K^- beams at the ^{12}C target. The insert is the magnification above 250 MeV/c with the background subtracted. The figure is taken from [4]

target nucleus. Studies of hypernuclei at CERN by using the (K^-, π^-) reaction were continued with low momentum K^- beams with limited intensities (10^4–10^5/s) and with rather large pion background in the beams.

Hypernuclear production with the inverse reaction of (K^-, π^-), the (π^+, K^+) reaction, was invented at the Brookhaven National Laboratory in 1984. The quality and the intensity of the π^+ beams is better and larger than the K^- beams since the mass of the π^+ meson is smaller than K^-. Momentum transfer from the π^+ beams to the produced hypernuclei is also larger than that in the (K^-, π^-) reaction, and therefore, the probability to produce excited hypernuclear states at higher spins is also larger. Hypernuclear experiments with the (π^+, K^+) reaction were continued at KEK until the beginning of the twenty-first century.

One of potential ways to study hypernuclei with the K^- and π^+ beams is to employ a missing mass method. The missing mass M, for example with the ^{89}Y target nucleus, is expressed as

$$M = \sqrt{(E_\pi + M(^{89}\text{Y}) - E_K)^2 - |\mathbf{p}_\pi - \mathbf{p}_K|^2}$$
$$= \sqrt{(E_\pi + M(^{89}\text{Y}) - E_K)^2 - (p_\pi^2 + p_K^2 - 2p_\pi p_K cos\theta)^2} \qquad (4.2)$$

where E and p are the total energy and three-momentum, respectively, and subscripts π and K are respectively for π^+ beams and out-going K^+. The angle between the in-coming π^+ beam and the out-going K^+ is denoted as θ, and $M(^{89}\text{Y})$ is the mass of the ^{89}Y target nucleus. These quantities are measured by dedicated spectrometers. As an examples, the Superconducting Kaon Spectrometer (SKS) with the Beam spectrometer (QQDQQ) [5] is shown in Fig. 4.3. Four-momenta of in-coming π^+ beams impinging in the target are measured by the Beam Spectrometer, and those of out-going K^+ mesons are measured by SKS. Then, the missing mass defined in Eq. (4.2) can be deduced. With the deduced missing mass, the Λ binding energy, $-B_\Lambda$, can be extracted by subtracting the mass of the core nucleus to which the Λ-hyperon is attaching as well as subtracting the mass of the Λ-hyperon. Figure 4.4 shows the deduced distribution of $-B_\Lambda$ in the $^{89}_\Lambda$Y hypernucleus [5]. The upper panel shows the deduced $-B_\Lambda$ distribution and the lower panel shows the one with fittings. Details of the analyses can be found in [5]. One of the striking results in this experiment is that the Λ-orbitals can be clearly seen even in the deep s-orbitals. It is because that the Λ-hyperon is a different Fermion in comparison with nucleons and because it does not obey the Pauli principle with the other nucleons in the observed hypernucleus. Therefore, Λ-hyperons can move freely among the Λ-orbitals. This phenomena can be seen only with hypernuclei. Such an observation has shown that a Λ-hyperon can be used as a probe to study the deep part of the potential of the core nucleus, and the investigation of the hypernucleus can also contribute to understand the nucleus without hyperons. Further experimental studies with the secondary meson beams including K^- and π^+ are continued at the J-PARC accelerator facility.

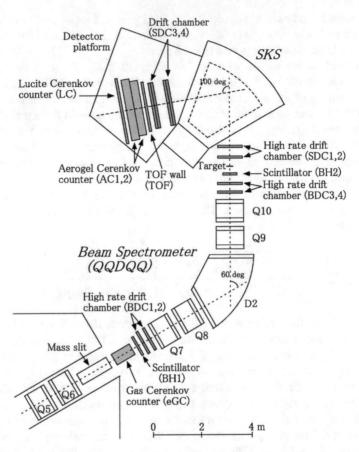

Fig. 4.3 Schematic view of the SKS spectrometer together with the Beam Spectrometer (QQDQQ). Four-momenta of π^+ beams and out-going K^+ mesons are respectively measured by the Beam Spectrometer and SKS. The figure is taken from [5]

The FINUDA collaboration at LNF-INFN invented another unique method to produce Λ-hypernuclei with low-energy K^- stopped at thin nuclear targets. Collisions of electrons and positrons at 0.51 GeV are produced by the DAΦNE e^+e^- collider and the production of $\phi(1020)$ mesons takes place in the center of the FINUDA spectrometer shown in the top part of Fig. 4.5. The collision point is also shown by "+" on the bottom part of the figure. At the collision point, the $\phi(1020)$ meson decays to $K^- + K^+$ with 127 MeV/c, as shown in the bottom left part of the figure. Because of the low momentum of the produced K^- from the decay, the K^- meson can be stopped in the thin target foils as shown in the bottom part of Fig. 4.5, and hypernuclei are produced with the stopped K^- by emitting π^- (conventional production) or π^+ (double-charge exchange reaction). The figure shows the case for the double-charge exchange reaction for searching for neutron rich hypernuclei. In the figure, a candidate of neutron rich nuclei are stopped in the Si-micro vertex

Fig. 4.4 The observed distribution of the binding energy of Λ-hyperons, $-B_\Lambda$, for $^{89}_{\Lambda}\text{Y}$ (top) and with fittings (bottom). The observed distribution is fitted with a single Gaussian for the s orbit or two Gaussians for the other orbits (F1), where the Gaussian widths were fixed to be the energy resolution. The whole bound region were fitted well by additionally introducing three Gaussians (F2) representing extra yields in between the bumps. The figure is taken from [5]

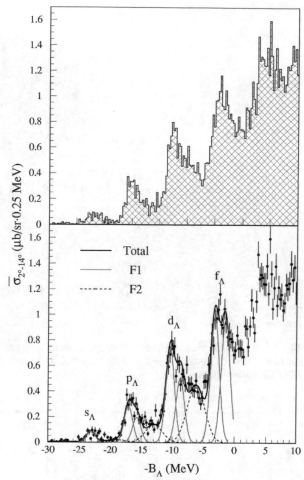

detectors inside the target barrels, and it emits a proton and π^-. By measuring four-momenta of K^- and π^- as well as identifying K^+, the missing mass of the produced hypernuclei are measured (for example, see [6]). Figure 4.6 shows the distribution of deduced $-B_\Lambda$ of the $^{12}_{\Lambda}\text{C}$ hypernucleus with different fittings [6]. They show that the resolution in the missing spectroscopy has been improved in comparison to the other experiments at CERN, BNL and KEK since the experiment can be performed with thin target foils to reduce the straggling of the out-going π^-. Though the series of the FINUDA experiments were very successful, the project was unfortunately terminated.

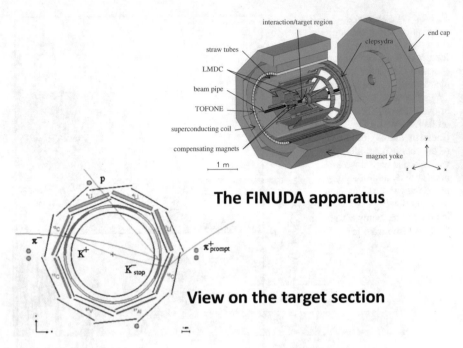

Fig. 4.5 Schematic view of the FINUDA spectrometer (top part) and the magnification around the target station. Description of the tracks are found in the text. The figure is the courtesy of the FINUDA collaboration

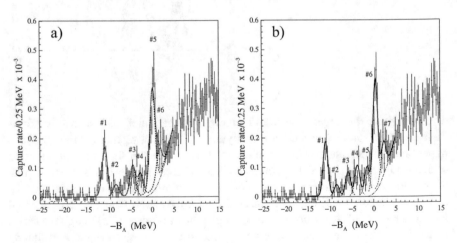

Fig. 4.6 Distributions of deduced $-B_\Lambda$ for the $^{12}_\Lambda$C hypernuclei studied by FINUDA. Two figures are with fitting by (**a**) six and (**b**) seven Gaussian functions. The figure is taken from [6]

4.1.4   Hypernuclear Experiments with Primary Electron Beams

Single Λ-hypernuclei can also be studied by employing primary electron beams. With electron beams, hypernuclei are produced by virtual photons involved in the $(e,e'K^+)$ reaction. One of the unique features of the hypernuclear production with $(e,e'K^+)$ is that the energy of the primary beam is well defined by the accelerator rigidity thus there is no ambiguity on the momentum of the in-coming beams in the missing mass deduction by using Eq. (4.2). The intensity of the primary electron beams is much larger than the secondary meson beams, and the hypernuclear spectroscopy can be conducted with a thin target foil even though the hypernuclear production in the $(e,e'K^+)$ reaction is small. Therefore, the resolution in the missing mass spectroscopy has been expected in the sub-MeV scale. Furthermore, hypernuclei can be produced with a large spin flip amplitude for the produced Λ-hyperon due to the unity spin of the photon involved in the hypernuclear production. The $(e,e'K^+)$ reaction produces a neutron-rich hypernucleus since a proton is converted to a Λ-hyperon. The challenge in this kind of experiment is to deal with a huge electromagnetic background.

One successful experiment with primary electron beams is the E01-011 experiment performed at Thomas Jefferson Laboratory (JLab) [7]. Figure 4.7 shows the experimental layout of the JLab E01-011 experiment. It is a double armed spectrometer. The out-going K^+ mesons were measured by the HKS spectrometer, which has a central momentum of 1.2 GeV/c and a 16 msr solid angle. The scattered electrons (central momentum of 0.35 GeV/c) were measured by the ENGE-type

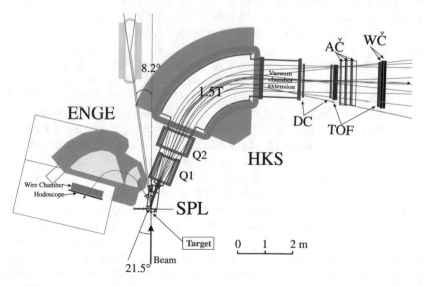

Fig. 4.7 Schematic layout of the JLab E01-011 experiment. Scattered electrons are measured by the ENGE spectrometer, and out-going K^+ mesons are measured by the HKS spectrometer. The figure is taken from [7]

Fig. 4.8 Spectra of the
deduced binding energy $-B_\Lambda$
of the $^7_\Lambda$He hypernucleus
measured by the emulsion
experiment [8] (top) and the
JLab E01-011 experiment [7]
after background subtraction
and acceptance corrections
(bottom). It is clearly shown
that the missing mass
resolution in the JLab
E01-011 experiment is
excellent compared to the
former emulsion experiment.
The figure is taken from [7]

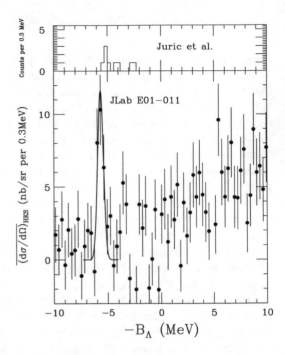

split-pole spectrometer that was vertically tilted by 8° from the dispersion plane and
shifted vertically to suppress electron backgrounds originating from bremsstrahlung
and Møller scattering that have very sharp forward distributions (tilt method). The
electron beam energy was set to 1.851 GeV, giving a virtual photon energy of about
1.5 GeV. The bottom panel of Fig. 4.8 shows the distribution of the deduced binding
energy $-B_\Lambda$ of the $^7_\Lambda$He by the JLab E01-011 experiment [7] after background
subtraction and acceptance corrections. It is compared with the result of a former
emulsion experiment reported in 1973 [8], which is shown in the top panel of the
figure. It is clearly shown that the missing mass resolution in the JLab E01-011
experiment is excellent in comparison with the former emulsion experiment. The
width of the observed peak by JLab E01-011 is 0.63 ± 0.12 MeV in FWHM.

The KaoS-A1 collaboration at the MAMI C accelerator facility of Johannes-
Gutenberg Mainz University in Germany also employed the $(e,e'K^+)$ reaction
to produce light single-Λ hypernuclei but measure discrete π^+ from the two-
body decay of the $^4_\Lambda$H hypernucleus. This method can give the best possible
pion momentum resolution from the two-body decay of hypernuclei with counter
techniques. Figure 4.9 shows the deduced π^- momentum distribution from the two-
body decay of the $^4_\Lambda$H hypernucleus measured by the KaoS-A1 experiment for true
coincidences (green) and accidental coincidences (blue) scaled by the ratio of the
time gate widths [9]. A monochromatic peak at 133 MeV/c was observed, which
is a unique signature for the two-body decay of the stopped $^4_\Lambda$H hypernucleus. The
top panel shows on the corresponding binding energy scale the distribution of data
on the Λ-hyperon binding energy in $^4_\Lambda$H from the former emulsion experiments

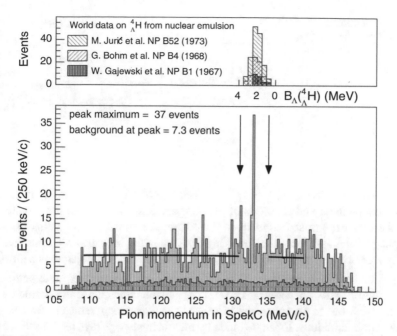

Fig. 4.9 Pion momentum distribution from the two body decay of the $_\Lambda^4$H hypernucleus measured by the KaoS-A1 experiment for true coincidences (green) and accidental coincidences (blue) scaled by the ratio of the time gate widths [9]. A monochromatic peak at 133 MeV/c was observed, which is the unique signature for the two-body decay of stopped $_\Lambda^4$H hypernucleus. The top panel shows on the corresponding binding energy scale the distribution of data on the Λ-hyperon binding energy in $_\Lambda^4$H from the former emulsion experiments [8, 10, 11]. The figure is taken from [9]

[8, 10, 11]. The clear peak has shown that the momentum resolution is comparable or better than the former emulsion experiments serving the best resolution in the past. It has shown that the decay pion spectroscopy from the hypernuclear two-body decay can be a very powerful tool to measure the hypernuclear binding energy with excellent mass resolution. With such a technique, the measured binding energy by the former old emulsion experiments can also be crosschecked by modern spectroscopy techniques, which has not been possible for almost a half century.

4.1.5   Gamma-Ray Spectroscopy on Hypernuclei

Though the resolution of hypernuclear spectroscopy with the induced reaction of secondary meson beams and primary electron beams has been improved by employing different techniques, as discussed above, the resolution has still to be better in order to study spin-dependent interactions in Λ-hypernuclei. Figure 4.10 shows schematically an example of low-lying states in Λ-hypernuclei. A Λ-hyperon is bound with a core nucleus, $^{A-1}$Z, thus a forming a hypernucleus, $_\Lambda^A$Z. Because of

Fig. 4.10 Schematic
explanation of the low-lying
level of Λ-hypernuclei

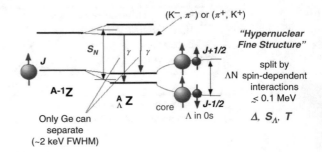

the spin of the Λ-hyperon, $s = 1/2$, and because of the spin dependent interaction between a nucleon and Λ, levels of the core nucleus, for example the ground state and the first excited state shown in the figure, are split to two levels due to the coupling of the Λ-hyperon to these levels. This is the so called *hypernuclear fine structure*. Splittings are induced by different spin-dependent interactions with $\vec{s}_N \cdot \vec{s}_\Lambda$, $\vec{l}_{N\Lambda} \cdot \vec{s}_\Lambda$ and the tensor part S_{12} [12, 13]. Since the spin-dependent interactions are expected to be small, the energy differences between the spilt levels should be also small, which has to be measured in order to deduce the parameters for the spin-dependent interactions. It can be made by measuring two γ-rays feeding to the split levels from the same state as shown in the figure by red arrows in the level scheme. Because of the small energy difference of the split levels, two γ-ray energies are very similar, and they can be resolved by measuring γ-rays only with Ge detectors that have a typical energy resolution of approximately 2 keV for 1 MeV γ-rays.

A pioneering experiment to measure hypernuclear γ-rays with Ge detectors was the KEK E336 experiment by the Hyperball collaboration with the $^7\text{Li}(\pi^+, K^+)$ reaction to study the $^7_\Lambda\text{Li}$ hypernucleus [14]. Figure 4.11 shows γ-ray spectra in coincidence with in-coming π^+ beams and out-going K^+ mesons with different conditions. The bound and unbound regions are selected by the (π^+, K^+) missing mass method. The panels (a) and (b) show γ-ray spectra respectively in the unbound and bound regions. A γ-transition at 2050 keV is assigned as a slow E2 transition which is emitted after the produced hypernucleus is stopped in the target. Fast M1 transitions are not observed in the panel (b). They are Doppler broadened since they are emitted while the produced hypernucleus is moving in the target. An event-by-event Doppler correction has been made, and the Doppler corrected γ-ray spectrum is shown in panel (c) in the figure. Peaks of three M1 transitions at 692, 3186 and 3877 keV are sharply observed in the Doppler corrected spectrum. With these observations, a level scheme of the $^7_\Lambda\text{Li}$ hypernucleus is built as shown in Fig. 4.12. After the pioneering experiment, the Hyperball collaboration also measured γ-rays of $^9_\Lambda\text{Be}$ [15], $^{11}_\Lambda\text{B}$ [16], $^{15}_\Lambda\text{N}$ [17] and $^{16}_\Lambda\text{O}$ [17, 18] hypernuclei.

Another striking result has been also reported recently by the Hyperball-J collaboration. They have measured γ-rays feeding from the 1^+ excited state to the 0^+ ground state in $^4_\Lambda\text{He}$ produced by the $^4\text{He}(K^-, \pi^-)$ reaction at J-PARC [19]. The four panels of Fig. 4.13 show γ-ray spectra measured in the $^4\text{He}(K^-, \pi^-)$ reaction, (a) and (b) with the unbound region in the missing mass, and (c) and (d) with the

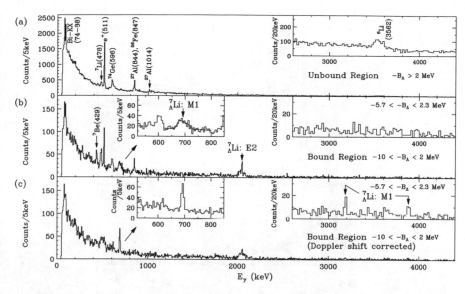

Fig. 4.11 Gamma-ray spectra measured in the $^7\mathrm{Li}(\pi^+,K^+)$ reaction for (**a**) the unbound region and for (**b**) the bound region. The two peaks at 692 and 2050 keV in (**b**) are assigned as M1 and E2 transitions, respectively, in $^7_\Lambda\mathrm{Li}$. The panel (**c**) is the same spectrum as (**b**) but the event-by-event Doppler-shift correction was applied. The figure is taken from [14]

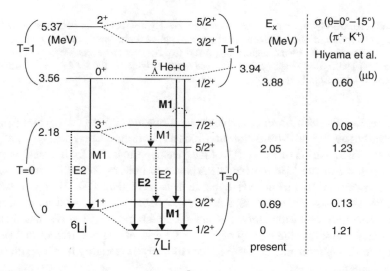

Fig. 4.12 Level scheme and γ-transitions of $^7_\Lambda\mathrm{Li}$. Thick arrows show transitions observed, and the column labeled "present" shows level energies measured in the experiment at KEK [14]. The calculated cross section values to populate those states are also shown in the left column. The figure is taken from [14]

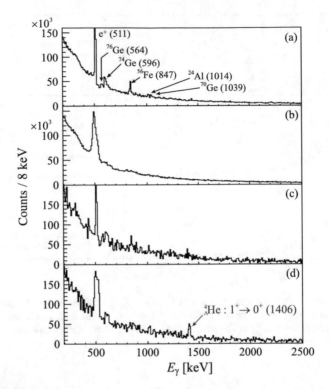

Fig. 4.13 Gamma-ray energy spectra measured by Hyper- ball-J in coincidence with the ^{4}He(K^-,π^-) reaction. Missing mass selections are applied to the highly unbound region for (**a**) and (**b**), and to the $^4_\Lambda$He bound region for (**c**) and (**d**). An event-by-event Doppler correction is applied for (**b**) and (**d**). A single peak is observed in (**d**) attributed to the M1 $1^+ \rightarrow 0^+$ transition. The figure is taken from [19]

bound region. Event-by-event Doppler corrections are applied for (b) and (d). One can clearly see that a γ-transition at 1406 keV is only observed in (d), and it is assigned as a transition of $1^+ \rightarrow 0^+$ of the $^4_\Lambda$He hypernucleus. By the measurement, a level scheme of the $^4_\Lambda$He hypernucleus has been built as shown in Fig. 4.14. The level is compared to that of $^4_\Lambda$H hypernucleus studied in [20, 21] (A. Kawachi, 1997, Ph. D. thesis, University of Tokyo, unpublished). The ground state energy of $^4_\Lambda$H has also been confirmed by the KaoS-A1 collaboration [9] as discussed above. Figure 4.14 show that the energy of $1^+ \rightarrow 0^+$ transitions is different in $^4_\Lambda$He and $^4_\Lambda$H. It is a strong evidence of the existence of charge symmetry breaking involving Λ and nucleons.

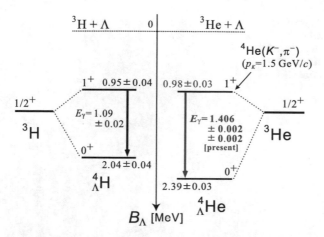

Fig. 4.14 Level schemes of the mirror hypernuclei, $^4_\Lambda$H and $^4_\Lambda$He. The Λ binding energies (*B_Λ) of the ground states of $^4_\Lambda$H and $^4_\Lambda$He are taken from past emulsion experiments [8]. The B_Λ values of the ground and the first excited 1^+ states are obtained using the present data and past γ-ray data [20, 21] (A. Kawachi, 1997, Ph. D. thesis, University of Tokyo, unpublished). The first error value of the measured energy of the γ-ray in $^4_\Lambda$He corresponds to the statistical error, and the second one is the systematic uncertainty. The B_Λ of the ground state of $^4_\Lambda$H was also confirmed by the KaoS-A1 experiment as discussed above [9]. The figure is taken from [19]

4.1.6 Spectroscopy on Double-Λ Hypernuclei

Studies on double-Λ hypernuclei are the only way to extract the information of the Λ-Λ interaction. They can be produced by the double-strangeness exchanging reaction (K^-, K^+) via a production of Ξ^--hyperons.[1] The study of double-Λ hypernuclei is an experimental challenge since the production cross section of double-Λ hypernuclei is expected to be small. Furthermore, twofold mesonic- or non mesonic-decay of the produced double-Λ hypernuclei has to be reconstructed with excellent precisions in order to extract the binding energy $-B_{\Lambda\Lambda}$ and $\Delta B_{\Lambda\Lambda}$ of two Λ-hyperons. Though the nuclear emulsion is a traditional experimental technique employed in the early era of the hypernuclear experiments, it still serves the best resolution in the reconstruction of the particle tracks and decays. The emulsion technique with a modern track scanning system can be employed together with a modern technique of counter experiments, and the pioneering experiment with this combination was the KEK E373 experiment with an emulsion/scintillating-fiber hybrid system [22]. In the KEK E373 experiment, the first uniquely identified double-Λ hypernuclear event, so called "Nagara event" was observed, and it is shown in Fig. 4.15.

---

[1]This particle is called "cascade" particle.

Fig. 4.15 Photograph and
schematic drawing of
NAGARA event. See text for
detailed explanation. The
figure is taken from [22]

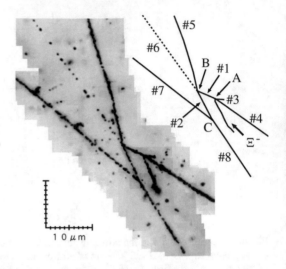

In Fig. 4.15, a $\varXi^-$-hyperon came to rest at point A, from which three charged
particles (tracks No. 1, No. 3, and No. 4) were emitted. One of them decayed into a
π^- meson (track No. 6) and two other charged particles (tracks No. 2 and No. 5) at
point B. The particle of track No. 2 decayed again to two charged particles (tracks
No. 7 and No. 8) at point C. The measured lengths and emission angles of these
tracks are measured. The particle of track No. 7 left the emulsion stack and entered
the downstream scintillating-fiber block detector. Track No. 5 ended in a 50-μm-
thick acrylic base film. The kinetic energy of each charged particle was calculated
from its range. The single hypernucleus (track No. 2) was identified from event
reconstruction of its decay at point C. Mesonic decay modes of single hypernuclei
were rejected because their Q values are too small. The decay mode of the single
hypernucleus is non-mesonic with neutron emission. If either track No. 7 or No.
8 has more than unit charge, the total kinetic energy of the two charged particles
is much larger than the Q value of any possible decay mode because of the long
ranges of tracks No. 7 and No. 8. Therefore, both tracks No. 7 and No. 8 are singly
charged, and only $_\Lambda$He isotopes are acceptable. The kinematics of all possible decay
modes of the double hypernucleus (track No. 1) which decays into $_\Lambda$He (track No.
2) and π^+ (track No. 6) were checked, and $B_{\Lambda\Lambda}$ and $\Delta B_{\Lambda\Lambda}$ were calculated. Since
track No. 5 ended in the base film, only the lower limit of its kinetic energy can
be determined. For the decay modes without neutron emission, the range of the
particle of track No. 5 was increased to minimize the missing momentum. If the
sum of the momenta of the three charged particles (tracks No. 2, No. 5, and No.
6) deviated from zero by more than three standard deviations even after the range
of track No. 5 was increased from the missing momentum, that decay mode was
rejected. For the decay modes with neutron emission, the upper limits of $B_{\Lambda\Lambda}$ and
$\Delta B_{\Lambda\Lambda}$ were obtained. Kinematical analysis of the production reaction was made
by assuming that the $\varXi^-$-hyperon was captured by a light nucleus in the emulsion

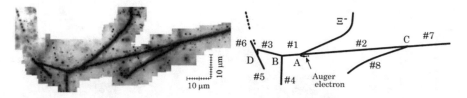

Fig. 4.16 A superimposed image from photographs and a schematic drawing of the KISO event. See [24] for the interpretation of the tracks. The figure is taken from [24]

(^{12}C, ^{14}N, or ^{16}O). This assumption is reasonable, taking into account the existence of the short track No. 3 and the Coulomb barrier of the target nucleus. For each of the modes without neutron emission, if the sum of momenta deviated from zero by more than three standard deviations, the mode was rejected. For the modes with one neutron emission, the momentum of the neutron was assigned to the missing momentum of the three charged particles (tracks No. 1, No. 3, and No. 4). For the modes with more than one neutron emission, the lower limits of the total kinetic energy of the neutrons were calculated from the missing momentum. The values of $B_{\Lambda\Lambda}$ and $\Delta B_{\Lambda\Lambda}$ were calculated with the Ξ^--hyperon binding energy B_{Ξ^-} set to zero. Hence these values are lower limits of $B_{\Lambda\Lambda}$ and $\Delta B_{\Lambda\Lambda}$, and their true values are larger, depending on the actual value of B_{Ξ^-}. A comparison of the values of $B_{\Lambda\Lambda}$ and $\Delta B_{\Lambda\Lambda}$ obtained from both points A and B was made. After rejecting the modes which have inconsistent values, only one interpretation remained,

$$^{12}C + \Xi^- \rightarrow {}^{6}_{\Lambda\Lambda}He + {}^4He + t$$

$$^{6}_{\Lambda\Lambda}He \rightarrow {}^{5}_{\Lambda}He + p + \pi^-.$$

It was the first uniquely identified double-Λ hypernuclear event.

The further analyses of the KEK E373 experiment reveals more double-Λ hypernuclear events, Mikage-, Demachiyanagi- and Hida-events [23]. Recently, a candidate of the production and decay of $^{14}_{\Xi}$N bound system was discovered by the analyses of the KEK E373 experiment, and the event was named as Kiso-event [24]. The photomicrograph of the Kiso-event and its track interpretation is shown in Fig. 4.16. Details of the Kiso-event can be found in [24].

4.2 Hypernuclear Spectroscopy with Heavy Ion Beams

In this and the following sections, an overview of the former hypernuclear experiment with heavy ion beams will briefly be given. The concept of the HypHI hypernuclear studies at GSI and FAIR will also be discussed.

As discussed above, hypernuclei have been studied already for more than six decades with different experimental techniques. However, because of the experimental difficulties and the very short lifetime of hyperons, the number of known and

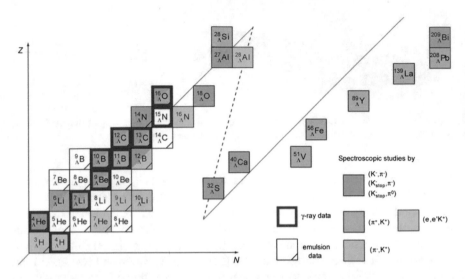

Fig. 4.17 Lambda-hypernuclear chart. The experimentally identified Λ- hypernuclei and the experimental methods used to study them are shown. The figure is the updated version from [25]

investigated hypernuclei has yet been limited. Figure 4.17 shows a chart of known single-Λ hypernuclei with symbols indicating the applied experimental methods. As already discussed, they were produced with induced reactions of meson- and electron-beams by converting one of the nucleons in the target nuclei to a Λ-hyperon. Therefore, the isospin of the produced hypernucleus in these reactions is similar to the target nucleus. It implies that it is difficult to study neutron- and proton-rich hypernuclei by the conventional hypernuclear production method. Studies of hypernuclei with different isospin values are important to investigate the isospin dependence of the interaction involving a Λ-hyperon, and the importance of the three-body force induced by the ΛN-ΣN coupling may become pronounced in hypernuclei with neutron or proton excess. Furthermore, with the meson induced reaction, the maximum number of Λ-hyperons bound in hypernuclei is limited only up to two. Neutron- or proton-rich hypernuclei as well as triple or more-multiple Λ hypernuclei can be produced by induced reaction of heavy ion beams at high energies, and the hypernuclear spectroscopy with heavy ion beams will be discussed in the following.

Hypernuclear production via heavy ion collisions was first discussed by Kerman and Weiss [26]. They suggested that it could be the only way to produce very exotic hypernuclei. Recently, Gaitanos et al. [27] studied theoretically hypernuclear production with heavy ion beams based on the GiBUU model [27]. In high energy heavy ion collisions, it is well known that the participant-spectator model explains the general feature of the reaction. The nucleons in the region where two nuclei overlap in collision can participate the reaction (they are called "participants"), while the nucleons in the non-overlapping region pass by each other without experiencing a large disturbance (they are called "spectators"). Hyperons such as Λ are produced in the participant region near mid-rapidity. Because of their wide rapidity distribution, one may produce a hypernucleus in coalescence of hyperon(s) in the projectile fragments. Thus the velocity of produced hypernuclei as projectile spectators is close to that of projectile. Because of the energy threshold of ~1.6 GeV for Λ production in an elementary process of $NN \rightarrow \Lambda KN$, the produced hypernuclei have a large velocity with $\beta > 0.9$, and their effective lifetime in the laboratory frame is longer than at rest due to a large Lorentz factor. Decays of hypernuclei can be studied in-flight, and most of their decay vertices are located a few tens of centimeters behind the target in which hypernuclei are produced.

The first attempt to produce and identify a hypernucleus as a projectile spectator with heavy ion beams was made in reaction of 2.1 A GeV ^{16}O projectiles impinged on a polyethylene target at Lawrence Berkeley National Laboratory (LBL) in the 70s [28]. However, the deduced hypernuclear cross section exceeded the theoretical prediction by far, which was explained with the experimental difficulties on the implemented kaon triggers. Later another attempt was undertaken in the late 80s at Joint Institute for Nuclear Research (JINR) by using streamer chambers [29, 30]. In the experiment at JINR, hypernuclei were produced with beams of 3.7 A GeV ^{4}He and 3.0 A GeV ^{7}Li, which impinged on a polyethylene target. Figure 4.18 shows one of the photographed indications of a hypernuclear decay taken in the experiment at JINR Dubna [29]. In this experiment, the particle identification of particles from hypernuclear decays was ambiguous, therefore, the identification of the produced hypernucleus was not performed. Nevertheless, a production cross section for $^4_\Lambda$H of ~0.3 μb was deduced, which was reproduced by the theoretical calculation based on a model of Λ coalescence in the projectiles [31, 32].

In the hypernuclear production with meson- or electron-beams as discussed in the previous section, hypernuclei are produced by the elementary processes such as $n(K^-, \pi^-)\Lambda$, $n(\pi^+, K^+)\Lambda$ and $p(e, e'K^+)\Lambda$, therefore, ground and excited states of produced hypernuclei can be studied by the missing mass method only by measuring in-coming beams and out-going particles from the reaction. On the other hand, in the hypernuclear production with heavy ion beams, hypernuclei are not produced by elementary processes. Therefore, the hypernuclear states must be reconstructed by the invariant mass method with measurements of all particles emitted from the hypernuclear decays. The invariant mass M is defined by the following relation;

$$(Mc^2)^2 = \left(\sum E\right)^2 - \left|\sum \vec{p}\,c\right|^2 \tag{4.3}$$

Fig. 4.18 One of the photographed indications of a hypernuclear decay taken in the experiment at JINR Dubna [29]

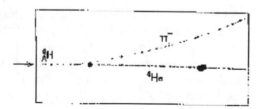

Fig. 4.19 Invariant mass spectrum with background subtraction for ^{3}He $+ \pi^-$. The solid histogram overlaid on the data is the simulated (Monte Carlo, MC) $^3_\Lambda$H signal normalized so that the peak bin matches the data. The figure is taken from [33]

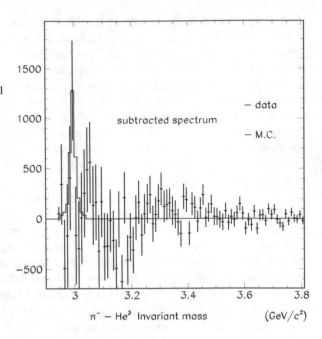

where $\sum E$ and $\sum \vec{p}$ are the sum of the energy and the three-vector sum, respectively, of all decay particles. In the experiments at LBL and JINR Dubna, the invariant mass of produced hypernuclei was not measured.

Hypernuclei have also been studied by central collisions of relativistic heavy ion beams, and they are produced at mid-rapidity. In the E468 experiment employing beams form the AGS accelerator at the Brookhaven National Laboratory (BNL) in the USA, a hypertriton, $^3_\Lambda$H, has been produced in central 11.5 A GeV/c Au + Pt collisions [33]. Figure 4.19 shows an invariant mass spectrum with background subtraction for ^{3}He $+ \pi^-$. The solid histogram overlaid on the data is the simulated (MC) and normalized $^3_\Lambda$H signal. This experiment was the first successful attempt to produce and identify a hypernucleus produced with heavy ions by using the invariant mass method. The STAR collaboration with the Relativistic Heavy Ion Collider (RHIC) at BNL used ultra-relativistic heavy ion collisions (Au + Au) to study hypertriton and anti-hypertriton [34]. Figure 4.20 shows invariant mass distributions for (a) ^{3}He $+ \pi^-$ and (b) $\overline{^3\text{He}} + \pi^+$ final states observed by the STAR collaboration, and signals from the hypertriton ($^3_\Lambda$H) and anti-hypertriton ($\overline{^3_\Lambda\text{H}}$) are clearly shown respectively in panels (a) and (b). The ALICE collaboration at the Large Hadron Collider (LHC) at CERN also observed anti-hypertritons by reconstructing the $\overline{^3\text{He}} + \pi^+$ final state [35].

In the experiments at AGS/BNL, RHIC/BNL and LHC/CERN [33–35], hypernuclei were produced by central collisions. On the other hand, the former two experiments at LBL and JINR produced hypernuclei by peripheral collisions. By means of the central collision, the mass number of produced hypernuclei is

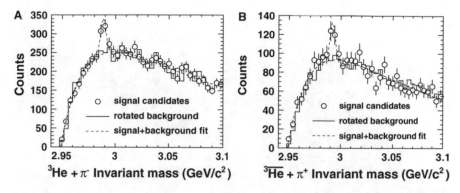

Fig. 4.20 Invariant mass distributions for (**a**) ^{3}He $+ \pi^-$ and (**b**) $\overline{^3\text{He}} + \pi^+$ final states. Clear signals of $^3_\Lambda$H and $^3_{\overline{\Lambda}}\overline{\text{H}}$ are shown respectively in panels (**a**) and (**b**). The figure is taken from [34]

practically limited to $A < 5$, however, such a limitation does not exist in the hypernuclear production with peripheral collisions of heavy ion beams.

The HypHI project [36] which the author of this article started and leads employs peripheral collisions of relativistic heavy ion beams in fixed nuclear targets to produce and identify hypernuclei in the projectile rapidity region, similar to the experiments at LBL and JINR. A hypernucleus is produced as a projectile fragment. In such reactions, a projectile fragment can capture a hyperon produced in the hot participant region to produce a hypernucleus. In this reaction, the energy of heavy ion beams should exceed the energy threshold for the hyperon production, and the velocity of the produced hypernucleus should be similar to that of the projectile. Thus, the produced hypernucleus has a large Lorentz factor ($\gamma > 0.9$), and the decay of the hypernucleus takes place well behind the production target. This makes it possible to study hypernuclei in flight. Since a hypernucleus is produced from a projectile fragment, isospin and mass values of the produced hypernuclei, unlike in other hypernuclear experiments, can be widely distributed, similar to projectile fragmentation reactions. As already mentioned, one of the unique features of the hypernuclear spectroscopy with projectile fragmentation reactions is that due to the large Lorentz factor of the produced hypernuclei the decay of the hypernuclei can be observed in flight behind the production target. Lifetime values of the observed hypernuclei can be deduced from the distribution of the *proper decay time*, obtained from the measured distance from the target, in the rest frame of the mother state of interest, thus making the deduced lifetime values independent of the detectors' time resolution. The lifetimes of hypernuclei are of interest since they are sensitive to the overall wave function of the hyperon located within the core nucleus. The lifetime of light Λ-hypernuclei has been conjectured to be similar to the lifetime of a free Λ hyperon if it is weakly bound to the core nucleus [37]. Deviations from the value of 263.2 ps of the Λ lifetime would possibly provide new information on Lambda's wavefunction inside hypernuclei. Studies of hypernuclei as a projectile spectator produced by peripheral collisions of heavy ion beams should be a uniquely good way to look into hypernuclear lifetime values and their decay properties.

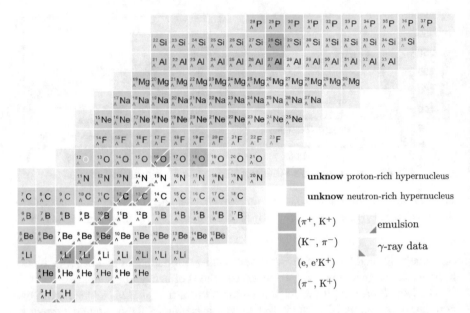

Fig. 4.21 Chart of hypernuclei that can be reached with heavy ion beams together with those already known. Unknown proton- and neutron-rich hypernuclei indicated by green and yellow colours, respectively, are expected to be studied by using heavy ion or rare-isotope beams

With projectile fragmentation reactions and coalescence of hyperons, hypernuclei heavier than those produced in central collisions of heavy ion beams can be produced, and hypernuclei with neutron or proton excess will also be produced. Therefore, the Λ-hypernuclear chart shown in Fig. 4.17 is expected to be expanded to exotic nuclei containing hyperons, i.e., exotic hypernuclei. Furthermore, hypernuclei with multi-strangeness content, i.e., with $S < -2$ can be produced. Figure 4.21 shows reachable hypernuclei with the HypHI project at GSI and FAIR. It is shown that hypernuclei with extreme isospin values can be well studied. The former two experiments with projectile fragmentation reactions at LBL and JINR has revealed a possibility to study hypernuclei with peripheral collisions of heavy ion beams on fix nuclear targets, however the feasibility of precise spectroscopy with the invariant mass method was never demonstrated. Therefore, the HypHI collaboration performed its first experiment at GSI in 2009 to demonstrate the feasibility. In this experiment, a reaction of ^{6}Li + ^{12}C at 2 A GeV was used to produce and identify light hypernuclei.

4.3   The HypHI Experiment

The HypHI experiment will be presented in this section. The experimental setup and the analysis methods will briefly be discussed. Results on the analyses on Λ-hyperon, $^3_\Lambda$H and $^4_\Lambda$H as well as on the $d + \pi^-$ and $t + \pi^-$ final states will also be discussed.

The HypHI collaboration has proposed a series of experiments at the GSI Helmholtz Centre for Heavy Ion Research, using induced reactions of stable heavy ion beams and rare-isotope beams, with the aim of producing and measuring hypernuclei with the invariant mass method [36]. In the proposed experiments, charged particles and neutrons from the mesonic or non-mesonic weak decay of hypernuclei are tracked and identified in order to reconstruct the hypernuclear mass values. The lifetime of produced hypernuclei can be extracted by measuring of the proper time in the rest frame of the hypernuclear decay. This methodology also allows investigating neutron- and proton-rich hypernuclei as well as hypernuclei with more than two units of strangeness, and several hypernuclei can be studied in a single experiment.

The first HypHI experiment to study $^3_\Lambda$H and $^4_\Lambda$H hypernuclei was performed by means of projectile fragmentation reactions of ^{6}Li projectiles at 2 A GeV delivered on a carbon target. It was performed to demonstrate the feasibility of the new experimental method by observing the Λ, $^3_\Lambda$H and $^4_\Lambda$H. It also aimed at measurements of the lifetime of $^3_\Lambda$H and $^4_\Lambda$H as well as the production cross sections of Λ, $^3_\Lambda$H and $^4_\Lambda$H in the ^{6}Li + ^{12}C reaction at 2 A GeV. Furthermore, all possible final states produced by the projectile fragmentation reaction with a capture of Λ in the ^{6}Li + ^{12}C reaction were investigated. The experimental setup, analyses and results will be summarized in the following subsection. The analyses on the HypHI experiment were already completed, and results are published in [38–42].

4.3.1   Experimental Setup

The first HypHI experiment took place at the GSI Helmholtz Centre for Heavy Ion Research. Figure 4.22 shows a schematic layout of the experiment. Projectiles of ^{6}Li at 2 A GeV with an average intensity of 3×10^6 beam particle per second bombarded at a carbon graphite target with a thickness of 8.84 g/cm^2. The ALADiN magnet [43] was used as a bending magnet for charged particles produced from the target and hypernuclear decay vertices, as shown in Fig. 4.22. A magnetic field of approximately 0.75 T was applied. The distance between the target and the center of the ALADiN magnet was 2.35 m. A small array of plastic finger hodoscopes labeled in the figure as TOF-start was used as a start counter for Time-of-Flight (TOF) measurements. In order to track charged particles, three layers of scintillating fiber detectors [44], identified as TR0, TR1 and TR2 in the figure, were set in front of the ALADiN magnet. TR0 was placed at 4.5 cm behind the target, and the

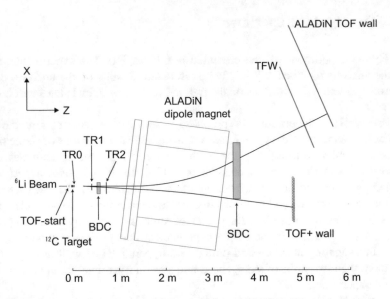

Fig. 4.22 Schematic layout of the experimental setup in the first HypHI experiment

distance of TR1 and TR2 from the target center was 40 and 70 cm respectively. A drift chamber BDC with six wire layers (xx', uu' and vv') to track charged particles was positioned between the two fiber detectors arrays, TR1 and TR2. Since it was installed around the beam axis, the wires around the beam region were made as insensitive by wrapping the wires by thin Teflon sheets. Behind the ALADiN magnet, two hodoscopes with plastic scintillating bars, TFW and ALADiN TOF wall, provided the stop signal for TOF measurements and the hit position information for π^- mesons. Positively charged particles and fragments were measured by another plastic hodoscope, labeled as TOF+ wall in the figure. An additional drift-chamber, labeled SDC, with four wire layers (xx' and yy'), was positioned behind the ALADiN magnet to measure hit positions of outgoing charged particles. Wires of SDC near the beam region were also made insensitive by shooting the wires to the ground.

The trigger system for the data acquisition electronics combined three trigger stages. The first stage was a tracking trigger which is generated by signals from the scintillating fiber tracking arrays with VUPROM2 [45, 46]. It checked for secondary vertex sites behind the target caused by free-Λ and hypernuclear decays. The second stage was linked to π^- detection by the TFW wall. The third stage required the detection of $Z = 2$ charged fragments in TOF+ wall. More details on the experimental setup and the preliminary results can be found in [38]. The efficiency of the triggers as well as the acceptance is discussed in [41]. The integrated luminosity was 0.066 pb^{-1}.

The tracking system with scintillating fiber detector arrays and the two drift chambers was used for reconstructing tracks and determining the secondary vertex.

The four scintillating hodoscope walls, used to perform time-of-flight measurements of charged particles, also worked as part of the tracking system. Track fitting was done by means of the Kalman filter algorithm, as summarized in [47].

4.3.2 Particle Identification

First, the invariant mass distributions of the final states of two-body mesonic weak decays of $\Lambda \to p + \pi^-$, $^3_\Lambda H \to {}^3He + \pi^-$, $^4_\Lambda H \to {}^4He + \pi^-$ were analyzed. In each event, daughter candidates for those decays were identified and used to reconstruct the secondary vertex, as corresponding to the mesonic weak decay of interest. In conjunction with the invariant mass calculation, the mother bound state candidates were selected after a series of geometrical considerations in order to find the most probable candidate per event.

After the track fitting procedure based on the Kalman Filter algorithm [47], the different track candidates are associated with their p-values, the goodness-of-fit criteria. The p-value approach involves determining "likely" or "unlikely" in statistical hypothesis testing by determining the probability, assuming the null hypothesis were true. It is often used in high energy physics as well as commonly used in economics, finance, political science, psychology, biology, criminal justice, criminology, and sociology. The p-value for each of the candidates is used to exclude poorly fitted tracks. The fitting procedure gives the most probable momentum vector of the particle or fragment. The particle identification is obtained by applying the additional information from the hodoscope walls, such as the time-of-flight and the energy deposit. The charge, Z, of the positively charged particles and fragments can be determined from the correlation between the energy deposit, ΔE, in the scintillating bars of the TOF+ wall and the momentum-to-charge ratio, P/Z, obtained by the track fitting.

The left panel of Fig. 4.23 shows the correlation of ΔE vs. P/Z. A clear separation is seen between the hydrogen, helium and lithium isotopes. In the case of the He isotopes, the species are determined by their momentum separation since they have a velocity close to that of the projectile: the time-of-flight measurement does not help in their identification. The right panel of Fig. 4.23 exhibits the separation between the ^3He and ^4He species in the momentum-to-charge ratio distribution. Each species' contribution is modeled by a Gaussian probability density function and allows determining intervals of momentum-to-charge [3.35 GeV/c, 4.5 GeV/c] and [4.6 GeV/c, 6.6 GeV/c] for the identification of ^3He and of ^4He respectively. The ^3He contribution to the contamination of the ^4He identification is estimated to be approximately 1.7%, while the contamination of ^4He into the identification of ^3He is approximately 1.8%.

In the case of hydrogen isotopes, the time-of-flight measurement is used to calculate the velocity, β, of each of the track candidates. The correlation of β vs. P/Z provides the identification of proton, deuteron and triton species. π^- mesons are also identified with this correlation obtained from the time-of-flight measurement

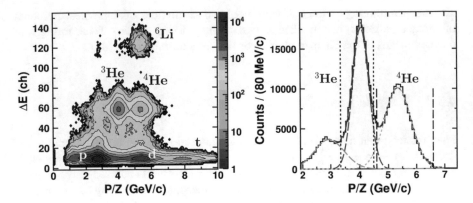

Fig. 4.23 Shown in the left panel is the correlation between the energy deposit of positively charged isotopes in the TOF+ wall detector and the momentum-to-charge ratio P/Z. The right panel depicts the projection of the momentum-to-charge ratio for the helium isotope identification in the right panel. The drawn-in vertical lines show the momentum intervals (3.35 GeV/c ≤ P/Z ≤ 4.5 GeV/c & 4.6 GeV/c ≤ P/Z ≤ 6.6 GeV/c) used to distinguish ^{3}He and ^{4}He. Each contribution is modeled by a Gaussian probability density function shown individually by blue dashed lines and obtained from the combined fit of the distribution, shown as a red solid line. The figure is taken from [39]

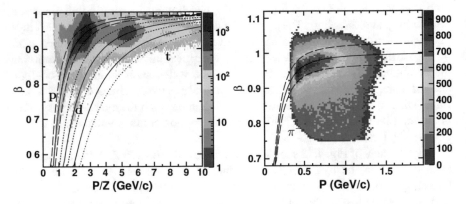

Fig. 4.24 Correlation between the velocity β and the momentum for the Z = 1 charge isotopes (left panel) and the π^- meson (right panel). The theoretical function $\beta = 1/\sqrt{(m/p)^2 + 1}$ is drawn as a black line for each hydrogen isotope (proton, deuteron and triton) and the π^- meson. Selection bands for the particle identification are drawn as dashed lines for each species of interest. The figure is taken from [39]

by the TFW detector. Figure 4.24 shows those correlations for the positively charged particles (left panel) and the negatively charged particles (right panel). The species of hydrogen isotopes and π^- meson should be distributed around the theoretical function of $\beta = 1/\sqrt{(m/p)^2 + 1}$, where m and p respectively are the mass of the species of interest and the reconstructed momentum. Each theoretical line is indicated by a solid line in Fig. 4.24. Each hydrogen species is identified by falling into the mass interval of $\pm 15\%$ from the nominal mass of the species of interest. In the left panel of Fig. 4.24, the selection bands used for the identification are indicated by dashed lines, and can be observed in the correlation plot. The identification of π^- mesons is achieved in a manner similar to hydrogen isotopes by using the time-of-flight information at the TFW wall. The right panel of Fig. 4.24 shows the corresponding correlation of β vs. P/Z, and the selection of π^- mesons is made by defining a band of ± 0.05 in β around the theoretical line.

4.3.3   Vertex Reconstruction and Invariant Mass of $p + \pi^-$, $^3He + \pi^-$ and $^4He + \pi^-$

Once the particle and fragment identification has been performed, reconstruction of the secondary vertex is necessary in order to obtain the information on the Λ hyperon and the hypernuclei of interest. The secondary vertex of interest corresponds to the two-body mesonic weak decay of the Λ hyperon and of the $^3_\Lambda H$ and $^4_\Lambda H$ hypernuclei. First, the following pair of track candidates is associated to match that two-body decay: $p + \pi^-$, $^3He + \pi^-$ and $^4He + \pi^-$. Next, several criteria are applied in selecting the best secondary vertex candidate per event among the vertex candidates of paired daughter particles. To begin with, the p-value from the track fitting for the daughter of interest has to be greater than 0.005 and 0.2 for the positively charged fragment and the π^- meson candidates, respectively.

Next, geometrical considerations are applied. The distance of closest approach for the track pair, $d_{2tracks}$, shown schematically in the left panel of Fig. 4.25, is calculated to define an estimated secondary vertex position. This distance has to be less than 4 mm for the secondary vertex candidate to be accepted. This is

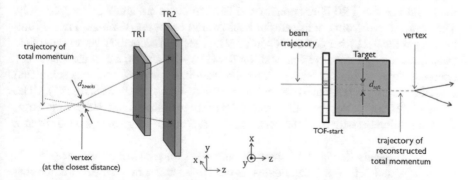

Fig. 4.25 Geometrical rules for vertex selection. The distance of closest approach of the track pair and estimation of the secondary vertex position are shown in the left panel. The right panel shows the distance between the extrapolated position of the reconstructed mother track candidate and the beam position as measured by the TOF-start detector at the target position in the horizontal plane of the laboratory frame. The figure is taken from [39]

Fig. 4.26 Longitudinal secondary vertex position in the laboratory frame. The tallest dashed black vertical lines correspond to the interval selection for the secondary vertex candidate vertex position (-10 cm $\leq Z$ vertex position ≤ 30 cm). The shorter red vertical line corresponds to the requirement for the lifetime measurements (Z vertex position > 6 cm). The figure is taken from [39]

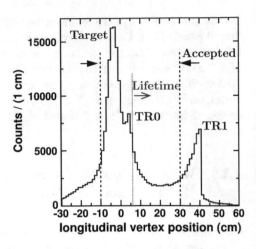

followed by evaluating the distance in the horizontal plane of the laboratory frame between the extrapolated position of the reconstructed mother track candidate at the target position and the beam position measured by the TOF-start detector. A schematic view of this distance d_{tofs} is shown in the right panel of Fig. 4.25. During the experiment, the beam trajectories were adjusted by the accelerator to be perpendicular to the surface layer of the TOF-start detector, allowing the hit position of the TOF-start detector to represent the position of the production vertex in the target.

The last rule applied in the vertex selection is based on the calculated longitudinal vertex position of the secondary vertex of interest in the laboratory frame. The secondary vertex is expected to be between the location of the production target and the first layer of the TR1 fiber detector. Figure 4.26 shows the distribution of the longitudinal Z positions of all secondary vertex candidates. The contribution from the target and the TR0 fiber detector can be seen at -4 cm and 2 cm, respectively. The contribution from combinatorial background becomes dominant after 30 cm, peaking at the TR1 fiber detector position. The accepted secondary vertex candidate must have a longitudinal vertex position between -10 and 30 cm in the laboratory frame. After applying the rules for vertex selection, the invariant mass is calculated from the reconstructed four-vector of the pair of daughter candidates, (p, π^-), (^{3}He, π^-) and (^{4}He, π^-). For the estimation of the lifetime of the mother states of interest, the longitudinal position of the secondary vertex must be greater than 6 cm at a minimum.

The top panels of Fig. 4.27 show the invariant mass distributions for p + π^-, ^{3}He + π^-, and ^{4}He + π^- candidates. It has been shown in each distribution that there is a signal peak at around the mass of Λ, $^3_\Lambda$H and $^4_\Lambda$H. For the estimation of the background contributions to the invariant mass distributions, the event mixing method was initially employed. Estimated background distributions are shown in Fig. 4.27 by the open symbols. After the first estimation of the background contributions by the mixed event method, the background contribution was then

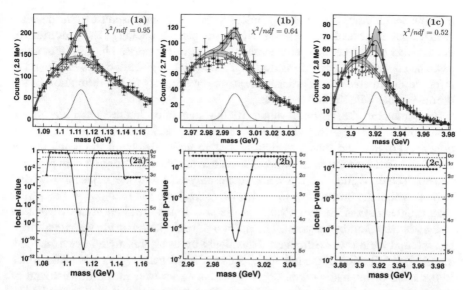

Fig. 4.27 Invariant mass distribution for candidates of Λ, $^3_\Lambda$H and $^4_\Lambda$H, are represented by the filled circles in panels (**1a**), (**1b**), and (**1c**), respectively. The shaded orange region represents one standard deviation of the fitted model centred at the solid blue line. The dotted lines show the separate contributions of the signal and the background with, respectively, black and coloured lines. The data represented by open triangles correspond to invariant mass distributions of the mixed event analysis. The local p-value distribution of the background-only hypothesis in full range of fit of Λ, $^3_\Lambda$H and $^4_\Lambda$H, are shown in panels (**2a**), (**2b**) and (**2c**), respectively. The red dashed lines illustrate the p-values corresponding to significances of 1σ, 2σ, 3σ, 4σ, 5σ and 6σ. The figures are taken from [39]

modelled by a Chebychev polynomial probability density function with a signal in the invariant mass distributions represented by a Gaussian probability density function. An extended maximum likelihood estimator provided the share of signal and background in each invariant mass spectrum. For estimating the significance of the signals, a hypothesis test was applied, and a profiled maximum likelihood ratio test gives us a significance of 6.7σ, 4.7σ, 4.9σ for Λ hyperon, $^3_\Lambda$H and $^4_\Lambda$H, respectively. Details of the analyses on the invariant mass were already discussed in [39].

4.3.4 Lifetime of $^3_\Lambda$H and $^4_\Lambda$H, and the Puzzle on the Hypertriton Lifetime

After reconstructing the invariant mass, the lifetime values of Λ, $^3_\Lambda$H and $^4_\Lambda$H were extracted. As described in the previous section, one of the merits in the present work is to deduce the lifetime value of the observed hypernuclei in a different way to the other conventional hypernuclear experiments since the deduction of the hypernuclear lifetime is completely independent of the time response of the detector systems but dependent on the sensitivity to the particle tracking. The procedure of the analyses to deduce the lifetime values was already discussed in [39]. It has also to be noted that an *unbinned* maximum likelihood fitting method was employed to infer the lifetime values in order to improve the accuracy of the estimation of the lifetime values with limited statistics.

For deducing the lifetime values of the signals of interest, first the signal contribution was determined by subtracting the background from the signal-plus-background, and thus two data sets were built. To obtain the amplitude of signal-plus-background contribution, the data in the fitted peak region with $\bar{m} \pm 2\sigma_m$ were selected. While, for the background-only data set, the adjacent sideband regions within the intervals of $[\bar{m} - 4\sigma_m, \bar{m} - 2\sigma_m]$ and $[\bar{m} + 2\sigma_m, \bar{m} + 4\sigma_m]$ were selected, where $\bar{m}$ is the mean value of the peak in the invariant mass distribution and σ_m is the width of the peak. The portion of background inside the signal-plus-background region was estimated deductively by using the mixed event method, and the normalisation of the integrated area to the sideband region was made. Since hypernuclei of interest as well as Λ-hyperons decay well in the downstream of the production target, a cut for the longitudinal vertex position was applied to be more than 6 cm distant from the target in order to minimise contamination from reaction events at the target and at the tracking detector immediately behind the target. With the conditions described above, the proper decay time $t = l/(\beta\gamma c)$ is calculated for the two data sets in the rest frame of the mother state of interest by the measured decay length l, where $\beta\gamma c = p/m$, p and m the momentum and the mass of the mother state of interest. Then signal contributions were obtained by subtracting the background-only data set from the signal-plus-background data set. The correction by the acceptance and reconstruction efficiency was made by means of full Monte Carlo simulations, as discussed in [39]. Because of the limited statistics, an *unbinned* maximum likelihood fitting method was employed to deduce the lifetime values. A model of an exponential probability density function with the background was chosen to represent the distribution of the proper decay time $l/(\beta\gamma c)$ with a lifetime τ [39]. Likelihood profiles for estimation of the mean decay length $c\tau$ of Λ hyperon, $^3_\Lambda H$ and $^4_\Lambda H$ hypernuclei are shown in the top panels of Fig. 4.28, and deduced lifetime values for Λ, $^3_\Lambda H$ and $^4_\Lambda H$ are 262^{+56}_{-43} ps, 183^{+42}_{-32} ps and 140^{+48}_{-33} ps, respectively, with the error values estimated by a standard deviation of $\pm 1\sigma$ (68.3% confidence level).

In order to check the quality of the determined lifetime values, a χ^2 test was performed with the *binned* data of decay length $l/\beta\gamma$ in the signal region. The fitted model of the signal region included the binned contribution of the exponential probability density function added to the *binned* data of the sideband regions. The binned fitted model at the bin i can be expressed as discussed in [39] as;

$$\text{ModelFit}_i = \left\{ \int_{bin_i} A/(c\tau) \exp(-l/(\beta\gamma \, c\tau)) dl \right\} \cdot 1/w_i' + B_l[l_i/(\beta_i\gamma_i)] + B_h[l_i/(\beta_i\gamma_i)]$$

$$(4.4)$$

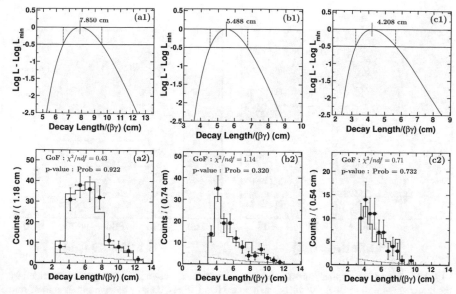

Fig. 4.28 Ratio of likelihood functions for interval estimation of Λ hyperon (**a1**), $^3_\Lambda$H (**b1**), and $^4_\Lambda$H hypernuclei (**c1**). Interval estimation for 1 standard deviation is shown on each profiled likelihood ratio. *Binned* decay length distributions of the signal region of Λ hyperon (**a2**), $^3_\Lambda$H (**b2**), and $^4_\Lambda$H hypernuclei (**c2**) with the fitted model, which included the exponential function resulting from the *unbinned* maximum likelihood fit and the background contribution estimated by the sidebands. The black line represents the fitted model, while the blue dotted line represents the contribution of the exponential function. The figures are taken from [39]

where A is the normalisation factor of the exponential probability density function, $B_l[l_i/(\beta_i\gamma_i)]$ and $B_h[l_i/(\beta_i\gamma_i)]$ are the two sideband data sets at bin i. In the bottom panels of Fig. 4.28, fitted model distributions are shown by solid lines together with binned data of the signal region of the signal-plus-background data shown by filled circles with error bars. By comparing the two distributions, reduced χ^2 values of the modelled proper decay time function over the *binned* data as $\chi^2 = \sum_{bin_i}(SB[l_i/(\beta_i\gamma_i)] - \text{ModelFit}_i)^2/(SB[l_i/(\beta_i\gamma_i)])$ was calculated. They are 0.43 for Λ, 1.14 for $^3_\Lambda$H and 0.71 for $^4_\Lambda$H. It has to be emphasised once more here that the deduced lifetime values have been obtained by the *unbinned* fitting, and the reduced χ^2 from the binned representation was used only to cross check the goodness-of-fit.

One of the striking results in the HypHI experiment is the observation of a significantly shorter lifetime value of $^3_\Lambda$H than that of the Λ-hyperon. With the former data [34, 48–53], it had been difficult to determine the lifetime of $^3_\Lambda$H since these experimental data are distributed widely with large error bars. Therefore, it had been concluded that due to its small Λ-binding energy the lifetime of $^3_\Lambda$H should be similar to that of the free Λ-hyperon without strong experimental evidences. However, as shown above, our result on the lifetime of $^3_\Lambda$H reveals 183^{+42}_{-32} ps. Thus, in order to summarise the former results for the lifetime of $^3_\Lambda$H together with our

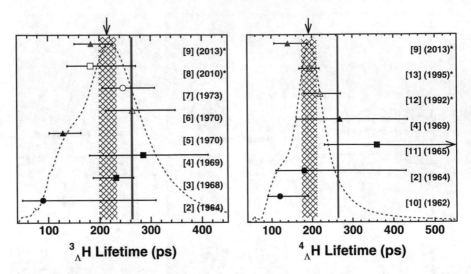

Fig. 4.29 World data comparison of $^3_\Lambda$H and $^4_\Lambda$H lifetimes. The combined average is represented by the arrow at the top, while the width of the hatched band corresponds to the one standard deviation of the average. The blue vertical line at 263.2 ps with a width of ± 2 ps shows the known lifetime of the free Λ-hyperon. References to counter experiments are marked by an asterisk. Figures are taken from [40] with reference numbers also according to [40]

result, all the existing data until 2014 for the lifetime observations of $^3_\Lambda$H were combined with a similar methodology of the Particle Data Group (PDG), and the results were already discussed in [40]. The same analyses were also employed for the lifetime of $^4_\Lambda$H [40]. For $^3_\Lambda$H, data are taken from [34, 39, 48–53] and for $^4_\Lambda$H from [30, 39, 48, 50, 54–56]. Results of the combined analyses are shown in Fig. 4.29. Details are also discussed in [57] including an alternative statistical analysis applying a Bayesian approach [40]. Figure 4.29 shows that the combined lifetime values of $^3_\Lambda$H and $^4_\Lambda$H are significantly smaller than that of the Λ-hyperon. The exclusion band at 95% confidence level, useful for discarding theoretical models, was also deduced. Theoretical values outside of the ranges between 186 and 254 ps for $^3_\Lambda$H and between 158 and 233 ps for $^4_\Lambda$H can be excluded with 95% confidence level [40]. In the analyses of the HypHI data with the Bayesian approach described in detail in [40], lifetime values of $^3_\Lambda$H and $^4_\Lambda$H were estimated to be approximately 217^{+19}_{-16} ps and 194^{+20}_{-18} ps, respectively, and the upper limit at 95% confidence level was also deduced to be 250 ps and 227 ps, respectively, excluding the possible theoretical prediction above this limit with 95% confidence level [40]. The most recent theoretical model predicts 256 ps [37] for $^3_\Lambda$H, and it has to be noted that the result of this theoretical calculation is above the upper limit at 95% confidence level obtained by both the combined analyses and the Bayesian approach.

Very recently, the STAR and ALICE collaborations also measured the $^3_\Lambda$H lifetime, and they observed 155^{+25}_{-22} ps [58] and 181^{+54}_{-39} ps [35], respectively. It has to be noted that there are no theoretical models which can reproduce the short lifetime of $^3_\Lambda$H so far, and the errors on the experimental lifetime value should be still minimized to come to conclusions about the lifetime of $^3_\Lambda$H.

As discussed above, the lifetime of the $^4_\Lambda$H hypernucleus is concluded to be significantly shorted than that of the free Λ-hyperon. It has not been fully understood though the Λ-binding energy in $^4_\Lambda$H is much larger than that in $^3_\Lambda$H, being approximately 2.2 MeV. It is not practical to study the $^4_\Lambda$H hypernucleus by experiments using the ultra-relativistic heavy ion collisions such as STAR and ALICE because of the extremely small production cross section, therefore, it has to be experimentally studied further by heavy ion beams with a fixed target.

4.3.5   Production Cross Section of Λ, $^3_\Lambda H$ and $^4_\Lambda H$, and Their Kinematics

Hypernuclear production cross sections for $^3_\Lambda$H and $^4_\Lambda$H as well as the production cross section of the Λ-hyperon in the reaction of ^{6}Li+^{12}C at 2 A GeV or $\sqrt{s_{NN}} = 2.70$ GeV have been deduced with the experimental data. Details of the analyses are given in [41]. A production cross section of $3.9 \pm 1.4 \,\mu$b for $^3_\Lambda$H and of $3.1 \pm 1.0 \,\mu$b for $^4_\Lambda$H in the projectile rapidity region was inferred. Also the yield of Λ-hyperons in the projectile rapidity region was deduced, and the total production cross section of the Λ-hyperon was extracted by employing UrQMD theoretical calculations [59, 60] that are considered to model the Λ phase space reasonably [61–67]. It is found to be equal to 1.7 ± 0.8 mb. A global fit based on a Bayesian approach was performed in order to include and propagate statistical and systematic uncertainties. Production ratios of $^3_\Lambda$H/$^4_\Lambda$H, $^3_\Lambda$H/Λ and $^4_\Lambda$H/Λ were included in the inference procedure. A summary of the cross sections as well as the production ratios is given in Table 4.1.

Table 4.1 Summary of the deduced cross sections and the yield ratios with the data of the HypHI experiment

	$\langle x \rangle$	σ_{stat}	σ_{sys}	σ_{prior}
Λ_{tot} (mb)	1.7	±0.7 (stat)	±0.4 (sys)	±0.2 (prior)
Λ_{obs} (mb)	0.3	±0.1 (stat)	±0.06 (sys)	±0.03 (prior)
$^3_\Lambda$H (μb)	3.9	±1.3 (stat)	±0.3 (sys)	±0.3 (prior)
$^4_\Lambda$H (μb)	3.1	±1.0 (stat)	±0.3 (sys)	±0.1 (prior)
$^3_\Lambda$H/$^4_\Lambda$H	1.4	±0.7 (stat)	±0.1 (sys)	±0.2 (prior)
$^3_\Lambda$H/Λ ($\times 10^{-3}$)	2.6	±1.4 (stat)	±0.3 (sys)	±0.2 (prior)
$^4_\Lambda$H/Λ ($\times 10^{-3}$)	2.1	±1.1 (stat)	±0.1 (sys)	±0.2 (prior)

$\langle x \rangle$ and σ_{stat} correspond to the expected value and its statistical standard deviation of the posterior probability density function. σ_{sys} and σ_{prior} stand for the systematic uncertainties and the prior sensitivity uncertainties [41]

The multiplicity distributions of $^3_\Lambda$H and $^4_\Lambda$H signal as a function of the rapidity in the center-of-mass system, $y0$, and transversal momentum, Pt, are also studied [41], and they are shown in the top panels of Fig. 4.30. The hypernuclear signal was extracted from the experimental data set and the mixed event data set of $y0 - Pt$ observables. The bins size of data sets were 40 MeV/c in Pt and 0.02 and 0.03 unit of $y0$ for $^3_\Lambda$H and $^4_\Lambda$H respectively. For each bin, the signal contribution was estimated by a maximum likelihood ratio method from the background contribution (mixed event) and the signal-plus-background contribution (experimental data). The panels (c) and (d) of Fig. 4.30 show the projected rapidity distribution of the data set with the following distributions: the extracted signal (*Smodel*), the background-only model from the mixed event analysis (*Bmodel*) and the signal-plus-background contribution (*SBmodel*). The last four panels of Fig. 4.30, show the projected rapidity $y0$ and Pt distribution of the extracted signal for $^3_\Lambda$H and $^4_\Lambda$H hypernuclei on the left- and right-hand side, respectively. The mean value and standard deviation of the rapidity distribution of $^3_\Lambda$H and $^4_\Lambda$H are respectively $\langle y0 \rangle = 0.98 \pm 0.01$, $\sigma_{y0} = 0.06 \pm 0.01$ and $\langle y0 \rangle = 1.00 \pm 0.01$, $\sigma_{y0} = 0.07 \pm 0.01$. One can remark that the experimental rapidity distribution $^3_\Lambda$H and $^4_\Lambda$H falls within the Monte Carlo experimental acceptance. Moreover the multiplicity density decrease in the rapidity region [0.8 ; 0.9] shown in the $^3_\Lambda$H and $^4_\Lambda$H rapidity distribution represents a physical limit since it is still within the experimental acceptance.

4.3.6 Invariant Mass and Lifetime of $d + \pi^-$ and $t + \pi^-$ Final States

In addition to the analyses discussed above, the invariant mass distributions of all other possible final states were studied with the data of the HypHI experiment. Surprisingly, signals in the $d + \pi^-$ and $t + \pi^-$ final states were observed. Details of the analyses were discussed in [42]. Figure 4.31 shows the invariant mass distributions of $d + \pi^-$ in panels (a1) and (a2) and $t + \pi^-$ in panels (b1) and (b2). The longitudinal decay vertex position (Z) was requested to be set between -10 cm $< Z < 30$ cm in panels (a1) and (b1) and between -2 cm $< Z < 30$ cm in (a2) and (b2). The production target was placed between -6 and -2 cm, thus the cut condition of -10 cm $< Z < 30$ cm includes vertices from the production target while the other condition excludes the target region. The deduced invariant mass distributions are represented by the filled-in circles. The mass values were calibrated by using the data for the reconstructed invariant mass peak positions of Λ, $^3_\Lambda$H and $^4_\Lambda$H. Fitting of the distributions of signal-plus-background from the data were performed in a similar fashion to $^3_\Lambda$H and $^4_\Lambda$H. By hypothesis testing via profiled likelihood ratio tests, the significance values of the observed peaks of $d + \pi^-$ for -10 cm $< Z < 30$ cm and -2 cm $< Z < 30$ cm were determined to be respectively 5.3 and 3.7 σ, and for $t+\pi^-$ they are 5.0 and 5.2 σ, respectively. Vertex distributions of the $d + \pi^-$ and $t + \pi^-$ were analysed to deduce the lifetime values of the initial

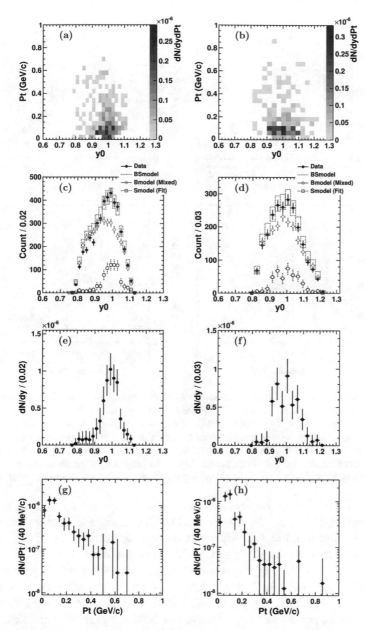

Fig. 4.30 Multiplicity distribution as a function of the rapidity observable y0 and of the transversal momentum Pt in the center-of-mass system for $^3_\Lambda$H in panel (**a**), $^4_\Lambda$H in panel (**b**), respectively. In panels (**c**) and (**d**) the projected rapidity distributions of the data set $^3_\Lambda$H and $^4_\Lambda$H respectively is shown in black full circle together with the extracted signal contribution *Smodel* (open box), the background-only distribution from the mixed event analysis *Bmodel* (open circle) and the signal-plus-background model *BSmodel* (dash box representing the 1-σ standard deviation interval). Panels (**e**), (**f**) and (**g**), (**h**) show the rapidity and Pt distribution of the extracted $^3_\Lambda$H and $^4_\Lambda$H signal, respectively. Figures are taken from [41]

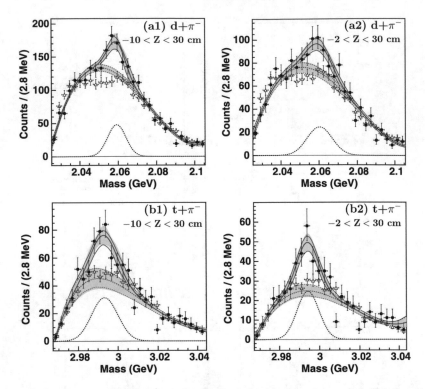

Fig. 4.31 Invariant mass distributions of d+π^- final state candidate in panels (**a1**) (**a2**), and of t + π^- in panels (**b1**) (**b2**). Panels (**a1**) and (**b1**) are for -10 cm $< Z < 30$ cm, and (**a2**) and (**b2**) are for -2 cm $< Z < 30$ cm. Observed distributions are represented by filled-in circles. The shaded orange region represents one standard deviation of the fitted model centred at the solid blue line of the total best fit. The black and coloured dotted-lines respectively show the separate contributions of the signal and the background. The open triangles represent the data corresponding to the invariant mass distribution of the mixed event analysis. The figures are taken from [42]

states decaying the d + π^- and t + π^- final states by using a similar manner for $^3_\Lambda$H and $^4_\Lambda$H. As shown in Fig. 4.32, the resulting lifetime values with the d + π^- and t + π^- final states are 181^{+30}_{-24} ps and 190^{+47}_{-35} ps, respectively.

As discussed in [42], systematic uncertainties and the possibility to produce peaks in the d+π^- and t+π^- invariant mass distributions by mis-reconstructing the other possible decay channels were studied. It was concluded that the other channels could not create peaks like those observed in the d + π^- and t + π^- final states. Therefore, a possible interpretation for the observed t + π^- and d + π^- final states might be the two- and three-body decays of an unknown bound state of two neutrons associated with Λ, $^3_\Lambda$n, via $^3_\Lambda$n $\rightarrow$ t + π^- and $^3_\Lambda$n $\rightarrow$ t* + π^- $\rightarrow$ d + n + π^-, respectively. With this interpretation, the production mechanism of $^3_\Lambda$n might be the Fermi break-up of excited heavier hyperfragments [68]. On the other hand, the direct coalescence of a Λ-hyperon and a di-neutron state is unlikely, because the di-neutron state is known to be unbound.

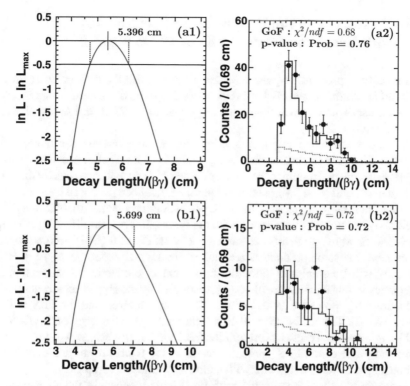

Fig. 4.32 Profiled likelihood ratio for interval estimation of the lifetime of $d + \pi^-$ (**a1**) and $t + \pi^-$ (**b1**) final states. Interval estimation for 1 standard deviation is shown on each profiled likelihood ratio. *Binned* decay length distributions of the signal region of $d + \pi^-$ (**a2**) and $t + \pi^-$ (**b2**) are shown with the fitted model. The model includes the exponential function that resulted from the *unbinned* maximum likelihood fit and the background contribution estimated by the sidebands. The fitted model is represented by the black line, while the contribution of the exponential function is represented by the dotted blue line. The figures are taken from [42]

The possibility of a bound state made of a Λ-hyperon and two neutrons, ${}^{3}_{\Lambda}$n, was studied recently in several theoretical works [69–72], all concluding that the $n - n - \Lambda$ system is unlikely to be bound. These calculations are based on quite general arguments relying on existing, well established hypernuclear data.

Since the results from the HypHI collaboration on ${}^{3}_{\Lambda}$n are not conclusive, the origin of the observed structures in the $d + \pi^-$ and $t + \pi^-$ final states must be experimentally verified and clarified. The ALICE collaboration studied $d + \pi^-$ finals states, and there is no signal observed [73]. Since ALICE and HypHI observe particles in different rapidity regions, the origin of the structures in the $d + \pi^-$ and $t + \pi^-$ final states should be experimentally studied in the projectile rapidity region like the HypHI experiment, which will be conducted at FAIR Phase 0 (GSI) and FAIR Phase 1. Newly proposed experiments at FAIR Phase 0 (GSI) to study the $d + \pi^-$ and $t + \pi^-$ structures as well as to measure lifetime values of ${}^{3}_{\Lambda}$H and ${}^{4}_{\Lambda}$H with better accuracy will be discussed in the following sections. Future plans for the hypernuclear spectroscopy at FAIR Phase 1 will also be discussed.

4.4 Perspective of the Hypernuclear Spectroscopy with Heavy Ion Beams at FAIR Phase 0 and 1

In this section, plans for hypernuclear experiments with the FRS at FAIR Phase 0 (GSI) and with the Super-FRS at FAIR Phase 1 will be discussed. A possibility of hypernuclear experiments at the high energy cave of NuSTAR at FAIR will also be briefly mentioned.

As discussed in the previous section, further experiments for hypernuclear physics with heavy ion beams should be conducted with a better precision than the HypHI experiment to confirm the short lifetime of $^3_\Lambda$H as well as to clarify the origin of the $d + \pi^-$ and $t + \pi^-$ signals. An experiment has been proposed by a part of the HypHI collaboration together with the Super-FRS Experiment Collaboration [74] to perform hypernuclear spectroscopy with the fragment separator FRS [75] as a high-resolution forward magnetic spectrometer at FAIR Phase 0 (GSI). In the proposed experiment, the same production reaction of the HypHI experiment, ^{6}Li + ^{12}C at 2 A GeV, will be employed since the hypernuclear production cross section and kinematics in this reaction are already investigated by the HypHI experiment.

Figure 4.33 shows the layout of the SIS 18 synchrotron and the FRS at FAIR Phase 0 (GSI). The ^{6}Li beams will be accelerated up to an energy of 2 A GeV and will be injected to the FRS. Through the FRS via S1, the beams will arrive at the mid-focal plane of the FRS, indicated as S2 in the figure. In the S2 area, a fixed carbon target is situated together with a pion spectrometer complex.

A proposed experimental setup with two dipole magnets in the S2 area of the FRS is shown in Fig. 4.34 schematically. The ^{6}Li projectiles with an intensity of about 10^6 particles per second are delivered from the left side of the figure, and they

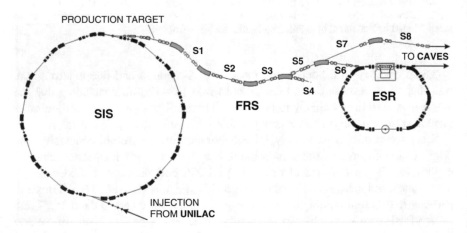

Fig. 4.33 Layout of the SIS-FRS-ESR at GSI. In the proposed experiment, a conventional production target located at the entrance of the FRS is not used, but the hypernuclear production target will be located at the mid-focal plane, S2. A system of the π^- measurements will be installed at the mid-focal plane S2, too

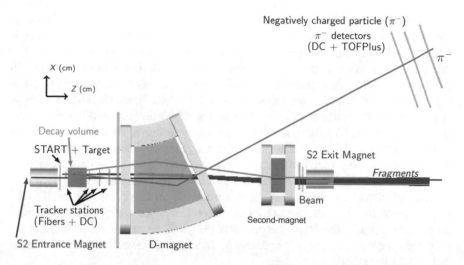

Fig. 4.34 Proposed setup for the future HypHI experiment with the Fragment separator FRS at GSI

are bombarded to the fixed carbon target with a thickness of 8 g/cm^2. In front of the target, a small plastic hodoscope which is similar to TOF-start employed in the HypHI experiment [39] will be mounted to measure timing and positions of beam particles to define start-time for time-of-flight measurements. It is shown together with the target as indicated as "START + TARGET" in the figure. Hypernuclei are produced in the target and fly to the forward direction. Then, hypernuclei decay in flight, and they emit particles. Particles from the decay also fly to the forward direction, and they are injecting to the dipole magnet indicated as D-magnet in the figure. Decays of the hypernuclei taking place inside "Decay volume" as shown in the figure will also be measured. Across the "Decay volume", several tracking detectors will be mounted, that consist of scintillating fibre detectors and the drift chamber (SDC) used in the HypHI experiment [39]. These tracking detectors will contribute to find decay vertices of hypernuclei. The π^- mesons from the hypernuclear decay are bent in the D-magnet and swept upward in the figure. Approximately 10% of π^- from the hypernuclear decay are registered in the detector system indicated as "π^- detectors" in the figure. These detectors will consist of a plastic hodoscope and a drift chamber. The plastic hodoscope will be implemented by rearranging the TOF+ detector used in the HypHI experiment, and the drift chamber will be SDC which was also used in HypHI experiment [39]. For π^-, four momenta will be obtained by tracking across the magnet, time-of-flight measurements and the information of ΔE in the plastic hodoscope. In the proposed system, a momentum resolution of π^- is expected to be 1–2%.

In the proposed experiment, positively charged decay residues of hypernuclear π^- decay, such as ^{2}H, ^{3}He and ^{4}He, will propagate to the second half of the FRS, and they should be in the aperture of the magnet located at the exit of S2 indicated as "S2 Exit Magnet" in the figure. However, decay residues are also bent inside the D-magnet as shown in the figure, therefore, the bending of the decay residues must be compensated in order to deliver them to the second half of the FRS. For this purpose, another dipole magnet located in the right side of the D-magnet will be installed. With this magnet, decay residues can be measured from S2 to S4 in Fig. 4.33. A momentum resolution of measured decay residues from S2 through S4 is expected to be better than 10^{-3}. One of the disadvantages in this setup is the small momentum and spatial acceptance for decay residues. However, it will make a counting rate at S4 drastically smaller (about three orders of magnitude lower than the rate of the projectiles), and one can make a trigger to the data acquisition system only by coincident measurements of particle registrations in the detectors at S4 and "π^- detectors" at S2. Because of a small trigger rate of an order of 10^2–10^3 per second with this coincidence trigger, a secondary vertex trigger, which played an important role in the HypHI experiment [39], will be unnecessary. Without the vertex trigger, the efficiency of the data accumulation will be at least ten times larger than that in the HypHI experiment.

With reconstructed π^- from the hypernuclear decay and decay residues together with finding of hypernuclear decay vertices in "Decay volume", an invariant mass of produced hypernuclei will be calculated. Despite the small momentum acceptance of decay residues due to the second half of the FRS, a resultant invariant mass resolution will not be narrow around the observed peak region since π^- will be measured with a large momentum range at S2 in the proposed setup. In order to observe $^3_\Lambda$H and $^4_\Lambda$H as well as the $d + \pi^-$ signal, only four magnetic settings of the FRS between S2 and S4 including the magnets at S2 will be required; (1) 12.5 Tm for measuring $^3_\Lambda$H, (2) 16.5 Tm for both $^4_\Lambda$H and the $d + \pi^-$ final state, (3) 11.0 Tm for $^4_\Lambda$H and (4) 14.0 Tm for all the three states. Monte Carlo simulations have been performed, and expected invariant mass distributions for $^3_\Lambda$H and $d + \pi^-$ final state are shown in the left and right panels, respectively, of Fig. 4.35. For $^3_\Lambda$H a

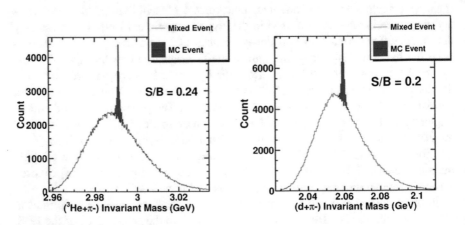

Fig. 4.35 Expected invariant mass distributions for $^3_\Lambda$H (left) and $d + \pi^-$ final state (right). In the panels, "S/B" indicates a ratio of the signal to the background. "MC" indicates Monte Carlo simulations

ratio of the signal to the background is assumed to be 0.24. For $d + \pi^-$, a smaller ratio of the signal to the background, 0.20, is assumed because the background for $d + \pi^-$ was observed larger than $^3_\Lambda$H in the analyses of the HypHI experiment. Resultant mass resolution for the both cases are approximately $800 \, \text{keV}/c^2$, which is much better than the results of the HypHI experiment ($\sim 5 \, \text{MeV}/c^2$). The expected mass resolution for the reconstruction of $^4_\Lambda$H will be similar to the case of $d + \pi^-$ because of the similarity of the magnetic rigidity of decay residues. Measurements with all the four magnetic setups mentioned above can be achieved within 2 weeks of beamtime.

The proposed experiment can also be performed without the second dipole magnet (right magnet in Fig. 4.34), but the acceptance for the decay residues will be reduced by 25%. Another option with a superconducting solenoid magnet at S2, instead of the dipole magnet(s), has been also considered, and Monte Carlo simulations to design an experimental apparatus with these configurations are currently in progress.

This method will be extended to measure other hypernuclei including those towards the proton drip-line with Super-FRS [76] at FAIR Phase 1. Figure 4.36 shows the schematic layout of the Super-FRS. Primary beams or proton-rich secondary beams will be delivered to the mid-focal plane of the Super-FRS, as indicated by a red circle in the figure, through the pre-separator and the first half of the main separator of Super-FRS. Hypernuclei will be produced at the mid-focal plane, and pions from the production and decays of hypernuclei will be measured there. Decay residues will be measured by the second half of the main separator of the Super-FRS, in a similar fashion to the experiment with FRS as described above. Because of a good separation power of the combination of the pre-separator and the first half of the main separator, proton rich beams will also be used to study proton-rich hypernuclei for the first time. Hypernuclear experiments can also be performed at the high energy cave of NuSTAR, indicated as HEC in Fig. 4.36

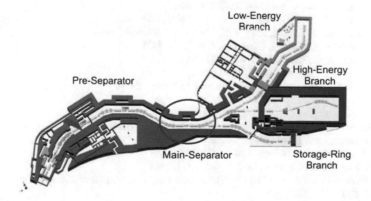

Fig. 4.36 Layout of Super-FRS at FAIR. The detector system for measuring mesons including π^- from the hypernuclear decay will be installed at the position indicated by the red circle

together with the GLAD magnet and the neutron detector system NeuLAND. Hypernuclear decay channels with neutron emissions will be measured, and some of neutron-rich hypernuclei will be studied.

The success of the proposed hypernuclear experiments with the FRS and the Super-FRS will open a new doorway to the hypernuclear studies with heavy ion beams.

References

1. P.B. Demorest et al., Nature **467**, 1081 (2010)
2. M. Danysz, J. Pniewski, Phil. Mag. **44**, 348 (1953)
3. D.H. Davis, Hypernuclei. Contemp. Phys. **27**(2), 91 (1986)
4. M.A. Faessler et al., Phys. Lett. B **46**, 468 (1973)
5. Hotchi et al., Phys. Rev. C **64**, 044302 (2001)
6. FINUDA Collaboration, Phys. Lett. B **622**, 35 (2005)
7. S.N. Nakamura et al., Phys. Rev. Lett. **110**, 012502 (2013)
8. M. Juric et al., Nucl. Phys. B **52**, 1 (1973)
9. A. Esser et al., Phys. Rev. Lett. **114**, 232501 (2015)
10. G. Bohm et al., Nucl. Phys. B **4**, 511 (1968)
11. W. Gajewski et al., Nucl. Phys. B **1**, 105 (1967)
12. R.H. Dalitz, A. Gal, Ann. Phys. (N.Y.) **116**, 167 (1978)
13. D.J. Millener et al., Phys. Rev. C **31**, 499 (1985)
14. H. Tamura et al., Phys. Rev. Lett. **84**, 5963 (2000)
15. H. Akikawa et al., Phys. Rev. Lett. **88**, 082501 (2002)
16. Y. Ma et al., Eur. Phys. J. A **33**, 243 (2007)
17. M. Ukai et al., Phys. Rev. C **77**, 054315 (2008)
18. M. Ukai et al., Phys. Rev. Lett. **93**, 23250 (2004)
19. T.O. Yamamoto et al., Phys. Rev. Lett. **115**, 222501 (2015)
20. M. Bedjidian et al., Phys. Lett. B **62**, 467 (1976)
21. M. Bedjidian et al., Phys. Lett. B **83**, 252 (1979)
22. H. Takahashi et al., Phys. Rev. Lett. **87**, 212502 (2001)
23. J.K. Ahn et al., Phys. Rev. C **88**, 014003 (2013)
24. K. Nakazawa et al., Prog. Theor. Exp. Phys. **2015**, 033D02 (2015)
25. O. Hashimoto, H. Tamura, Prog. Part. Nucl. Phys. **57**, 564 (2006)
26. A.K. Kerman, M.S. Weiss, Phys. Rev. C **8**, 408 (1973)
27. T. Gaitanos et al., Phys. Lett. B **675**, 297 (2009)
28. K. Nield et al., Phys. Rev. C **13**, 1263 (1976)
29. A. Abdurakhimov et al., Nuovo Cimento A **102**, 645 (1989)
30. S. Avramenko et al., Nucl. Phys. A **547**, 95c (1992)
31. M. Wakai et al., Phys. Rev. C **38**, 748 (1988)
32. M. Sano, M. Wakai, Prog. Theor. Phys. Suppl. **117**, 99 (1994)
33. T. Armstrong et al., Phys. Rev. C **70**, 024902 (2004)
34. STAR Collaboration, Science **328**, 58 (2010)
35. ALICE Collaboration, Phys. Lett. B **754**, 360 (2016)
36. T. Saito et al., Letter of intent (2006). http://www.gsi-schwerionenforschung.org/documents/ DOC-2005-Feb-432-1.ps
37. H. Kamada et al., Phys. Rev. C **57**, 1595 (1998)
38. T. Saito et al., Nucl. Phys. A **881**, 218 (2012)
39. C. Rappold et al., Nucl. Phys. A **913**, 170 (2013)
40. C. Rappold et al., Phys. Lett. B **728**, 543 (2014)

41. C. Rappold et al., Phys. Lett. B **747**, 129 (2015)
42. C. Rappold et al., Phys. Rev. C **88**, 041001(R) (2013)
43. ALADiN Collaboration, Proposal for a forward spectrometer at the 4π detector, GSI Report 88-08, GSI Darmstadt (1988)
44. D. Nakajima et al., Nucl. Instrum. Methods A **608**, 287 (2009)
45. S. Minami et al., GSI Scientific Report (2007), p. 223
46. S. Minami et al., GSI Scientific Report (2008), p. 52
47. C. Rappold et al., Nucl. Instrum. Methods A **622**, 231 (2010)
48. R.J. Prem, P.H. Steinberg, Phys. Rev. **136**, B1803 (1964)
49. G. Keyes et al., Phys. Rev. Lett. **20**, 819 (1968)
50. R.E. Phillips, J. Schneps, Phys. Rev. **180**, 1307 (1969)
51. G. Bohm et al., Nucl. Phys. B **16**, 46 (1970)
52. G. Keyes et al., Phys. Rev. D **1**, 66 (1970)
53. G. Keyes et al., Nucl. Phys. B **67**, 269 (1973)
54. N. Crayton et al., in *Proceedings, 11th International Conference on High Energy Physics*, 1962, p. 460
55. Y.W. Kang et al., Phys. Rev. **139**, B401 (1965)
56. H. Outa et al., Nucl. Phys. A **585**, 109 (1995)
57. C. Rappold et al., in *Proceedings of HYP2015* (2015)
58. Y. Xu for the STAR Collaboration, *Proceedings of the 12th International Conference on Hypernuclear and Strange Particle Physics (HYP2015), JPS Conference Proceedings*, p. 021005 (2017)
59. S. Bass et. al., Prog. Part. Nucl. Phys. **41**, 255 (1998)
60. M. Bleicher et al., J. Phys. G Nucl. Part. **25**, 1859 (1999)
61. M. Merschmeyer, Ph.D. thesis, University of Heidelberg (2004), 104 pp
62. H. Petersen et al., arXiv:0805.0567 (2008)
63. E. Bratkovskaya et al., Prog. Part. Nucl. Phys. **53**, 225 (2003)
64. E.L. Bratkovskaya et al., Phys. Rev. C **69**, 054907 (2004)
65. K. Dey, B. Bhattacharjee, Phys. Rev. C **89**, 054910 (2014)
66. N. Abgrall et al., NA61/SHINE Collaboration, Phys. Rev. C **89**, 025205 (2014)
67. G. Agakishiev et al., Eur. Phys. J. A (ISSN1434-6001) **50** (2014)
68. A. Sanchez Lorente et al., Phys. Lett. B **697**, 222 (2011)
69. A. Gal, H. Garcilazo, Phys. Lett. B **736**, 93 (2014)
70. E. Hiyama et al., Phys. Rev. C **89**, 061302(R) (2014)
71. H. Garcilazo, A. Valcarce, Phys. Rev. C **89**, 057001 (2014)
72. J.M. Richard et al., Phys. Rev. C **91**, 014003 (2015)
73. S. Piano on Behalf of ALICE Collaboration, *Proceedings of the 12th International Conference on Hypernuclear and Strange Particle Physics (HYP2015), JPS Conference Proceedings*, p. 021004 (2017)
74. J. Aysto et al., The Super-FRS Collaboration, in *Proceeding Conference Advances in Radioactive Isotope Science (ARIS2014)*, JPS Conf. Proc. **6**, 020035 (2015)
75. H. Geissel et al., Nucl. Instrum. Methods B **70**, 286 (1992)
76. H. Geissel et al., Nucl. Instrum. Methods B **204**, 71 (2003)

Chapter 5
Hyperons and Resonances in Nuclear Matter

Horst Lenske and Madhumita Dhar

Abstract Theoretical approaches to interactions of hyperons and resonances in nuclear matter and their production in elementary hadronic reactions and heavy ion collisions are discussed. The focus is on baryons in the lowest SU(3) flavor octet and states from the SU(3) flavor decuplet. Approaches using the SU(3) formalism for interactions of mesons and baryons and effective field theory for hyperons are discussed. An overview of application to free space and in-medium baryon-baryon interactions is given and the relation to a density functional theory is indicated. SU(3) symmetry breaking is discussed for the Lambda hyperon. The symmetry conserving Lambda-Sigma mixing is investigated. In asymmetric nuclear matter a mixing potential, driven by the rho- and delta-meson mean-fields, is obtained. The excitation of subnuclear degrees of freedom in peripheral heavy ion collisions at relativistic energies is reviewed. The status of in-medium resonance physics is discussed.

5.1 Introduction

In 1947, Rochester and Butler observed a strange pattern of tracks on a photographic emulsion plate which was exposed in a high altitude balloon mission to cosmic rays [1]. That event marks the inauguration of strangeness physics, indicating that there might be matter beyond nucleons and nuclei. Seven years later, that conjecture was confirmed by Danysz and Pniewski with their first observation of a hypernucleus [2], produced also in a cosmic ray event. These observations had and are still

H. Lenske (✉)
Institut für Theoretische Physik, JLU Giessen, Gießen, Germany
e-mail: horst.lenske@physik.uni-giessen.de

M. Dhar
Balurghat College, Balurghat, India

© Springer International Publishing AG, part of Springer Nature 2018
C. Scheidenberger, M. Pfützner (eds.), *The Euroschool on Exotic Beams - Vol. 5*,
Lecture Notes in Physics 948, https://doi.org/10.1007/978-3-319-74878-8_5

having a large impact on elementary particle and nuclear physics. In recent years, a series of spectacular observations on hypernuclear systems were made, giving new momentum to hypernuclear research activities, see e.g. [3–7].

The group-theoretical approach introduced independently by Murray Gell-Mann and Yuval Ne'eman in the beginning of the sixties of the last century was the long awaited for breakthrough towards a new understanding of hadrons in terms of a few elementary degrees of freedom given by quarks and gluon gauge fields as the force carrier of strong interactions. One of the central predictions of early QCD was the parton structure of hadrons. Once that conjecture was confirmed by experiment in the early 1970s [9], Quantum Chromo Dynamics (QCD) has evolved into the nowadays accepted standard model of strong interaction physics. Since long, QCD theory has become part of the solid foundations of modern science. As indispensable part of the scientific narrative QCD gauge theory has become a central topic in particle and nuclear physics text books, for instance the one by Cheng and Li [8]. The $\frac{1}{2}^{+}$ baryon octet and the $\frac{3}{2}^{+}$ baryon decuplet together with their valence quark structures are displayed in Fig. 5.1.

Lambda-hypernuclei are being studied already for decades. They are the major source of information on the $S = -1$ sector of nuclear many-body physics. The status of the field was comprehensively reviewed quite recently by Gal et al. [10]. The "hyperonization puzzle" heavily discussed for neutrons stars [11] is another aspect of the revived strong interest in in-medium strangeness physics. In the past, (π, K) experiments were a major source of hypernuclear spectroscopy. More recently, those studies were complemented by electro-production experiments at JLab and, at present, at MAMI at Mainz. The FINUDA collaboration at the Frascati ϕ-meson factory observed for the first time the exotic superstrange system $^{6}_{\Lambda}H$ [3].

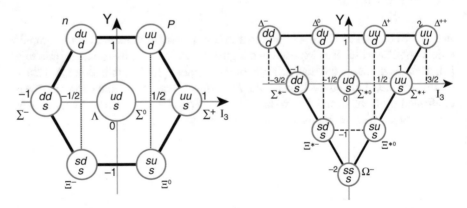

Fig. 5.1 The first two baryon SU(3) flavor multiplets are given by an octet (left) and a decuplet (right). The valence quark content of the baryons is indicated explicitly. The vertical axes are representing the hypercharge $Y = S + B$, given by the strangeness S and baryon number B; the horizontal axes indicate $I_3 = Q - \frac{1}{2}Y$ the third component of the isospin I, which includes also the charge number Q. The group theoretical background and construction of these kind of diagrams is discussed in depth in textbooks, see e.g. [8]

The STAR experiment at RHIC has filtered out of their data samples exciting results for a totally unexpected reduced lifetime of the Lambda-hyperon bound in $^3_\Lambda H$. Soon after, that result was observed also by the HypHI-experiment at the FRS@GSI [5]. Observations by the ALICE collaboration at the LHC confirm independently this unexpected—and yet to be explained—result. Recent observations on light hypernuclei and their antimatter counterparts at RHIC [6] and the LHC [7], respectively, seem to confirm the surprising life-time reduction and, moreover, point to a not yet understood reaction mechanism. The HypHI group also found strong indications for a $nn\Lambda$ bound state [4] which—if confirmed—would be a spectacular discovery of the first and hitherto only charge-neutral system bound by strong interactions.

Resonance studies with peripheral light and heavy ion reactions were initiated in the late 1970s and thereafter continued at SATURNE and later at the Synchrophasotron at Dubna and at KEK. The major achievements were the observation of an apparent huge mass shift of the Delta-resonance by up to $\Delta M \sim -70$ MeV. Detailed theoretical investigations, however, have shown that the observed shifts were in fact due to distortions of the shape of the spectral distribution induced mainly by reaction dynamics and residual interactions. The new experiments at GSI on the FRS have shown already a large potential for resonance studies under well controlled conditions and with hitherto unreached high energy resolution. Once the Super-FRS will come into operation resonance physics with beams of exotic nuclei will be possible, thus probing resonances in charge-asymmetric matter.

Viewed from another perspective, investigations of nucleon resonances in nuclear matter are a natural extension of nuclear physics. The Delta-resonance, for example, appears as the natural partner of the corresponding spin-isospin changing $\Delta S = 1, \Delta I = 1$ nuclear excitation, well known as Gamow-Teller resonance (GTR). GTR charge exchange excitations and the Delta-resonance are both related to the action of $\sigma \tau_\pm$ spin-isospin transition operators, in the one case on the nuclear medium, in the other case on the nucleon. Actually, a long standing problem of nuclear structure physics is to understand the coupling of the nuclear GTR and the nucleonic Δ_{33} modes: The notorious (and never satisfactorily solved) problem of the quenching of the Gamow-Teller strength is related to the redistribution of transition strength due to the coupling of Δ *particle-nucleon hole* (ΔN^{-1}) and purely nucleonic double excitations of particle-hole (NN^{-1}) states. Similar mechanisms, although not that clearly seen, are present in the Fermi-type spectral sections, i.e. the non-spin flip charge exchange excitations mediated by the $\tau_\pm$ operator alone. In that case the nucleonic particle-hole states may couple to the $P_{11}(1440)$ Roper resonance. Since that $J^\pi = \frac{1}{2}^+, I = \frac{1}{2}$ state falls out of the group theoretical systematics it is a good example for a dynamically generated resonance. From a nuclear structure point of view the $\Delta_{33}(1232)$ and other resonances are of central interest because they are important sources of induced three-body interactions among nucleons. Moreover, the same type of operators are also acting in weak interactions leading to nuclear beta-decay or by repeated action to the rather exotic double-beta decay with or, still hypothetical, without neutrino emission. These investigations will possibly also serve to add highly needed information on interactions of high energy neutrinos with matter and may give hints on nucleon resonances in neutron star matter. On the experimental side, concrete steps towards a new approach to resonance physics are under way. The production of nucleon resonances in peripheral heavy ion collisions, their implementation into stable and exotic nuclei in peripheral charge exchange reactions, and studies of their in-medium decay spectroscopy are the topics of the baryon resonance collaboration [12]. The strangeness physics, hypernuclear research, and resonance studies belong to the experimental program envisioned for the FAIR facility.

In this article we intend to point out common aspects of hypernuclear and nucleon in-medium resonance physics. Both are allowing to investigate the connections, cross-talk, and dependencies of nuclear many-body dynamics and sub-nuclear degrees of freedom. Such interrelations can be expected to become increasingly important for a broader understanding of nuclear systems under the emerging results of effective field theory and lattice QCD. The unexpected observations of neutron stars heavier than two solar masses are a signal for the need to change the paradigm of nuclear physics. Already in low energy nuclear physics we have encountered ample signals for the entanglement of nuclear and sub-nuclear scales, e.g. in three-body forces and quenching phenomena. In Sect. 5.2 we introduce the concepts of flavor SU(3) physics and discuss the application to in-medium physics of the baryons for a covariant Lagrangian approach. The theoretical results are recast into a density functional theory with dressed in-medium meson-baryon vertices in Sect. 5.3. Results for hypermatter and hypernuclei are discussed in Sect. 5.4, addressing also the investigations by modern effective field theories. In Sect. 5.5 we derive hyperon interactions in nuclear matter by exploiting the constraints imposed by SU(3) symmetry. Investigations of baryon resonances in nuclei are discussed in Sects. 5.6 and 5.7. In Sect. 5.8 the article is summarized and conclusions are drawn before closing with an outlook.

5.2 Interactions of SU(3) Flavor Octet Baryons

5.2.1 General Aspects of Nuclear Strangeness Physics

Hypernuclear and strangeness physics in general are of high actuality as seen from the many experiments in operation or preparation, respectively, and the increasing amount of theoretical work in that field. There are a number of excellent review articles available addressing the experimental and theoretical status, ranging from the review of Hashimoto and Tamura [13], the very useful collection of papers in [14] to the more recent review on experimental work by Feliciello et al. [15] to the article by Gal et al. [10]. The latest activities are also recorded in two topical issues: in Ref. [16] strangeness (and charm) physics are highlighted and in Ref. [17] the contributions of strangeness physics with respect to neutron star physics is discussed. In [18] we have reviewed the status of in-medium baryon and baryon resonance physics, also covering production reactions on the free nucleon and on nuclei.

On free space interactions of nucleons a wealth of experimental data exists which are supplemented by the large amount of data on nuclear spectroscopy and reactions. Taken together, they allow to define rather narrow constraints on interactions and, with appropriate theoretical methods, to predict their modifications in nuclear matter. Nuclear reaction data have provided important information on the energy and momentum dependence on the one hand and the density dependence on the other hand of interactions in nuclear matter. For the hyperons, however, the situation is much less well settled. All attempts to derive hyperon-nucleon (YN) interactions in the strangeness $S = -1$ channel are relying, in fact, on a small sample of data points obtained mainly in the 1960s. By obvious reasons, direct experimental information on hyperon-hyperon (YY) interactions is completely lacking. A way out of that dilemma is expected to be given by studies of hypernuclei. Until now only single-Lambda hypernuclei are known as bound systems, supplemented by a few cases of $S = -2$ double-Lambda nuclei. While a considerable number of Λ-hypernuclei is known, no safe signal for a particle-stable Σ or a $S = -2$ Cascade hypernucleus has been recorded, see e.g. [10, 13].

On the theoretical side, large efforts are made to incorporate strange baryons into the nuclear agenda. The conventional non-relativistic single particle potential models, the involved few-body methods for light hypernuclei, and the many-body shell model descriptions of hypernuclei were reviewed recently in the literature cited above and will not be repeated here. Approaches to nucleon and hyperon interactions based on the meson-exchange picture of nuclear forces have a long tradition. They are describing baryon-baryon interactions by one-boson exchange (OBE) potentials like the well known Nijmegen Soft Core model (NSC) [19], later improved to the Extended Soft Core (ESC) model [20, 21], for which over the years a number of parameter sets were evaluated [22–24]. The Jülich model [25–27] and also the more recently formulated Giessen Boson Exchange model (GiBE) [28, 29] belongs to the OBE-class of approaches. In the Jülich model the $J^P = 0^+$ scalar interaction channel is generated dynamically by treating those mesonic states explicitly as correlations of pseudo-scalar mesons. In the GiBE and the early NSC models scalar mesons are considered as effective mesons with sharp masses. The extended Nijmegen soft-core model ESC04 [20, 21] and ESC08 [22, 24] includes two pseudo-scalar meson exchange and meson-pair exchange, in addition to the standard one-boson exchange and short-range diffractive Pomeron exchange potentials. The Niigata group is promoting by their fss- and $fss2$-approaches a quark-meson coupling model which is being updated regularly [30, 31]. The resonating group model (RGM) formalism is applied to the baryon–baryon interactions using the SU(6) quark model (QM) augmented by modifications like peripheral mesonic or ($q\bar{q}$) exchange effects.

Chiral effective field theory (χ EFT) for NN and NY interactions are a more recent development in baryon-baryon interaction. The review by Epelbaum et al. [32] on these subjects is still highly recommendable. The connection to the principles of QCD are inherent. A very attractive feature of χ EFT is the built-in order scheme allowing in principle to solve the complexities of baryon-baryon interactions systematically by a perturbative expansion in terms of well-ordered and properly defined classes of diagrams. In this way, higher order interaction diagrams are generated systematically, controlling and extending the convergence of the calculations over an increasingly larger energy range.

In addition to interactions, studies of nuclei require appropriate few-body or many-body methods. In light nuclei Faddeev-methods allow an ab initio description by using free space baryon-baryon (BB) potentials directly as practiced e.g. in [33]. Stochastic methods like the Green's function Monte Carlo approach are successful for light and medium mass nuclei [34]. A successful approach up to oxygen mass region, the so-called p-shell nuclei, is the hypernuclear shell model of Millener and collaborators [35, 36]. Over the years, a high degree of sophistication and predictive power for hypernuclear spectroscopy has been achieved by these methods for light nuclei. For heavier nuclei, density functional theory (DFT) is the method of choice because of its applicability over wide ranges of nuclear masses. The development of an universal nuclear energy density functional is the aim of the UNEDF initiative [37–39]. Already some time ago we have made first steps in such a direction [40] within the Giessen Density Dependent Hadron Field (DDRH) theory. The DDRH approach incorporates Dirac-Brueckner Hartree-Fock (DBHF) theory into covariant density functional theory [41–46]. Since then, the approach is being used widely on a purely phenomenological level as e.g. in [47–50]. In the non-relativistic sector comparable attempts are being made, ranging from Brueckner theory for hypermatter [51, 52] to phenomenological density functional theory extending the Skyrme-approach to hypernuclei [53, 54]. In recent works energy density functionals have been derived also for the Nijmegen model [24] and the Jülich χ EFT [55]. Relativistic mean-field (RMF) approaches have been used rather early for hypernuclear investigation, see e.g. [56, 57]. A covariant DFT approach to hypernuclei and neutron star matter was used in a phenomenological RMF approach in Ref. [58] where constraints on the scalar coupling constants were derived by imposing the constraint of neutron star masses above two solar masses. In Refs. [59–61] and also [62, 63] hyperons and nucleon resonances are included into the RMF treatment of infinite neutron star matter, also with the objective to obtain neutron stars heavier than two solar masses. A yet unexplained additional repulsive interaction at high densities is under debate. A covariant mean-field approach, including a non-linear realization of chiral symmetry, has been proposed by the Frankfurt group [64] and is being used mainly for neutron star studies.

5.2.2 Interactions in the Baryon Flavor Octet

In Fig. 5.2 the SU(3) multiplets are shown which are considered in the following. The lowest $J^P = \frac{1}{2}^+$ baryon octet, representing the baryonic ground state multiplet, is taken into account together with three meson multiplets, namely the pseudo-scalar nonet ($\mathcal{P}$) with $J^P = 0^-$, the scalar nonet ($\mathcal{S}$) with $J^P = 0^+$, and the vector nonet ($\mathcal{V}$) with $J^P = 1^-$, which, in fact, consists of four subsets, $\mathcal{V} = \{\mathcal{V}^\mu\}_{|\mu=0...3}$, according to the four components of a Lorentz-vector. In addition, we include also the corresponding meson singlet states, represented physically by the η', ϕ, σ' states but not shown in Fig. 5.2. Masses and lifetimes of baryons are displayed in Table 5.1.

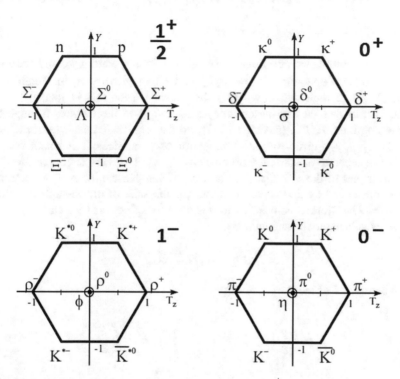

Fig. 5.2 The SU(3) multiplets considered in this section: the $\frac{1}{2}^+$ baryon octet (upper left), the scalar 0^+ (upper right), the vector 1^- (lower left), and the pseudo-scalar 0^- (lower right). The octets have to be combined with the corresponding meson-singlets, which are not displayed, thus giving rise to meson nonets

Table 5.1 Mass, lifetime, and valence quark configuration of the $J^P = \frac{1}{2}^+$ octet baryons

State	Mass (MeV)	Lifetime (s)	Configuration
p	938.27	$> 6.62 \times 10^{+36}$	[uud]
n	939.57	880.2 ± 1	[udd]
Λ	1115.68	2.63×10^{-10}	[uds]
Σ^+	1189.37	0.80×10^{-10}	[uus]
Σ^0	1192.64	7.4×10^{-20}	[uds]
Σ^-	1197.45	1.48×10^{-10}	[dds]
Ξ^0	1314.86	2.90×10^{-10}	[uss]
Ξ^-	1321.71	1.64×10^{-10}	[dss]

The data are taken from Ref. [65]

The mesons of each multiplet couple to the baryons by vertices of a typical, generic operator structure characterized the Lagrangian densities

$$\mathcal{L}_{\mathcal{M}}^{BB'}(0^-) = -\ g_{BB'\mathcal{M}}\bar{\psi}_{B'}\frac{1}{m_\pi}\gamma_5\gamma_\mu\psi_B\partial^\mu\Phi_{\mathcal{M}} \tag{5.1}$$

$$\mathcal{L}_{\mathcal{M}}^{BB'}(1^-) = -\ g_{BB'\mathcal{M}}\bar{\psi}_{B'}\gamma_\mu\psi_B V_{\mathcal{M}}^\mu + \frac{f_{BB'\mathcal{M}}}{2(M_B+M_{B'})}\bar{\psi}_{B'}\sigma_{\mu\nu}\psi_B F_{\mathcal{M}}^{\mu\nu} \tag{5.2}$$

$$\mathcal{L}_{\mathcal{M}}^{BB'}(0^+) =\ g_{BB'\mathcal{M}}\bar{\psi}_{B'}\mathbb{1}\psi_B\Phi_{\mathcal{M}} \tag{5.3}$$

given by Lorentz-covariant bilinears of baryon field operators ψ_B and the Dirac-conjugated field operators $\bar{\psi}_B = \gamma_0\psi_B^\dagger$ and Dirac γ-matrices, to which meson fields $(\mathcal{V}, \mathcal{S})$ or their derivatives $(\mathcal{P})$ are attached, such that in total a Lorentz-scalar is obtained. Vector mesons may also couple to the baryons through their field strength tensor $F_{\mathcal{M}}^{\mu\nu}$ (see Eq. (5.11)) and the relativistic rank-2 tensor operator $\sigma_{\mu\nu} = \frac{i}{2}[\gamma_\mu, \gamma_\nu]$, given by the commutator of γ matrices. The tensor coupling involves a separate tensor coupling constant $f_{BB'\mathcal{M}}$. The mass factors in the $\mathcal{P}$ and the tensor part of the $\mathcal{V}$ vertices are serving to compensate the energy-momentum scales introduced by derivative operators for the sake of dimensionless coupling constants. The Bjorken-convention [66] is used for spinors and γ-matrices.

We introduce the flavor spinor Ψ_B

$$\Psi_B = (N, \Lambda, \Sigma, \Xi)^\mathsf{T} \tag{5.4}$$

being composed of the isospin multiplets

$$N = \begin{pmatrix} p \\ n \end{pmatrix}, \quad \Sigma = \begin{pmatrix} \Sigma^+ \\ \Sigma^0 \\ \Sigma^- \end{pmatrix}, \quad \Xi = \begin{pmatrix} \Xi^0 \\ \Xi^- \end{pmatrix} \tag{5.5}$$

and the iso-singlet Λ. Each baryon entry is given by a Dirac spinor. For later use, we also introduce the mesonic isospin doublets

$$K = \begin{pmatrix} K^+ \\ K^0 \end{pmatrix}, \quad K_c = \begin{pmatrix} \overline{K^0} \\ -K^- \end{pmatrix}, \tag{5.6}$$

and corresponding structures are defined in the vector and scalar sector involving the K^* and the κ mesons, respectively. The Σ hyperon and the π isovector-triplets are expressed in the basis defined by the spherical unit vectors $e_{\pm,0}$ which leads to

$$\Sigma \cdot \pi = \Sigma^+\pi^- + \Sigma^0\pi^0 + \Sigma^-\pi^+, \tag{5.7}$$

also serving to fix phases [67].

The full Lagrangian is given by

$$\mathcal{L} = \mathcal{L}_B + \mathcal{L}_M + \mathcal{L}_{int} \tag{5.8}$$

accounting for the free motion of baryons with mass matrix $\hat{M}$,

$$\mathcal{L}_B = \overline{\Psi}_B \left[i\gamma_\mu \partial^\mu - \hat{M} \right] \Psi_B \quad , \tag{5.9}$$

and the Lagrangian density of massive mesons,

$$\mathcal{L}_M = \frac{1}{2} \sum_{i \in \{\mathcal{P}, \mathcal{S}, \mathcal{V}\}} \left(\partial_\mu \Phi_i \partial^\mu \Phi_i - m_{\Phi_i}^2 \Phi_i^2 \right) - \frac{1}{2} \sum_{\lambda \in \mathcal{V}} \left(\frac{1}{2} F_\lambda^2 - m_\lambda^2 V_\lambda^2 \right) \tag{5.10}$$

where $\mathcal{P}, \mathcal{S}, \mathcal{V}$ denote summations over the lowest nonet pseudo-scalar, scalar, and vector mesons. The field strength tensor of the vector meson fields V_λ^μ, $\lambda \in \{\omega, \rho, K^*, \phi, \gamma\}$ is defined by

$$F_\lambda^{\mu\nu} = \partial^\mu V_\lambda^\nu - \partial^\nu V_\lambda^\mu \quad . \tag{5.11}$$

For the scattering of charged particles and in finite nuclei, the electromagnetic vector field V_γ^μ of the photon is included.

Of special interest for nuclear matter and nuclear structure research are the mean-field producing meson fields, given by the isoscalar-scalar mesons σ, σ', the isovector-scalar δ meson, physically observed as the $a_0(980)$ meson, and their isoscalar-vector counterparts ω, ϕ and the isovector-vector ρ meson, respectively. While the pseudo-scalar and the vector mesons are well identified as stable particles or as well located poles in the complex plane, the situation is less clear for the scalar nonet. We identify $\sigma = f_0(500)$, $\delta = a_0(980)$, and $\kappa = K_0^*(800)$ as found in the compilations of the Particle Data Group (PDG) [65]. The so-called κ-meson is of particular uncertainty. It has been observed only rather recently as resonance-like structures in charmonium decay spectroscopy. A two-bump structure with maxima at about 640 and 800 MeV has been detected. In the recent PDG compilation a mean mass $m_\kappa = 682 \pm 29$ MeV is recommended [65]. We use $m_\kappa = 700$ MeV. The octet baryons are listed in Table 5.1, the meson parameters are summarized in Table 5.2.

Last but not least we consider the interaction Lagrangian $\mathcal{L}_{int}$. SU(3) octet physics is based on treating the eight baryons on equal footing as the genuine mass-carrying fields of the theory. Although SU(3) flavor symmetry is broken on the baryon mass scale by about $\pm 20\%$, it is still meaningful to exploit the relations among coupling constants imposed by that symmetry, thus defining a guideline and reducing the number of free parameters considerably. The eight $J^P = \frac{1}{2}^+$ baryons are collected into a traceless matrix $\mathcal{B}$, which is given by a superposition of the eight Gell-Mann matrices λ_i combined with the eight baryons $B_i \in \{N, \Lambda, \Sigma, \Xi\}$, leading

Table 5.2 Intrinsic quantum numbers and the measured masses [65] of the mesons used in the calculations

Channel	Meson	Mass (MeV)	Cut-off (MeV/c)
0^-	π	138.03	1300
0^-	η	547.86	1300
0^-	$K^{0,+}$	497.64	1300
0^+	σ	500.00	1850
0^+	δ	983.00	2000
0^+	κ	700.00	2000
1^-	ω	782.65	1700
1^-	ρ	775.26	1700
1^-	K^*	891.66	1700

Also the cut-off momenta used to regularize the high-momentum part of the tree-level interactions are shown. The mass of the isoscalar-scalar σ meson is chosen at the center of the $f_0(500)$ spectral distribution [65]

to the familiar form

$$
\mathcal{B} = \sum_{i=1\cdots8} \lambda_i B_i =
\begin{pmatrix}
\dfrac{\Sigma^0}{\sqrt{2}} + \dfrac{\Lambda}{\sqrt{6}} & \Sigma^+ & p \\[2ex]
\Sigma^- & -\dfrac{\Sigma^0}{\sqrt{2}} + \dfrac{\Lambda}{\sqrt{6}} & n \\[2ex]
-\Xi^- & \Xi^0 & -\dfrac{2\Lambda}{\sqrt{6}}
\end{pmatrix},
\tag{5.12}
$$

which is invariant under SU(3) transformations. The pseudo-scalar ($\mathcal{P}$), vector ($\mathcal{V}$), and the scalar ($\mathcal{S}$) meson octet matrices are constructed correspondingly by replacing the baryons B_i by the appropriate mesons M_i. Taking the $J^P = 0^-$ pseudo-scalar mesons as an example we obtain the (traceless) octet matrix

$$
\mathcal{P}_8 =
\begin{pmatrix}
\dfrac{\pi^0}{\sqrt{2}} + \dfrac{\eta_8}{\sqrt{6}} & \pi^+ & K^+ \\[2ex]
\pi^- & -\dfrac{\pi^0}{\sqrt{2}} + \dfrac{\eta_8}{\sqrt{6}} & K^0 \\[2ex]
K^- & \overline{K^0} & -\dfrac{2\eta_8}{\sqrt{6}}
\end{pmatrix}.
\tag{5.13}
$$

which, for the full nonet, has to be completed by the singlet matrix $\mathcal{P}_1$, given by the 3×3 unit matrix multiplied by $\eta_1/\sqrt{3}$. Thus, the full pseudo-scalar nonet is described by $\mathcal{P} = \mathcal{P}_8 + \mathcal{P}_1$. We define the SU(3)-invariant baryon-baryon-meson vertex combinations

$$
\left[\overline{\mathcal{B}}\mathcal{B}\mathcal{P}\right]_D = \mathrm{Tr}\left(\{\overline{\mathcal{B}}, \mathcal{B}\}\,\mathcal{P}_8\right) \quad, \quad \left[\overline{\mathcal{B}}\mathcal{B}\mathcal{P}\right]_F = \mathrm{Tr}\left([\overline{\mathcal{B}}, \mathcal{B}]\,\mathcal{P}_8\right) \quad,
$$

$$
\left[\overline{\mathcal{B}}\mathcal{B}\mathcal{P}\right]_S = \mathrm{Tr}(\overline{\mathcal{B}}\mathcal{B})\mathrm{Tr}(\mathcal{P}_1)
\tag{5.14}
$$

where F and D couplings correspond to anti-symmetric combinations, given by anti-commutators, $\{X, Y\}$, and symmetric combinations, given by commutators $[X, Y]$, respectively. The singlet interaction term is indexed by S. With these relations we obtain the pseudo-scalar interaction Lagrangian

$$\mathcal{L}_{int}^{\mathcal{P}} = -\sqrt{2}\left\{g_D\left[\overline{\mathcal{B}}\mathcal{B}\mathcal{P}_8\right]_D + g_F\left[\overline{\mathcal{B}}\mathcal{B}\mathcal{P}_8\right]_F\right\} - g_S\frac{1}{\sqrt{3}}\left[\overline{\mathcal{B}}\mathcal{B}\mathcal{P}_1\right]_S, \quad (5.15)$$

with the generic SU(3) coupling constants $\{g_D, g_F, g_S\}$. The *de Swart* convention [67], underlying the Nijmegen and the Jülich approaches, is given by using $g_8 \equiv g_D + g_F$, $\alpha \equiv g_F/(g_D + g_F)$, and $g_1 \equiv g_S$ leading to the equivalent representation

$$\mathcal{L}_{int}^{\mathcal{P}} = -g_8\sqrt{2}\left\{\alpha\left[\overline{\mathcal{B}}\mathcal{B}\mathcal{P}_8\right]_F + (1-\alpha)\left[\overline{\mathcal{B}}\mathcal{B}\mathcal{P}_8\right]_D\right\} - g_1\frac{1}{\sqrt{3}}\left[\overline{\mathcal{B}}\mathcal{B}\mathcal{P}_1\right]_S, \quad (5.16)$$

In order to evaluate the couplings we define the pseudo-vector derivative vertex operator $m_\pi \Gamma_\mathcal{P} = \gamma_5\gamma_\mu\partial^\mu$, following Eq. (5.1). From the F- and D-type couplings, Eq. (5.15), we obtain the pseudo-scalar octet-meson interaction Lagrangian in an obvious, condensed short-hand notation, going back to de Swart [67],

$$\begin{aligned}
\mathcal{L}_{int}^{\mathcal{P}} = &- g_{NN\pi}(\overline{N}\Gamma_\mathcal{P}\tau N)\cdot\boldsymbol{\pi} + ig_{\Sigma\Sigma\pi}(\overline{\boldsymbol{\Sigma}}\times\Gamma_\mathcal{P}\boldsymbol{\Sigma})\cdot\boldsymbol{\pi}\\
&- g_{\Lambda\Sigma\pi}(\overline{\Lambda}\Gamma_\mathcal{P}\boldsymbol{\Sigma} + \overline{\boldsymbol{\Sigma}}\Gamma_\mathcal{P}\Lambda)\cdot\boldsymbol{\pi} - g_{\Xi\Xi\pi}(\overline{\Xi}\Gamma_\mathcal{P}\tau\Xi)\cdot\boldsymbol{\pi}\\
&- g_{\Lambda NK}\left[(\overline{N}\Gamma_\mathcal{P}K)\Lambda + \overline{\Lambda}\Gamma_\mathcal{P}(\overline{K}N)\right]\\
&- g_{\Xi\Lambda K}\left[(\overline{\Xi}\Gamma_\mathcal{P}K_c)\Lambda + \overline{\Lambda}\Gamma_\mathcal{P}(\overline{K_c}\Xi)\right]\\
&- g_{\Sigma NK}\left[\boldsymbol{\Sigma}\cdot\Gamma_\mathcal{P}(\overline{K}\tau N) + (\overline{N}\Gamma_\mathcal{P}\tau K)\cdot\boldsymbol{\Sigma}\right]\\
&- g_{\Xi\Sigma K}\left[\boldsymbol{\Sigma}\cdot\Gamma_\mathcal{P}(\overline{K_c}\tau\Xi) + (\overline{\Xi}\Gamma_\mathcal{P}\tau K_c)\cdot\boldsymbol{\Sigma}\right]\\
&- g_{NN\eta_8}(\overline{N}\Gamma_\mathcal{P}N)\eta_8 - g_{\Lambda\Lambda\eta_8}(\overline{\Lambda}\Gamma_\mathcal{P}\Lambda)\eta_8\\
&- g_{\Sigma\Sigma\eta_8}(\overline{\boldsymbol{\Sigma}}\cdot\Gamma_\mathcal{P}\boldsymbol{\Sigma})\eta_8 - g_{\Xi\Xi\eta_8}(\overline{\Xi}\Gamma_\mathcal{P}\Xi)\eta_8. \quad (5.17)
\end{aligned}$$

The—in total 16—pseudo-scalar $\mathcal{B}\mathcal{B}'$-meson vertices are completely fixed by the three nonet coupling constants (g_D, g_F, g_S) or, likewise, by (g_8, g_1, α).

Corresponding relations exist also for interactions induced by the vector and the scalar meson nonets. As in the pseudo-scalar case, they are given in terms of octet $(\mathcal{V}_8^\mu, \mathcal{S}_8)$ and singlet multiplets $(\mathcal{V}_1^\mu, \mathcal{S}_1)$, resulting in $\mathcal{V}^\mu = \mathcal{V}_8^\mu + \mathcal{V}_1^\mu$ and $\mathcal{S} = \mathcal{S}_8 + \mathcal{S}_1$, respectively, and having their own sets of respective coupling constants $(g_D, g_F, g_S)_{\mathcal{S}, \mathcal{V}}$. The $\mathcal{B}\mathcal{B}\mathcal{V}$ coupling constants are obtained in analogy to Eq. (5.17) by the mapping $\{K, \pi, \eta_8, \eta_1\} \rightarrow \{K^*, \rho, \omega_8, \phi_1\}$. Correspondingly, the scalar couplings $\mathcal{B}\mathcal{B}\mathcal{S}$ are obtained from Eq. (5.17) by replacing $\{K, \pi, \eta_8, \eta_1\} \rightarrow \{\kappa, \delta, \sigma_8, \sigma_1\}$. Thus, the baryon-baryon interactions as given by the exchange of particles from the three meson multiplets $\mathcal{M} \in \{\mathcal{P}, \mathcal{V}, \mathcal{S}\}$ are of a common structure which allows to express the $\mathcal{B}\mathcal{B}\mathcal{M}$ coupling constants in generic manner. For that purpose, we denote the isoscalar octet meson by $f \in \{\eta_8, \omega_8, \sigma_8\}$, the isovector octet meson by $a \in \{\pi, \rho, \delta\}$, and the iso-doublet mesons by $K \in \{K^{0,+}, K^{*0,+}\}$ and $\kappa = \{K_0^{*,0}, K_0^{*,+}\}$. Irrespective of the particular interaction channel, the coupling constants are then given by the relations [67]

$$
\begin{aligned}
g_{NNa} &= g_D + g_F, & g_{\Lambda NK} &= -\sqrt{\tfrac{1}{3}}(g_D + 2g_F), & g_{NNf} &= \tfrac{1}{\sqrt{3}}(3g_F - g_D), \\
g_{\Sigma\Sigma a} &= 2g_F, & g_{\Xi\Lambda K} &= \tfrac{1}{\sqrt{3}}(3g_F - g_D), & g_{\Lambda\Lambda f} &= -\tfrac{2}{\sqrt{3}}g_D, \\
g_{\Lambda\Sigma a} &= \tfrac{2}{\sqrt{3}}g_D, & g_{\Sigma NK} &= (g_D - g_F), & g_{\Sigma\Sigma f} &= \tfrac{2}{\sqrt{3}}g_D, \\
g_{\Xi\Xi a} &= -(g_D - g_F), & g_{\Xi\Sigma K} &= -(g_D + g_F), & g_{\Xi\Xi f} &= -\tfrac{1}{\sqrt{3}}(3g_F + g_D).
\end{aligned}
$$

$$(5.18)$$

where $g_{NNa} = g_8$ and, depending on the case, $\{g_D, g_F, g_S\}$ denote either the pseudo-scalar, vector, or scalar set of basic SU(3) couplings, respectively. The interactions due to the exchange of the isoscalar-singlet mesons $f' \in \{\eta_1, \phi_1, \sigma_1\}$ are treated accordingly with the result

$$
g_{NNf'} = g_{\Lambda\Lambda f'} = g_{\Sigma\Sigma f'} = g_{\Xi\Xi f'} = g_S = g_1. \tag{5.19}
$$

where again the proper $g_S \equiv g_1$ coupling constant for the $\{\mathcal{P}, \mathcal{V}, \mathcal{S}\}$ multiplet under consideration has to be inserted. The complete interaction Lagrangian is given by the sum over the partial interaction components

$$
\mathcal{L}_{int} = \sum_{\mathcal{M} \in \{\mathcal{P}, \mathcal{S}, \mathcal{V}\}} \mathcal{L}_{int}^{\mathcal{M}}. \tag{5.20}
$$

The coupling constants above define the tree-level interactions entering into calculations of T-matrices. The corresponding diagrams contributing to the $N\Lambda$ and $N\Sigma$ interactions in the $S = -1$ sector are shown in Fig. 5.3.

The advantage of referring to SU(3) symmetry is obvious: For each type of interaction (pseudo-scalar, vector, scalar) only four independent parameters are required to characterize the respective interaction strengths with all possible baryons. These are the singlet coupling constant g_S, the octet coupling constants g_D, g_F, g_S, and eventually the three mixing angles $\theta_{P,V,S}$, one for each meson multiplet, which relate the physical, dressed isoscalar mesons to their bare octet and singlet counterparts.

SU(3) symmetry, however, is broken at several levels and SU(3) relations will not be satisfied exactly. An obvious one is the non-degeneracy of the physical baryon and meson masses within the multiplets, also reflecting the non-degeneracy of (u, d, s) quark masses. As discussed below, this splitting leads to additional complex structure in the set of coupled equations for the scattering amplitudes because the various baryon-baryon (BB′) channels open at different threshold energies $\sqrt{s_{BB'}} = m_B + m_{B'}$. At energies $s < s_{BB'}$ a given BB′-channel does not contain asymptotic flux but contributes as a virtual state. Thus, $N\Lambda$ scattering, for example, will be modified

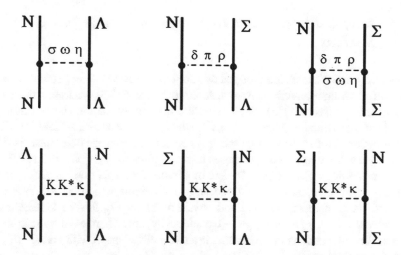

Fig. 5.3 Hyperon-nucleon OBE interactions in the $S = -1$ sector. Interactions without and with strangeness-exchange are displayed in the upper and lower row, respectively. The interactions from $S = 0$ scalar (σ, δ) and vector mesons (ω, ρ) mesons will contribute to the hypernuclear mean-field self-energies

Table 5.3 Baryon-baryon channels for fixed strangeness S and total charge Q

	$Q = -2$	$Q = -1$	$Q = 0$	$Q = 1$	$Q = 2$
$S = 0$			nn	np	pp
$S = -1$		$\Sigma^- n$	$\Lambda n \ \Sigma^0 n$ $\Sigma^- p$	$\Lambda p \ \Sigma^+ n$ $\Sigma^0 p$	$\Sigma^+ p$
$S = -2$	$\Sigma^- \Sigma^-$	$\Xi^- n \ \Sigma^- \Lambda$ $\Sigma^- \Sigma^0$	$\Lambda\Lambda \ \Xi^0 n$ $\Xi^- p \ \Sigma^+ \Lambda$ $\Sigma^+ \Sigma^0$	$\Xi^0 p \ \Sigma^+ \Lambda$ $\Sigma^+ \Sigma^0$	$\Sigma^+ \Sigma^+$
$S = -3$	$\Xi^- \Sigma^-$	$\Xi^- \Lambda \ \Xi^0 \Sigma^-$ $\Xi^- \Sigma^0$	$\Xi^0 \Lambda \ \Xi^0 \Sigma^0$ $\Xi^- \Sigma^+$	$\Xi^0 \Sigma^+$	
$S = -4$	$\Xi^- \Xi^-$	$\Xi^- \Xi^0$	$\Xi^0 \Xi^0$		

at any energy by virtual or real admixtures of $N\Sigma$ channels. For pratical purposes, a scheme taking into account the physical hadron masses is of clear advantage, if not a necessity. Irrespective of the mass symmetry breaking, total strangeness and charge are always conserved. That is assured by using the particle basis, which at the same time allows to use physical hadron masses. The corresponding charge-strangeness conserving multiplets are listed in Table 5.3.

5.2.3 Baryon-Baryon Scattering Amplitudes and Cross Sections

In practice, the BB' states are grouped into substructures which reflect the conservation laws of strong interaction physics. Assuming strict $SU(3)$ symmetry, this can be done in terms of flavor $SU(3) \otimes SU(3)$ irreducible representations (*irreps*), as e.g. in [8] and for a practical application in [24]. However, as seen from Table 5.1, $SU(3)$ symmetry is obviously broken on the mass scale: The average octet mass is $\bar{m}_8 = 1166.42$ MeV, the octet mass splitting is found as $\Delta m_8 = \bar{m}_\Xi - \bar{m}_N = 379.19$ MeV and, thus, the mass symmetry is broken by about 32%. This implies differences in the kinematical channel threshold and an order scheme which takes into account those effects is of more practical and physical use. In Fig. 5.4 the kinematics for the nucleonic $S = 0$ and the $S = -1$ nucleon-hyperon channels are illustrated in terms of the invariant channel momenta in a two-body channel with masses $m_{1,2}$ and Mandelstam total energy s:

$$q^2(s) = \frac{1}{4s} \left((s - (m_1 - m_2)^2)(s - (m_1 + m_2)^2) \right), \qquad (5.21)$$

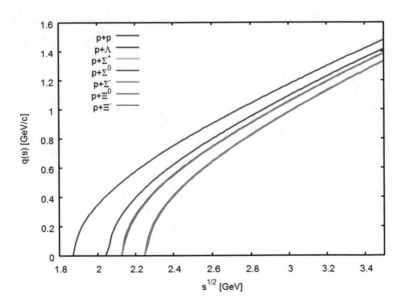

Fig. 5.4 The channel momentum $q(s)$ is shown as a function of the Mandelstam variable s for the nucleon-nucleon and the $S = -1$ nucleon-hyperon two-body channels. The threshold is defined by $q(s) = 0$

where at the channel threshold $s = s_{thres} = (m_1 + m_2)^2$ and $q^2(s_{thres}) = 0$. In observables like cross sections, the coupling to a channel opening at a certain energy produces typically a kink-like structure, seen as a sudden jump in both the data and numerical results. An example is found in Fig. 5.5. The spike showing up in the elastic Λp cross section at about $p_{lab} \sim 640$ MeV/c is due to the coupling to the $\Sigma^0 p$ channel which crosses the kinematical threshold at the $\sqrt{s} = M_{\Sigma^0} + M_p$ reached that channel momentum, namely at $p_{lab} \simeq 642$ MeV/c for a proton incident on a Λ hyperon. In scattering phase shifts such a channel coupling effect is typically seen as a sudden jump.

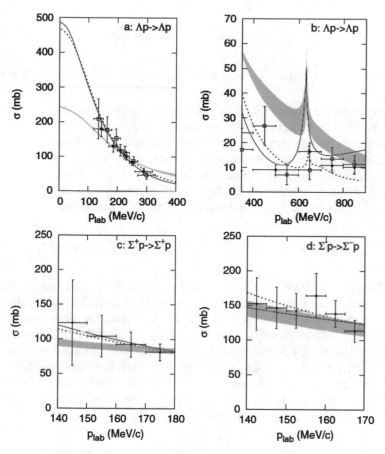

Fig. 5.5 Total cross sections as a function of p_{lab} for Λp and $\Sigma^{\pm} p$ elastic scattering. The shaded band is the LO EFT result for varying the cutoff as $\Lambda = 550 \ldots 700$ MeV/c. For comparison results are shown of the Jülich 04 model [25] (dashed), and the Nijmegen NSC97f model [19] (solid curve) (from Ref. [32])

The Lagrangian densities serve to define the tree-level interactions of the BB' configurations built from the octet baryons. The derived potentials include in addition also vertex form factors. Formally, they are used to regulate momentum integrals, physically they define the momentum range for which the theory is supposed to be meaningful. The OBE models typically use hard cut-offs in the range of 1–$2\,\mathrm{GeV/c}$. The χ EFT cut-offs are much softer with values around $600\,\mathrm{MeV/c}$.

In the $N\Lambda/N\Sigma$-system, for example, the tree-level interactions are given by

$$V_{N\Lambda,N\Lambda} = V^{(\eta)}_{N\Lambda,N\Lambda} + V^{(\sigma)}_{N\Lambda,N\Lambda} + V^{(\omega)}_{N\Lambda,N\Lambda} \tag{5.22}$$

$$V^{(I)}_{N\Sigma,N\Sigma} = V^{(\eta)}_{N\Sigma,N\Sigma} + V^{(\sigma)}_{N\Sigma,N\Sigma} + V^{(\omega)}_{N\Sigma,N\Sigma}$$

$$+ \left(V^{(\pi)}_{N\Sigma,N\Sigma} + V^{(\delta)}_{N\Sigma,N\Sigma} + V^{(\rho)}_{N\Sigma,N\Sigma} \right) \langle I || \tau_N \cdot \tau_\Sigma || I \rangle \tag{5.23}$$

$$V^{(I)}_{N\Lambda,N\Sigma} = \left(V^{(\pi)}_{N\Lambda,N\Sigma} + V^{(\delta)}_{N\Lambda,N\Sigma} + V^{(\rho)}_{N\Lambda,N\Sigma} \right) \left\langle I = \frac{1}{2} || \tau_N \cdot \tau_\Sigma || I = \frac{1}{2} \right\rangle, \tag{5.24}$$

where $\tau_{N,\Sigma}$ denote the isospin operators acting on the nucleon and the Σ hyperon, respectively. The strengths of the interactions inhibits a perturbative treatment. Rather, the scattering series must be summed to all orders. For that purpose, the bare interactions are entering as reaction kernels into a set of coupled Bethe-Salpeter equations, connecting BB' channels of the same total charge and strangeness. In a matrix notation, $\mathcal{V}$ indicates the Born terms, $\mathcal{G}$ denotes the diagonal matrix of channel Green's functions, and the resulting T-matrix is defined by

$$\mathcal{T}(q',q|P) = \mathcal{V}(q',q|P) + \int \frac{d^4k}{(2\pi)^4} \mathcal{V}(q',k|P)\mathcal{G}(k,P)\mathcal{T}(k,q|P) \tag{5.25}$$

describing the transition of the system for the (off-shell) four-momenta $q' \to q$ and fixed center-of-mass four-momentum P with $s = P^2$. A numerically solvable system of equations is obtained by projection to a three-dimensional sub-space. Such a reduction depends necessarily on the choice of projection method as discussed in the literature, e.g. [68, 69]. A widely used scheme is the so-called Blankenbecler-Sugar (BbS) reduction [70] consisting in projection of the intermediate energy variable to a fixed value, typically chosen as $k_0 = 0$. In this way, the full Bethe-Salpeter equation is reduced to an effective Lippmann-Schwinger equation in three spatial variables, but still obeying Lorentz invariance:

$$\mathcal{T}(\mathbf{q}',\mathbf{q}|s) = \mathcal{V}(\mathbf{q}',\mathbf{q}|s) + \int \frac{d^3k}{(2\pi)^3} \mathcal{V}(\mathbf{q}',\mathbf{k}|s)g_{BbS}(k,s)\mathcal{T}(\mathbf{k},\mathbf{q}|s) \tag{5.26}$$

where the propagator is now replaced by the Blankenbecler-Sugar propagator with (diagonal) elements

$$g_{BbS}(k, s) = \frac{1}{s - (E_1(k) + E_2(k))^2 + i\eta} \tag{5.27}$$

where $E_{1,2}(k) = \sqrt{m_{1,2}^2 + k^2}$ is the relativistic energy of the particles in that given channel. The T-matrix may be expressed by the K-matrix:

$$\mathcal{T} = (1 - i\mathcal{K})^{-1}\mathcal{K} \tag{5.28}$$

and very often the scattering problem is solved in terms of the (real-valued) K-matrix [68],

$$\mathcal{K}(\mathbf{q}', \mathbf{q}|s) = \mathcal{V}(\mathbf{q}', \mathbf{q}|s) + \int \frac{d^3k}{(2\pi)^3} \mathcal{V}(\mathbf{q}', \mathbf{k}|s) \frac{P}{s - (E_1(k) + E_2(k))^2} \mathcal{T}(\mathbf{k}, \mathbf{q}|s) \tag{5.29}$$

given by Cauchy principal value integral. In practice, moreover, a decomposition into invariants and partial wave matrix elements is performed which reduces the problem to a set of linear integral equations in a single variable, namely the modulus of the three-momentum involved. Here, we refrain from going into these mathematically very involved details. They have been subject of many well written standard text books and review papers, e.g. [68].

An interesting question is to what extent the higher order terms of the scattering series are contributing to the scattering amplitude. That is quantified in Fig. 5.6 where the s-wave matrix elements for the $(S = 0, I = 1)$ NN singlet-even channel are shown in Born approximation and for the fully summed K-matrix result. The higher order correlation effects are seen to be of overwhelming importance especially close to threshold and at low (on-shell) momenta.

Representative results for nucleon-hyperon scattering are shown in Fig. 5.5, comparing total cross sections obtained by EFT and the Jülich and Nijmegen OBE approaches, respectively. The latest χ EFT account for interactions up to next-to-leading-order (NLO) [71]. The resulting phase shifts in the spin-singlet channel are shown in Fig. 5.7 for a few baryon-baryon channels: the pp $(S = 0)$, $p\Sigma^+$ $(S = -1)$, and $\Sigma^+\Sigma^+$ $(S = -2)$. From the pp results it is seen that the NLO calculations are trustable up to about $p_{lab} \sim 300\,\mathrm{MeV/c}$.

In nuclear matter, the scattering equations are changed by the fact that part of the intermediate channel space is unavailable by Pauli-blocking. Formally, that is taken into account by the Pauli-projector $Q_F(\mathbf{p}_1, \mathbf{p}_2) = 1 - P_F(\mathbf{p}_1, \mathbf{p}_2)$, projecting to the

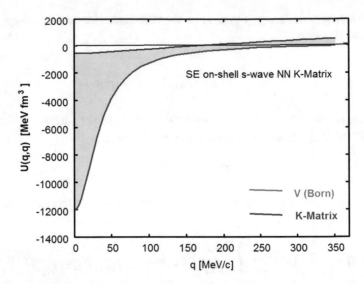

Fig. 5.6 The np s-wave interactions in the $(S = 0, I = 1)$ singlet-even channel in Born approximation, for which $U(q, q)$ denotes the s-wave component of $\mathcal{V}(q, q)$ (upper curve, red), and for the full K-matrix result, for which $U(q, q)$ corresponds to the s-wave component of $\mathcal{K}(q, q)$, (lower curve, blue). The difference between the Born-term and the fully summed K-matrix solution of the Lippmann-Schwinger is indicated by the shaded (blue coloured) area. The K-matrix results reproduce the measured scattering phase shift

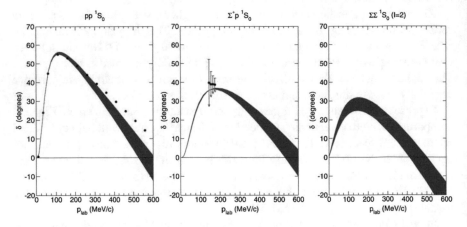

Fig. 5.7 The pp, $\Sigma^+ p$, and $\Sigma^+ \Sigma^+$ phase shifts in the $^1S_0$ partial wave. The filled bands represent the results at NLO [71]. The pp phase shifts of the GWU SAID-analysis [72] (circles) are shown for comparison. In the $\Sigma^+ p$ case the circles indicate upper limits for the phase shifts as deduced from the measured sections (from Ref. [71])

momentum region outside of the respective Fermi-spheres:

$$Q_F(\mathbf{p}_1, \mathbf{p}_2) = \Theta(p_1^2 - k_{F1}^2)\Theta(p_2^2 - k_{F2}^2). \tag{5.30}$$

The particle momenta $\mathbf{p}_{1,2}$ are given in the nuclear matter rest frame. Together with the time-like energy variables p_0 they define the four-momenta $p_{1,2} = (p_{1,2}^0, \mathbf{p}_{1,2})^T$ which are related the total two-particle four-momentum P and the relative momentum k by the Lorentz-invariant transformation

$$p_{1,2} = \pm k + x_{1,2}P \tag{5.31}$$

with

$$x_1 = \frac{s - m_2^2 + m_1^2}{2s} \quad ; \quad x_2 = \frac{s - m_1^2 + m_2^2}{2s} \tag{5.32}$$

obeying $x_1 + x_2 = 1$ and which in the non-relativistic limit reduce to $x_{1,2} = m_{1,2}/(m_1 + m_2)$. Furthermore, in-medium self-energies must be taken into account in propagators and vertices. In total, the in-medium K-matrix for a reaction $B_1 + B_2 \rightarrow B_3 + B_4$ is determined by a modified in-medium Lippmann-Schwinger equation, known as Brueckner G-matrix equation, given in full coupled channels form as

$$K_{B_1B_2,B_3B_4}(q, q') = \tilde{V}_{B_1B_2,B_3B_4}(q, q')$$

$$+ \sum_{B_5B_6} P \int \frac{d^3k}{(2\pi)^3} V_{B_1B_2,B_5B_6}(q, k) G_{B_5B_6}(k, q_s) Q_F(p_5^2, p_6^2) K_{B_5B_6,B_3B_4}(k, q') \tag{5.33}$$

where the integration has to be performed as a Cauchy Principal Value integral and the propagator $G_{B_5B_6}(k, q_s)$ includes now vector and scalar baryon self-energies. If the background medium consists only of protons and neutrons, the Pauli projection affects only the nucleonic part of the intermediate states $|B_5B_6\rangle$, i.e. if $B_5 = N$ or $B_6 = N$. The coupled equations have to be set up carefully with proper account of flavor exchange and antisymmetrization effects for the interactions in the channels of higher total strangeness, as discussed e.g. in [19, 73].

The calculation of in-medium interactions is in fact a very involved self-consistency problem, see e.g. [45, 74]. Interactions are entering in a nested manner to all orders into propagators, G-matrix equations, and self-energies. The (Dirac-) Brueckner-Hartree-Fock ((D)BHF) self-consistency cycle is indicated in Fig. 5.8 where the nested structure is recognized. However, as easily found, self-energies contribute to the Brueckner G-matrix only in second and higher order. Thus, a perturbative approach becomes possible for the channels which are weakly coupled to the medium: To a good approximation it is sufficient to solve the Brueckner equations with bare intermediate propagators, but including the Pauli-projector [29].

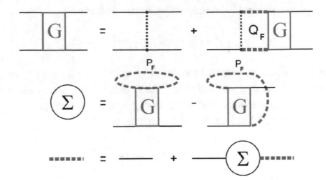

Fig. 5.8 The (D)BHF self-consistency problem for the G-matrix (upper row), connecting the single particle self-energy (middle row) and the one-body propagator (lower row). The two-body Pauli-project is denoted by Q_F, Eq. (5.30), while P_F is the projector onto the single particle Fermi sphere

Table 5.4 Low energy parameters for the indicated YN system

Channel	a (fm)	r (fm)	Model
ΣN $(I = \frac{1}{2})$	+0.90	−4.38	Jülich 04
ΣN $(I = \frac{3}{2})$	−4.71	3.31	Jülich 04
	−4.35	3.16	NSC97f 04
	−1.80	1.76	χ EFT LO
	−1.44	5.18	GiBE
ΛN $(I = \frac{1}{2})$	−2.56	2.75	Jülich 04
	−2.60	2.74	NSC97f
	−1.01	1.40	χ EFT LO
	−2.41	2.34	GiBE

Results from the OBE-oriented Jülich 04 [25], NSC97f [19], the recent Giessen Boson Exchange (GiBE) [29] models are compared to the leading order chiral EFT results [76]

For nuclear structure work, the low-energy behaviour of scattering amplitudes is of particular importance. The low momentum behaviour of the s-wave K-matrix is described by the effective range expansion for vanishing momentum q,

$$q \cot \delta_{|q \to 0} = -\frac{1}{a} + \frac{1}{2}q^2 r + \mathcal{O}(q^4) \tag{5.34}$$

where δ is the s-wave scattering phase shift and the expansion parameters a and r denote the s-wave scattering length and effective range, see e.g. [75]. In Table 5.4 the low energy parameters are given for several OBE approaches and compared to leading order (LO) χ EFT results.

In Fig. 5.9 the medium effects are illustrated for the total cross sections of NN-scattering and $\Sigma^+ p$-scattering. Both for NN and $\Sigma^+ p$ scattering one observes a

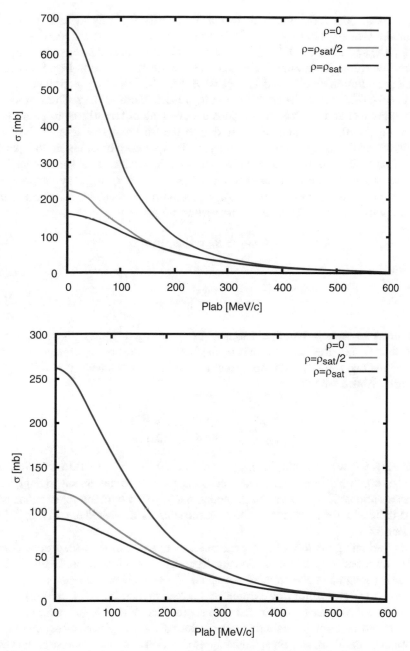

Fig. 5.9 Free-space and in-medium total cross sections for *pp* (up) and $\Sigma^+ p$ (down) elastic scattering. As indicated, the cross sections are evaluated in free space ($\rho = 0$), at half ($\rho = \rho_{sat}/2$) and at full nuclear matter density $\rho = \rho_{sat} = 0.16/\mathrm{fm}^3$, respectively (from Ref. [29])

rapid set in of a reduction with increasing density. Already at half the nuclear saturation density, i.e. $k_F = 209\,\text{MeV/c}$ the cross sections are reduced by a factor of $\frac{1}{3}$ (NN) and a factor of $\frac{1}{2}$ ($\Sigma^+ p$), respectively. At higher densities the reduction factors converge to an almost constant value close to the one in Fig. 5.9 seen for nuclear saturation density and $k_F = 263\,\text{MeV/c}$: the NN cross section is down by a factor of 0.25, the $\Sigma^+ p$ cross section by a factor close to 0.35. Thus, in nuclear matter the NN and the YN interactions are evolving differently, as to be expected from Eq. (5.33). The main effect is due to the different structure of the Pauli-projector. While the intermediate NN channels experience blocking in both particle momenta, in the intermediate YN channels only the nucleon states are blocked.

The elastic scattering amplitudes for $B_1 B_2 \rightarrow B_1 B_2$ we may separate into effective coupling constants $\tilde{g}_{B_1 B_1} \tilde{g}_{B_2 B_2}$ and a matrix element $\bar{M}_{B_1 B_2}$, amputated by all coupling constants. Thus, the cross section are given as

$$\sigma_{B_1 B_2} \sim \tilde{g}^2_{B_1 B_1} \tilde{g}^2_{B_2 B_2} \bar{\sigma}_{B_1 B_2} \tag{5.35}$$

where

$$\bar{\sigma}_{B_1 B_2} = \frac{4\pi}{q^2} |\bar{M}_{B_1 B_2}|^2 \tag{5.36}$$

The cross sections provide an estimate for the interaction strength of the iterated interactions, i.e. after the solution of the Bethe-Salpeter or Lippmann-Schwinger equations, respectively. Assuming that $\bar{\sigma}$ is an universal quantity within a given flavor multiplet, we find

$$R^2_{B_2 B_3} = \frac{\sigma_{B_1 B_2}}{\sigma_{B_1 B_2}} = \frac{\tilde{g}^2_{B_2 B_2}}{\tilde{g}^2_{B_3 B_3}} \tag{5.37}$$

From Fig. 5.9 we then obtain $R_{p\Sigma^+} \sim 0.62 \ldots 0.77$ for $\rho = 0$ to $\rho = \rho_{sat}$ and similar results are found also for other channels. These values are surprisingly close to assumptions of the naïve quark model, namely that hyperon interactions scale essentially with the number of u and d valence quarks, discussed e.g. in [56, 77] and in Sect. 5.4.

By evaluating the low-energy parameters as function of the background density we gain further insight into the density dependence of interactions. In Fig. 5.10 the scattering lengths of the coupled $n\Lambda$ and $n\Sigma^0$ channels are shown as a functions of the Fermi momentum of symmetric nuclear matter, $k_F = (3\pi^2 \rho/2)^{1/3}$. For $k_F \rightarrow 0$ the free space scattering length is approached. With increasing density of the background medium the scattering lengths decrease and approach an asymptotically constant value. Thus, at least in ladder approximation a sudden increase of the vector repulsion as discussed for a solution of the hyperon-problem in neutron stars is not in sight.

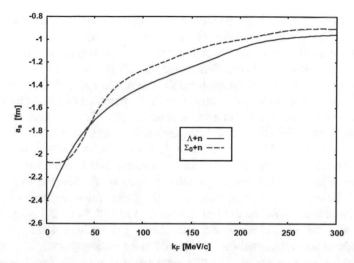

Fig. 5.10 The in-medium scattering lengths a_s for the coupled $n\Lambda$ and $n\Sigma^0$ channels are shown as a function of the Fermi momentum k_{F_N} of symmetric nuclear matter (from Ref. [29])

5.2.4 In-Medium Baryon-Baryon Vertices

A Lagrangian of the type as defined above leads to a ladder kernel $V^{BB'}(q', q)$ given in momentum representation by the superposition of one boson exchange (OBE) potentials $V^{BB'\alpha}(q', q)$. The solution of the coupled equations, Eq. (5.33) is tedious and sometimes numerically cumbersome by occasionally occurring instabilities. For certain parameter sets, unphysical YN and YY deeply bound states may show up. While for free space interactions the problems may be overcome, an approach avoiding the necessity to repeat indefinitely many times the (D)BHF calculations may be of advantage for applications in nuclear matter, neutron star matter, and especially in medium and heavy mass finite nuclei. Since systematic applications of (D)BHF theory in finite nuclei is in fact still not feasible, despite longstanding attempts, see e.g. the article of Müther and Sauer in [78], effective interactions in medium and heavy mass nuclei strongly rely on results from infinite nuclear matter calculations. Density functional theory (DFT) provides in principle the appropriate alternative, as known from many applications to atomic, molecular, and nuclear systems. However, DFT does not include a method to derive the appropriate interaction energy density which is a particular problem for baryonic matter. In the nuclear sector Skyrme-type energy density functionals (EDF), e.g. [79] for a hypernuclear Skyrme EDF, are a standard tool for nuclear structure research. The UNEDF initiative is trying to derive the universal nuclear EDF [39]. Finelli et al. started work on an EDF based on χ EFT. In [24] the ESC08 G-matrix was used to define an EDF.

Relativistic mean-field (RMF) theory relies on a covariant formulation of DFT and the respective relativistic EDF (REDF) [80–83]. Similar to the Skyrme-case, many different REDF versions are on the market, without and with non-linear self-interactions of the meson fields. One of the first RMF studies of hypernuclei was our work in [57] and many others have followed, see e.g. [84–86]. The Giessen DDRH theory is a microscopic approach with the potential of a true ab initio DFT description of nuclear systems. In a series of papers [40–43, 87] a covariant DFT was formulated with an REDF derived from DBHF G-matrix interactions. The density dependence of meson-baryon vertices as given by the ladder approximation are accounted for. Before turning to the discussion of the DDRH approach, we consider first a presentation of scattering amplitudes in terms of effective vertices, including the correlations generated by the solution of the Bethe-Salpeter equations.

For formal reasons we prefer to work with the full BB' T-matrix $T_{BB'}$. The ladder summation is done in the BB'-rest frame, but for calculations of self-energies and other observables the interactions are required in the nuclear matter rest frame rather than in the 2-body c.m. system. For that purpose the standard approach is to project the (on-shell) scattering amplitudes on the standard set of scalar (S), vector (V), tensor (T), axial vector (A) and pseudo scalar (P) Lorentz invariants, see e.g. [74, 88–90]. A more convenient representation, allowing also for at least an approximate treatment of off-shell effects, is obtained by representing the T-matrices in terms of matrix elements of OBE-type interactions, similar to the construction of the tree-level interactions $V^{BB'}$ but now using energy and/or density dependent effective vertex functionals Γ_a and propagators D_a for bosons with masses m_a. A natural choice is to use the same boson masses as in the construction of the tree-level kernels.

In the following, the expansion of a fully resummed interaction in terms of dressed vertices and boson propagators is sketched. For the reaction amplitude of the process $\mathcal{B} = (B_1B_2) \rightarrow \mathcal{B}' = (B_3B_4)$ we use a the *ansatz*

$$T_{BB'}(\mathbf{q}, \mathbf{q}'|q_s k_F) = \sum_a \Gamma_{B_1B_3}^{(a)\dagger}(q_s, k_F) U_{BB'}^{(a)}(\mathbf{q}, \mathbf{q}') \Gamma_{B_2B_4}^{(a)}(q_s, k_F) \tag{5.38}$$

where

$$U_{BB'}^{(a)}(\mathbf{q}, \mathbf{q}') = \mathcal{M}_{BB'}^{(a)} D_a(\mathbf{q}, \mathbf{q}') \tag{5.39}$$

is given by the invariant matrix element

$$\mathcal{M}_{BB'}^{(a)} = \langle B_4B_3|\mathcal{O}_a^{\dagger}(1) \cdot \mathcal{O}_a(2)|B_1B_2\rangle = \langle B_4|\mathcal{O}_a^{\dagger}|B_2\rangle \cdot \langle B_3|\mathcal{O}_a|B_1\rangle \tag{5.40}$$

containing the vertex operators $\mathcal{O}_a$ belonging to the interaction of type $a \in \{A, P, S, T, V\}$ defined above, but without coupling constants. The scalar product of the two operators is indicated by the dot-product. Non-relativistically, the operator set is given by $\mathcal{O}_a = \{1_\sigma, \boldsymbol{\sigma} \cdot \mathbf{q}, \boldsymbol{\sigma}, \boldsymbol{\sigma} \times \mathbf{q}\} \otimes \{1_\tau, \boldsymbol{\tau}\}$ for scalar, pseudo-scalar, vector, and pseudo-vector interactions of isoscalar and isovector character, respectively. The propagator $D_a(\mathbf{q}, \mathbf{q}')$ describes the exchange of the boson a, representing the specific interaction channel. Since the vertices are attached to the matrix element $\mathcal{M}^{(a)}_{BB'}$, the propagators are in fact of simple Yukawa-form,

$$D_a(\mathbf{q}, \mathbf{q}') = \frac{1}{(\mathbf{q} - \mathbf{q}')^2 + m_a^2},\tag{5.41}$$

thus depending only on the momentum transfer if meson self-energies are neglected, as above. Accordingly, the Born-terms are given by

$$V_{BB'a} = U^{(a)}_{BB'} g^2_{BB'a}\tag{5.42}$$

where $g^2_{BB'a}$ denotes the bare coupling constants in free space. Leaving out for simplicity the baryon indices and momentum arguments, we find the important relations

$$\frac{\delta V}{\delta U^{(a)}} = g_a^2\tag{5.43}$$

$$\frac{\delta T}{\delta U^{(a)}} = \Gamma^{(a)\dagger} \Gamma^{(a)}.\tag{5.44}$$

Formally, the Lippmann-Schwinger equation is solved by

$$T = \left(1 - \int dk V \mathcal{G}^* Q_F\right)^{-1} V\tag{5.45}$$

and applying Eqs. (5.43) and (5.44), we obtain

$$\Gamma^{(a)\dagger} \Gamma^{(a)} = \left(1 - \int dk V \mathcal{G}^* Q_F\right)^{-2} g_a^2 + \dots\tag{5.46}$$

where the dots indicate terms given by the variational derivatives of baryon (and meson) self-energies which, in general, are contained in the intermediate baryon-baryon propagator $\mathcal{G}^*$. Thus, the coupling constants of the dressed vertices are determined in leading order by

$$\Gamma^{(a)} \simeq \left(1 - \int dk V \mathcal{G}^* Q_F\right)^{-1} g_a.\tag{5.47}$$

Hence, we have recovered the well known integral equation for dressed vertices, generalizing Migdal's theorem for dressed vertices in interacting many-body systems [91, 92] to interactions in infinite nuclear matter. Restoring the full index

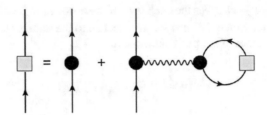

Fig. 5.11 Diagrammatic structure of the dressed vertex functionals $\Gamma_{BB'a}$ (filled square) in terms of the bare coupling constant $g_{BB'a}$ (filled circle) and the interaction $U_{BB'a}$, indicated by a wavy line. The integral over the complement of the combined Fermi spheres of the intermediate baryons is shown as a loop (see text)

structure, we find

$$\Gamma_{BB'a}(q_s, k_F) \simeq \frac{1}{1 - \int dq' V_a \mathcal{G}^* Q_F} \bigg|_{BB'} g_{BB'a}. \tag{5.48}$$

The diagrammatic structure is depicted in Fig. 5.11. Inspecting Eq. (5.48), one finds that dressed and bare vertices are related by a susceptibility matrix

$$\chi_{BB'a}(q_s, k_F) = \left(1 - \int dq' V_a \mathcal{G}^* Q_F\right)^{-1}_{\bigg|_{BB'}} \tag{5.49}$$

depending on the center-of-mass energy $\sqrt{s}$ through q_s and the set of Fermi momenta $k_F = \{k_{F_B}\}_{|B=n,p...}$.

Two limiting cases are of particular interest. At vanishing density where $Q_F \to 1$, and since $V_a \sim g_a^2$ we find that in free-space the dressed vertices retain their general structure as a fully summed series of tree-level coupling constants. At density $\rho \to \infty$, where $Q_F \to 0$ over the full integration range, we find $\Gamma_a \simeq g_a$. Albeit for another reason similar result is found in the high energy limit: The most important contribution to the integral is coming from the region around the Green's function pole. With increasing $\sqrt{s}$ the pole is shifted into the tail of the form factors regularizing the high momentum part of U_a. Thus, at large energies the residues are increasingly suppressed, as indicated by the decline of the matrix element shown in Fig. 5.6. As a conclusion, the dressing effects are the strongest for low energies and densities.

5.2.5 Vertex Functionals and Self-Energies

It is interesting to notice that a similar approach, but on a purely phenomenological level, was used long ago by Love and Franey to parameterize the NN T-matrix over a large energy range, $T_{lab} = 50\ldots 1000\,\text{MeV}$. Also the widely used M3Y G-matrix parametrization of Bertsch et al. [93] is using comparable techniques, as also the work in [94]. In order to generalize the approach a field-theoretical formulation is of advantage which is the line followed in DDRH theory. For that purpose, the dependence on $k_F \sim \rho_B^{1/3}$ is replaced by a functional dependence on the baryon four-current by means of the Lorentz-invariant operator relation $\rho_B^2 = j_{B\mu}j_B^\mu$ leading to vertex functionals $\hat{\Gamma}_a(\bar{\Psi}_B\Psi_B)$. The C-numbered vertices are recovered as expectation values, $\Gamma_a(\sqrt{s}, k_F) = \langle P, k_F|\hat{\Gamma}_a(\bar{\Psi}_B\Psi_B)|P, k_F\rangle$, under given kinematical conditions P and for a baryon configuration defined by k_F.

For studies of single particle properties it is sufficient to extract the vertices directly on the mean-field level. An efficient way is to use the baryon self-energies. For our present illustrating purposes it is enough to consider the Hartree tadpole-term. The self-energy $\Sigma_{aB}(k_F)$ due to the exchange of the boson a felt by the baryon B is given by

$$\Sigma_{aB}(k_F) = \Gamma_{BBa}(k_F)\frac{1}{m_a^2}\sum_{B'} \Gamma_{B'B'a}(k_F)\rho_{B'}^{(a)} \tag{5.50}$$

where on the right side we have inserted the above decomposition. Thus, a set of quadratic forms is obtained, bilinear in the vertex functions $\Gamma_{BBa}(k_F)$. $\rho_B^{(a)}$ denotes either a scalar ($a = s$) or a vector ($a = v$) ground state density of baryons of type B. Using on the left hand side the microscopic self-energies, the quadratic form can be evaluated. The vertices are fixed in their (relative) phases and magnitudes by the solutions

$$\frac{\Gamma_{B_1B_1a}(k_F)}{\Gamma_{B_2B_2a}(k_F)} = \frac{\Sigma_{aB_1}(k_F)}{\Sigma_{aB_2}(k_F)} \tag{5.51}$$

$$\Gamma_{aB}^2(k_F) = \frac{\Sigma_{aB}^2(k_F)}{\sum_{B'} \rho_{aB'}\Sigma_{aB'}(k_F)}. \tag{5.52}$$

Using DBHF self-energies as input the dressed vertices are obtained in ladder approximation, being appropriate for use in RMF theory. Although nuclear matter (D)BHF self-energies depend on the particle momentum [74], those dependencies are cancelling in the above expressions to a large extent. As discussed in [87] the mild remaining state dependence can be included by a simple correction factor, ensuring the proper reproduction of the nuclear matter equation of state and binding energies of nuclei.

Results of such a calculation using DBHF self-energies as input [74] are shown in Fig. 5.12 and applications to stable and exotic nuclei and neutrons stars are found e.g. in [41–43, 46, 87]. Hypernuclear results are discussed below.

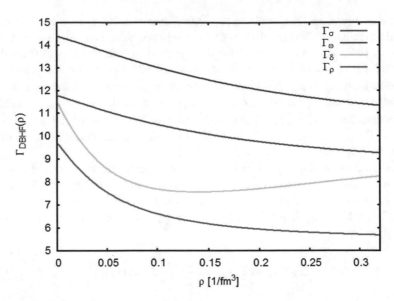

Fig. 5.12 In-medium $NN\alpha$ isoscalar ($\alpha = \sigma, \omega$) and isovector ($\alpha = \delta, \rho$) vertices determined from DBHF self-energies in asymmetric nuclear matter as explained in [74]

5.3 Covariant DFT Approach to Nuclear and Hypernuclear Physics

5.3.1 Achievements of the Microscopic DDRH Nuclear DFT

The RMF scheme for the extraction of the vertices within the DBHF ladder approximation has proven to lead to a quite successful description of nuclear matter and nuclear properties. Out of that scheme, DDRH theory has emerged, which later gave rise to the development of a purely phenomenological approach [47] by fixing the density dependence of the vertex functionals by fits to data. Relativistic DFT approaches based on density dependent vertices, e.g. by the Beijing group [95, 96], are now a standard tool for covariant nuclear structure mean-field theory, describing successfully nuclear ground states and excitations. While the DBHF realizations of DDRH theory are of a clear diagrammatic structure, namely including interactions in the ladder level only, the parameters of the phenomenological models are containing unavoidably already higher order effects from core polarization self-energies and other induced many-body interactions.

For obvious reasons, in nuclear structure research the meson fields giving rise to condensed classical fields are of primary importance. The isoscalar and isovector self-energies produced by scalar ($J^P = 0^+$) and vector mesons ($J^P = 1^-$) are dominating nuclear mean-field dynamics. In this section, we therefore set the focus on the mean-field producing scalar and vector mesons. The formulation is kept general in the sense that the concepts are not necessarily relying on the use of a DBHF description of vertices as discussed in the previous section. For clear distinction, here vertex functionals will be denoted by $g^*BB'M(\rho)$. The concepts have been developed earlier in the context of the Giessen Density Dependent Relativistic Hadron (DDRH) theory. By a proper choice of vertex functionals an ab initio approach is obtained. Binding energies and root-mean-square radii of stable and unstable nuclei are well described within a few percent [41, 42, 87]. The equations of state for neutron matter, neutron star matter, and neutron stars are obtained without adjustments of parameters. In [46] the temperature dependence of the vertices was studied and the thermodynamical properties of nuclear matter up to about a temperature of $T = 50\,\text{MeV}$ were investigated, looking for the first time into the phase diagram of asymmetric nuclear matter as encountered in heavy ion collisions and neutron stars by microscopically derived interactions. For that purpose, the full control of the isovector interaction channel—which is not well under control in phenomenological approaches—was of decisive importance. Results for hypermatter and hypernuclei will be discussed below.

5.3.2 Covariant Lagrangian Approach to In-Medium Baryon Interactions

A quantum field theory is typically formulated in terms of bare coupling constants which are plain numbers and the field theoretical degrees of freedom are contained completely in the hadronic fields. The preferred choice is a Lagrangian of the simplest structure, e.g. bilinears of matter field for the interaction vertices. That apparent simplicity on the Lagrangian level, however, requires additional theoretical and numerical efforts for a theory with coupling constants which are too large for a perturbative treatment. Hence, the theoretical complexities are only shifted to the treatment of the complete resummation of scattering series. DDRH theory attempts to incorporate the resummed higher order effects already into the Lagrangian with the advantage of a much simpler treatment of interactions. This is achieved by replacing the coupling constants $g_{BB'M}$ by meson-baryon vertices $g_{BB'M}^*(\Psi_{\mathcal{F}})$ which are Lorentz-invariant functionals of bilinears of the matter field operators Ψ_B. The derivation of those structures from Dirac-Brueckner theory has been discussed intensively in the literature [40–42, 45, 74]. The mapping of a complex system of coupled equations to a field theory with density dependent vertex functionals is leading to a formally highly non-linear theory. However, the inherent complexities are of a similar nature as known from quantum many-body theory in general. In practice, approximations are necessary. Here, we discuss the mean-field limit.

The essential difference of the Lagrangian discussed here to the one of Eq. (5.8) lies in the definition of the interaction vertices. Overall, their structure is constrained by the requirements of maintaining the relevant symmetries. Thus, at least the functionals must be Lorentz-scalars and scalars under SU(3) flavor transformations.

The simplest choice fulfilling these constraints is to postulate a dependence on the invariant density operator $\hat{\rho} = j_\mu j^\mu$ of the baryon 4-current operator $j^\mu = \bar{\Psi}_B \gamma^\mu \Psi_B$. Thus, we use $g^*_{BB'\mathcal{M}}(\Psi_B) = g^*_{BB'\mathcal{M}}(\hat{\rho})$. Applying the SU(3) relations, this implies immediately a corresponding structure for the fundamental interaction vertices, $\{g_D, g_F, g_S\} \rightarrow \{g_D^*(\hat{\rho}), g_F^*(\hat{\rho}), g_S^*(\hat{\rho})\}$. Thus, the interactions with density functional (DF) vertices are described by

$$\mathcal{L}_{int}^{DF} = -\sqrt{2} \sum_{\mathcal{M} \in \{\mathcal{P}, \mathcal{S}, \mathcal{V}\}} \left\{ g_D^{*(\mathcal{M})} \left[\bar{B}B P_8\right]_D + g_F^{*(\mathcal{M})} \left[\bar{B}B P_8\right]_F - g_S^{*(\mathcal{M})} \frac{1}{\sqrt{6}} \left[\bar{B}B P_1\right]_S \right\}. \tag{5.53}$$

By the same relations as in Eq. (5.17) we obtain baryon-meson vertices $g^*_{BB'M}(\hat{\rho})$ but with the complexity of intrinsic functional structures.

With the standard methods of field theory we obtain the meson field equations which are of the form

$$\left(\partial_\mu \partial^\mu + m_\mathcal{M}^2\right) \Phi_\mathcal{M}^s = \sum_{BB'} g^*_{BB'\mathcal{M}}(\hat{\rho}) \rho^{BB's} \tag{5.54}$$

$$\left(\partial_\mu \partial^\mu + m_\mathcal{M}^2\right) V_\mathcal{M}^\lambda = \sum_{BB'} g^*_{BB'\mathcal{M}}(\hat{\rho}) \rho^{BB'\lambda} \tag{5.55}$$

where Lorentz-scalar and Lorentz-vector fields are described by the first and second equation, respectively. The baryons are obeying the field equations

$$\left(\gamma_\mu \left(p^\mu - \Sigma_B^\mu(\hat{\rho})\right) - M_\mathcal{F} + \Sigma_B^{(s)}(\hat{\rho})\right) \Psi_B = 0 \tag{5.56}$$

The general structure of the self-energies is given by a folding part, defined as the sum over meson field multiplied by coupling constants, and an additional rearrangement terms resulting from the variation of the intrinsic functional structure of the vertices [41, 42]. As discussed above, the density dependence of the in-medium meson-baryon vertices is chosen by a functional structures given by $\hat{\rho}^2 = j_\mu j_B^\mu$ which is the Lorentz-invariant baryon density operator. Since higher order interaction effects are accounted for by the density-dependent vertex functionals, they must be used in a first order treatment only, i.e. on the level of effective Born-diagrams.

The rearrangement self-energies account for the static polarization of the surrounding medium due to the presence of the baryon. As was proven in [42], the rearrangement self-energies ensure the thermodynamical consistency of the theory. Their neglection would violate the Hugenholtz-van Hove theorem. Similar conclusion have been drawn before by Negele for non-relativistic BHF theory [97]. In mean-field approximation, the self-energies will become Hartree type mean-fields. The diagrammatic structure of the mean-field self-energies is depicted in Fig. 5.13. The structure of the rearrangement self-energies is discussed in detail in [45, 97].

Fig. 5.13 Derivation of the DFT mean-field self-energies by first variation of the energy density with respect to the density, leading to tadpole and rearrangement contributions, indicated by the first and second graph on the right hand side of the figure, respectively. The density dependent vertices are indicated by shaded squares, the derivative of the vertices by a triangle. The DFT tadpole self-energy accounts for the full Hartree-Fock self-energy shown in Fig. 5.8

With our choice of density dependence, only the vector self-energies are modified by rearrangement effects. Thus, the scalar self-energy $\Sigma_{\mathcal{F}}^s$ is obtained as a diagonal matrix with the elements of pure Hartree-structure

$$\Sigma_B^{(s)}(\hat{\rho}) = \sum_{\mathcal{M} \in S} \Phi_{\mathcal{M}}(\hat{\rho}) g_{BB\mathcal{M}}^*(\hat{\rho}). \tag{5.57}$$

The vector self-energies, contained in the matrix $\Sigma_{\mathcal{F}}^\mu$, are consisting of the direct meson field contributions

$$\Sigma_B^{(d)\mu}(\hat{\rho}) = \sum_{\mathcal{M} \in V} V_{\mathcal{M}}^\mu(\hat{\rho}) g_{BB\mathcal{M}}^*(\hat{\rho}) \tag{5.58}$$

and the rearrangement self-energies

$$\Sigma_B^{(r)\mu}(\hat{\rho}) = \sum_{B'B''\mathcal{M}} \frac{\partial g_{B'B''\mathcal{M}}^*(\hat{\rho})}{\partial \rho_\mu^{BB}} \frac{\delta}{\delta g_{B'B'f'\mathcal{M}}^*} \mathcal{L}_{int}^{DF} \quad, \tag{5.59}$$

where the latter are generic for a field theory with functional vertices. Physically, the $\Sigma^{(r)}$ are accounting for changes of the interactions under variation of the density of the system as resulting from static polarization effects. They are indispensable for the thermodynamical consistency of the theory [41, 42, 46].

A substantial simplification is obtained in mean-field approximation. As discussed in [46], the vertex functionals are replaced by ordinary functions of the expectation value of their argument. The meson fields are replaced by static classical fields and the rearrangement self-energies reduce to derivatives of the vertex functions with respect to the density. The mean-field constraint also implies that in a uniform system we can neglect in the average the space-like vector self-energy components. Thus, we are left with the scalar and the time-like vector self-energies. The scalar and the direct, Hartree-type vector mean-field self-energies, respectively, are obtained as

$$\Sigma_B^{(s)}(\rho_B) = \sum_{B',\mathcal{M}\in S} \frac{1}{m_{\mathcal{M}}^2} g_{BB\mathcal{M}}^*(\rho_B) g_{B'B'\mathcal{M}}^*(\rho_B)\rho_{B'}^{(s)} \tag{5.60}$$

$$\Sigma_B^{(d)}(\rho_B) = \sum_{B',\mathcal{M}\in V} \frac{1}{m_{\mathcal{M}}^2} g_{BB\mathcal{M}}^*(\rho_B) g_{B'B'\mathcal{M}}^*(\rho_B)\rho_{B'} \tag{5.61}$$

and the time-like vector rearrangement self-energies are given by the derivatives of the vertex functions with respect to the total baryon density of nucleons N and hyperons Y, $\rho_B = \rho_N + \rho_Y$,

$$\Sigma_B^{(r)}(\rho_B) = \sum_{B',\mathcal{M}} \frac{\partial g_{B'B'\mathcal{M}}^*(\rho_B)}{\partial \rho_B} \frac{\partial \mathcal{L}_{int}}{\partial g_{B'B'\mathcal{M}}^*}, \tag{5.62}$$

where the usual number and scalar densities are denoted by ρ_B and $\rho_B^{(s)}$, respectively. For later use we define the total time-like baryon vector self-energies

$$\Sigma^{(v)}(\rho_B) = \Sigma_B^{(d)}(\rho_B) + \Sigma_B^{(r)}(\rho_B). \tag{5.63}$$

By means of the SU(3) relations also the in-medium vertices in the other baryon-meson interaction channels can be derived [98], leading to a multitude of vertices which are not shown here. In magnitude (and sign) the BB' vertices are of course different form the one shown in Fig. 5.12. But a common feature is that the relative variation with density is similar to the NN-vertices in Fig. 5.12.

5.4 DBHF Investigations of Λ Hypernuclei and Hypermatter

5.4.1 Global Properties of Single-Λ Hypernuclei

An important result of the following investigations is that at densities found in the nuclear interior the Λ and nucleon vertices are to a good approximation related by scaling factors, $g_{\Lambda\Lambda\sigma,\omega}^* \simeq R_{\sigma,\omega} g_{NN\sigma,\omega}^*$. Microscopic calculations show that these ratios vary in the density regions of interest for nuclear structure investigations only slightly. Averaged over the densities up to the nuclear saturation density $\rho_{eq} = 0.16\,\mathrm{fm}^{-3}$, we find $\bar{R}_\sigma = 0.42$ and $\bar{R}_\omega = 0.66$ with variation on the level of 5%. Hence, as far as interactions of the Λ hyperon are concerned it is possible to describe their properties with interactions following the scaling approach. In fact, the scaling approach was used widely before in an ad hoc manner, e.g. in [56, 57]. Occasionally, the quark model is used as a justification by arguing that non-strange mesons couple only to the non-strange valence quarks of a baryon which for the Λ

gives scaling factors $R_{\sigma,\omega} = 2/3$. Surprisingly, the cited value of $\bar{R}_\omega$ is extremely close to the one expected by the quark counting hypothesis.

In this section, we investigate the scaling hypothesis by using self-consistent Hartree-type calculations for single-Lambda hypernuclei. In view of the persisting uncertainties on YN interactions, we treat $R_{\sigma,\omega}$ as free constants which are adjusted in fits to Λ separation energies. Since wave functions, their densities, and energies are calculated self-consistently we account simultaneously for effects also in the nucleonic sector induced by the presence of hyperons.

The nucleons ($B = n, p$) and the Lambda-hyperon ($B = \Lambda$) single particle states are described by stationary Dirac equations

$$\left(\boldsymbol{\alpha} \cdot \hat{\boldsymbol{p}} + \Sigma_B^{(v)}(r) + \gamma^0 M_B^*(r) - \varepsilon_{n\ell j}\right) \psi_{n\ell jm} = 0 \tag{5.64}$$

with $\hat{\mathbf{p}} = -i\nabla$ and $\boldsymbol{\alpha} = \gamma^0\boldsymbol{\gamma}$. We assume spherical symmetry. The Dirac spinors $\psi_{n\ell jm}$ with radial quantum number n and total angular momentum j and projection m are eigenfunctions to the eigenenergy $\varepsilon_{n\ell j}$. The orbital angular momentum of the upper component is indicated by ℓ. The direct vector self-energies are

$$\Sigma_B^{(d)}(r) = g_{BB\omega}^*(\rho_B)V_\omega^0(r) + \langle\tau_3\rangle_B g_{BB\rho}^*(\rho_B)V_\rho^0(r) + q_B V_\gamma^0(r). \tag{5.65}$$

$\rho_B(r) = \rho_p(r) + \rho_n(r) + \rho_\Lambda(r)$ is the radial dependent total baryon density with a volume integral normalized to the total baryon number $A_B = A_n + A_p + A_\lambda$ where in a single Λ nucleus $A_\Lambda = 1$. q_B denotes the electric charge of the baryon. We use $\langle\tau_3\rangle_B = \pm 1$ for the proton and neutron, respectively, and $\langle\tau_3\rangle_B = 0$ for the Λ. The scalar self-energies, contained in the relativistic effective mass M_B^*, are given by

$$\Sigma_B^{(s)}(r) = g_{BB\sigma}^*(\rho_B)\Phi_\sigma(r) + \langle\tau_3\rangle_B g_{BB\delta}^*(\rho_B)\Phi_\delta(r). \tag{5.66}$$

The condensed meson fields are determined by the classical field equations

$$\left(-\nabla^2 + m_\omega^2\right)V_\omega^0 = g_{NN\omega}^*(\rho_B)\left(\rho_p + \rho_n\right) + g_{\Lambda\Lambda\omega}^*(\rho_B)\rho_\Lambda \tag{5.67}$$

$$\left(-\nabla^2 + m_\rho^2\right)V_\rho^0 = g_{NN\omega}^*(\rho_B)\left(\rho_p - \rho_n\right) \tag{5.68}$$

$$-\nabla^2 V_\gamma^0 = e_p\rho_p^{(c)} \tag{5.69}$$

$$\left(-\nabla^2 + m_\sigma^2\right)\Phi_\sigma = g_{NN\sigma}^*(\rho_B)\left(\rho_p^{(s)} + \rho_n^{(s)}\right) + g_{\Lambda\Lambda\sigma}^*(\rho_B)\rho_\Lambda^{(s)} \tag{5.70}$$

$$\left(-\nabla^2 + m_\delta^2\right)\Phi_\delta = g_{NN\delta}^*(\rho_B)\left(\rho_p^{(s)} - \rho_n^{(s)}\right), \tag{5.71}$$

where e_p is the proton charge and $\rho_p^{(c)}$ denotes the proton charge density, including the proton charge form factor. The vector and scalar densities ρ_B and $\rho_B^{(s)}$ are given in terms of the sum and the difference, respectively, of the densities of upper and lower components of the wave functions of occupied states. For further details we refer to our previous work [42, 43, 45]. From the field-theoretical energy-momentum tensor the energy density is obtained as the ground state expectation value $\mathcal{E}_A(\mathbf{r}; Z, N, Y) = \langle T^{00}\rangle$ for a nucleus with A_p protons, A_n neutrons and A_Λ hyperons [40, 42].

Of particular interest are the nuclear binding energies defined by

$$B_A = \frac{1}{A}\left[\int d^3r\,\mathcal{E}_A(\mathbf{r}) - A_p M_p - A_n M_n - A_\Lambda M_\Lambda\right], \tag{5.72}$$

from which the baryon separation energies are obtained as usual. We treat the Λ interaction vertices in the scaling approach and determine the renormalization constants $R_{\sigma,\omega}$ by a fit to the available single-Λ hypernuclear data, as discussed in [40]. The lightest nuclei were left out of the fitting procedure because they are not well suited for a mean-field description. From a χ^2 procedure, we find that the two scaling constants are correlated linearly, leading to a χ^2 distribution with a steep, but long stretched valley. Two solutions of comparable quality are $(R_{1\sigma}, R_{1\omega}) = (0.506, 0.518)$ and $(R_{2\sigma}, R_{2\omega}) = (0.490, 0.553)$. On a 20% level both sets are compatible with the quark model hypothesis and with the SU(3)-symmetry based results [98]. Taking into account also the light nuclei, the values show a considerably large spread. In Fig. 5.14 relativistic mean-field the DDRH results are compared to known measured single-Λ separation energies. It is seen that

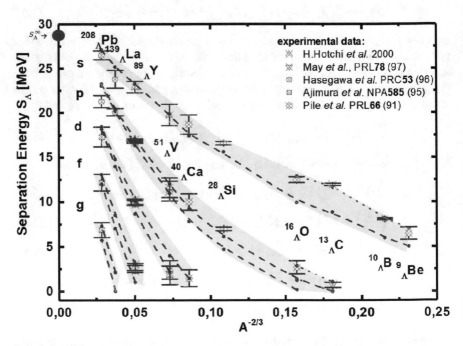

Fig. 5.14 Separation energies of the known $S = -1$ single Λ hypernuclei as a function of the mass number A to power $\gamma = -\frac{2}{3}$. Results of two sets of scaling parameter sets are compared to measured separation energies. For the $\ell > 0$ levels the spin-orbit splitting is indicated. The theoretical uncertainties are marked by colored/shaded bands. For $A \rightarrow \infty$ the limiting value $S_\Lambda^\infty \simeq 28$ MeV is asymptotically approached as indicated in the figure. Thus, we predict for the in-medium Λ-potential in ordinary nuclear matter a depth of 28 MeV. The data are from refs. [99–103]

Fig. 5.15 Energy spectra of single-Λ hypernuclei ($^{13}_{\Lambda}C$, $^{28}_{\Lambda}Si$, $^{51}_{\Lambda}V$, $^{139}_{\Lambda}La$, $^{208}_{\Lambda}Pb$) derived by the multi-pomeron exchange interaction MPa (solid lines) and the bare ESC08 interaction (dotted lines). Details of the Nijmegen approach are found in [23]. The theoretical results are compared to experimental values, marked by open circles (from [23])

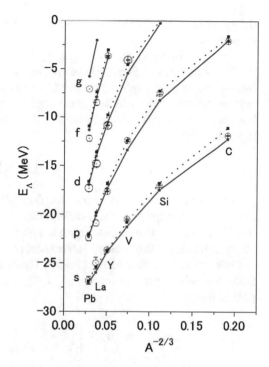

the global two-parameter fit leads to a surprisingly good description of the observed Λ separation energies. The remaining theoretical uncertainties are indicated.

Extrapolating the separation energies shown in Fig. 5.14 to (physical unaccessible) large mass number, the limiting value $S^{\infty}_{\Lambda} \simeq 28$ MeV is asymptotically approached for $A \to \infty$ which we identify with the separation energy of a single Λ-hyperon in infinite nuclear matter. Thus, we predict for the in-medium Λ-potential in ordinary nuclear matter a value of $U^{\infty}_{\Lambda} \sim -28$ MeV.

In Fig. 5.15 results of the Nijmegen group are shown for comparison. The ESC08 interaction with and without pomeron-exchange (see [23]) was used as input for non-relativistic Brueckner G-matrix calculations which then was used in a folding approach to generate the Lambda mean-field potentials. The relativistic DDRH and the non-relativistic ESC08-results agree rather well for the Lambda separation energies over the full known mass range of core nuclei. The agreement may be taken as an indication that a common understanding of single particle dynamics for single-Λ hypernuclei is obtained, at least within the presently available data base.

5.4.2 Spectroscopic Details of Single-Λ Hypernuclei

For the results shown in Fig. 5.14 the KEK-data of Hotchi et al. [99] for $^{41}_{\Lambda}V$ and $^{89}_{\Lambda}Y$ have been especially important because of their good energy resolution and the resolution of a large number of Λ bound states. This wealth of spectroscopic information did help to constrain further the dynamics of medium and heavy hypernuclei.

A caveat for those nuclei is that the hyperon is attached to a high-spin core, $^{40}V(6^-)$ and $^{88}Y(4^-)$. Hence, the Λ spectral distributions are additionally broadened by core-particle spin-spin interactions. A consistent description of the spectra could only be achieved by including those interactions into the analysis. A phenomenological approach was chosen by adding the core-particle spin-spin energy to the Λ eigenenergies

$$e_{(jJ_C)J\Lambda} = \varepsilon_{j\Lambda}^{RMF} + E_{jJ_c}\langle(jJ_C)J|\mathbf{j}_\lambda \cdot \mathbf{J}_C|(jJ_C)J\rangle \quad . \tag{5.73}$$

giving rise to a multiplet of states. The multiplet-spreading is found to account for about half of the spectral line widths. Hence, if neglected, badly wrong conclusions would be drawn on an extraordinary large spin-orbit splitting, too large by about a factor 2. Including the spin-spin effect leads to a spin-orbit energy fully compatible with the values known from light nuclei. The analysis includes also the contributions from the relativistic tensor vertex [104], modifying the effective Λ-spin-orbit potential to

$$U_\Lambda^{(so)} = \frac{1}{r}\mathbf{r} \cdot \nabla\left[\left(2\frac{M_B^*}{M_B}\frac{f_{\Lambda\Lambda\omega}}{g_{\Lambda\Lambda\omega}^*} + 1\right)\Sigma_\omega^\Lambda + \Sigma_\sigma^\Lambda\right]. \tag{5.74}$$

Here, the tensor strength f appears as an additional parameter. For the NN case, $f_{NN\omega}$ is known to be weak and usually it is set to zero. The small spin-orbit splitting observed in hypernuclei have led to speculations that the tensor part may be non-zero, partly cancelling the conventional spin-orbit potential, given by the sum of vector and scalar self-energies. This should happen for $f/g \sim -1$ as seen by considering that U_Λ^{so} is a nuclear surface effect where $M^* \sim M$ and also the self-energies are about the same. The KEK-spectra are described the best with vanishing Λ tensor coupling, $f_{\Lambda\lambda\omega}/g_{\Lambda\lambda\omega} = 0$, thus agreeing with the NN-case. Our results for the Λ single particle spectra in the two nuclei are found in Table 5.5. The averaged spin-orbit splitting is about 223 and 283 keV and the spin-spin interaction amounts to $E_{jJ_c} = 106$ keV and $E_{jJ_c} = 61.3$ keV in Vanadium and Yttrium, respectively. The experimentally obtained and the theoretical spectra of $_\Lambda^{89}Y$ are compared in Fig. 5.16.

5.4.3 Interactions in Multiple-Strangeness Nuclei

In contrast to the YN data—scarce as they are—for hyperon-hyperon systems like $\Lambda\Lambda$ no direct scattering data are available. The only source of (indirect) experimental information at hand comes from double-Lambda hypernuclei. The first observation of a double-Lambda hypernuclear event, assigned in an emulsion experiment as either $_{\Lambda\Lambda}^{10}Be$ or $_{\Lambda\Lambda}^{11}Be$, was reported as early as 1963 by Danysz et al. [105]. The probably best recorded case is the so-called Nagara event [106], a

Table 5.5 DDRH results for Λ single particle energies

Level	$^{89}_{\Lambda}Y$ (MeV)	$^{41}_{\Lambda}V$ (MeV)
$1s_{1/2}$	-22.94 ± 0.64	-19.8 ± 1.4
$1p_{3/2}$	-17.02 ± 0.07	-11.8 ± 1.3
$1p_{1/2}$	-16.68 ± 0.07	-11.4 ± 1.3
$1d_{5/2}$	-10.26 ± 0.07	-2.7 ± 1.2
$1d_{3/2}$	-9.71 ± 0.07	-1.9 ± 1.2
$1f_{7/2}$	-3.04 ± 0.11	–
$1f_{5/2}$	-3.04 ± 0.11	–

The experimentally unresolved fine structure due to the residual core-particle spin-spin interactions inhibits a precise determination of the genuine, reduced Λ single particle energies, defined by subtracting the core-particle spin-spin interaction energies. The shown errors are taking into account the resulting uncertainties in the reduced DDRH single-particle energies

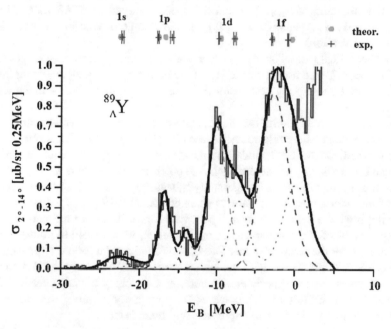

Fig. 5.16 Single Λ binding energies in $^{89}_{\Lambda}Y$. The DDRH results (dotted, dashed, and full lines) are compared to the data of Hotchi et al. [99], shown as histogram. The extracted single particle levels, obtained after defolding the spectrum and taken into account core polarization effects (see text), are indicated at the top of the figure

safely identified $^{6}_{\Lambda\Lambda}He$ hypernucleus produced at KEK by a (K^-, K^+) reaction at $p_{lab} = 1.66$ GeV/c on a ^{12}C target. The KEK-E373 hybrid emulsion experiment [106] traced the stopping of an initially produced Ξ^- hyperon, captured by a second carbon nucleus, which then was decaying into $^{6}_{\Lambda\Lambda}He$ plus an ^4He nucleus and a triton. The $^{6}_{\Lambda\Lambda}He$ nucleus was identified by it's decay into the known $^5_{\Lambda}He$ and a proton and a π^-. The data were used to deduce the total two-Lambda separation energy $B_{\Lambda\Lambda}$ and the Lambda-Lambda interaction energy $\Delta B_{\Lambda\Lambda}$ which is a particular highly wanted quantity. A re-analysis in 2013 [107] led to the nowadays accepted values $B_{\Lambda\Lambda} = 6.91 \pm 0.16$ MeV and $\Delta B_{\Lambda\Lambda} = 0.67 \pm 0.17$ MeV while the original value was larger by about 50% [106].

These data are an important proof that double-Lambda hypernuclei are indeed highly useful for putting firm constraints on the $\Lambda\Lambda$ $^1S_0$ scattering length. Theoretical studies for the $^6_{\Lambda\Lambda}He$ hypernucleus have been performed by a variety of approaches such as three-body Faddeev cluster model, in Brueckner theory, or with stochastic variational methods. In order to reproduce the separation energies obtained from the Nagara event the theoretical results suggest a $\Lambda\Lambda$ scattering length $a_{\Lambda\Lambda} = -1.3 \ldots -0.5$ fm, including cluster-type descriptions [21, 108–110], calculations with the NSC interactions [111, 112], and variational results [113]. Analyses of hyperon final state interactions in strangeness production reactions also allow to estimate $a_{\Lambda\Lambda}$. A recent theoretical analysis of STAR data [114] led to $a_{\Lambda\Lambda} = -1.25 \ldots -0.56$ fm [115] (note the change of sign in order to comply with our convention).

The theoretical results on $\Delta B_{\Lambda\Lambda}$ agree within the cited uncertainty ranges which should be considered an optimistic signal indicating a basic understanding of such a complicated many-body system for the sake of the extraction of a much wanted data as $\Delta B_{\Lambda\Lambda}$.

The experimental results on the $\Lambda\Lambda$ interaction energy have initiated on the theoretical side considerable activities, see e.g. [10, 13]. In [24] recent ESC results are discussed. In the ESC model the attraction in the $\Lambda\Lambda$ channel can only be changed by modifying the scalar exchange potential. The authors argue, that if the scalar mesons are viewed as being mainly $q\bar{q}$ states, the (attractive) scalar-exchange part of the interaction in the various channels satisfies $|V_{\Lambda\Lambda}| < |V_{N\Lambda}| < |V_{NN}|$, implying indeed a rather weak $\Lambda\Lambda$ potential. The ESC fits to the NY scattering data give values for the scalar-meson mixing angle, which seem to point to almost ideal mixing for the scalar mesons. This is also found for the former Nijmegen OBE models NSC89/NSC97. In these models an increased attraction in the $\Lambda\Lambda$ channel, however, gives rise to (experimentally unobserved) deeply bound states in the $N\Lambda$ channel. In the ESC08c model, however, the apparently required $\Lambda\Lambda$ attraction is obtained without giving rise to unphysical $N\Lambda$ bound states.

5.4.4 Hyperon Interactions and Hypernuclei by Effective Field Theory

As mentioned afore, a promising and successful approach to nuclear forces is chiral effective field theory which describes the few available NY scattering data quite well, see Fig. 5.5 and Ref. [32]. Already rather early the Munich and the Jülich groups have applied χ EFT also to hypernuclei. The early applications as in [116] were based on the leading order (LO) and next-to-leading order (NLO) diagrams shown in Fig. 5.17.

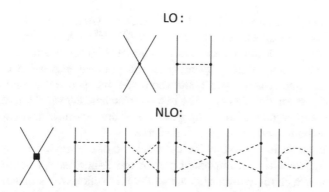

Fig. 5.17 Leading order (LO) and next-to-leading order (NLO) diagrams used in the χ EFT descriptions of hypermatter and hypernuclei. The full lines indicate either a nucleon or a hyperon, preferentially a Λ-hyperon. The LO contact interaction accounts for unresolved short range interactions. NLO interactions by pions are indicated by dashed lines

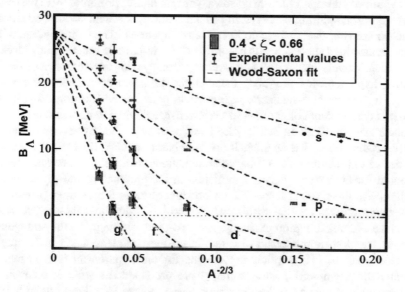

Fig. 5.18 χ EFT results for Λ separation energies over the hypernuclear mass table shown as a function of $A^{-2/3}$. The shaded areas indicate the theoretical uncertainties for the range of hyperon vertex scaling parameters ζ as indicated in the box. Also shown are results (dashed lines) obtained in calculations where the relativistic mean-field self-energies were fitted by potentials with Wood-Saxon form factors. The particle threshold is indicated by a dotted line (from Ref. [116])

The results obtained by Finelli et al. in [116] are in fact quite close to the OBE-oriented approaches of the covariant DDRH-theory and the non-relativistic ESC-model. The covariant FKVW density functional was used, incorporating SU(3) flavor symmetry, supplemented by constraints from QCD sum rules serving to estimate the scalar and vector coupling constants. In Fig. 5.18 the single-Λ

separation energies are shown. The LO and NLO nucleon-hyperon interactions were derived by fits to the spectra of $^{13}_\Lambda C$, $^{16}_\Lambda O$, $^{40}_\Lambda Ca$, $^{89}_\Lambda Y$, $^{139}_\Lambda La$, and $^{208}_\Lambda Pb$. As discussed above, a scaling description was used to adjust the hyperon interaction vertices. The Λ-nuclear surface term, appearing in the gradient expansion of a density functional for finite systems, was generated model-independently from in-medium chiral SU(3) perturbation theory at the two-pion exchange level. The authors found that term to be important in obtaining good overall agreement with Λ single particle spectra throughout the hypernuclear mass table.

It is quite interesting to follow their explanation of the small spin-orbit splitting seen in Λ hypernuclei. An important part of the Λ-nuclear spin-orbit force was obtained from the chiral two-pion exchange ΛN interaction which in the presence of the nuclear core generates a (genuinely non-relativistic, model-independent) contribution. This longer range contribution counterbalances the short-distance spin-orbit terms that emerge from scalar and vector mean fields, in exactly such a way that the resulting spin-orbit splitting of Λ single particle orbits is extremely small. A three-body spin-orbit term of Fujita-Miyazawa type that figures prominently in the overall large spin-orbit splitting observed in ordinary nuclei, is absent for a Λ attached to a nuclear core because there is no Fermi sea of hyperons. The confrontation of that highly constrained approach with empirical Λ single-particle spectroscopy turns out to be quantitatively successful, at a level of accuracy comparable to that of the best existing hypernuclear many-body calculations discussed before. Also the χ EFT approach predicts a Λ-nuclear single-particle potential with a dominant Hartree term of a central depth of about -30 MeV, consistent with phenomenology.

Since then, the work on SU(3)-χ EFT was been intensified by several groups and in several directions. The Munich-Jülich collaboration [55, 117–119], for example, has derived in-medium baryon-baryon interactions. A density-dependent effective potential for the baryon–baryon interaction in the presence of the (hyper)nuclear medium has been constructed. That work incorporates the leading (irreducible) three-baryon forces derived within SU(3) chiral effective field theory, accounting for contact terms, one-pion exchange and two-pion exchange. In the strangeness-zero sector the known result for the in-medium nucleon–nucleon interaction are recovered. In [55] explicit expressions for the hyperon-nucleon in-medium potential in (asymmetric) nuclear matter are presented. In order to estimate the low-energy constants of the leading three-baryon forces also the decuplet baryons were introduced as explicit degrees of freedom. That allowed to construct the relevant terms in the minimal non-relativistic Lagrangian and the constants could be estimated through decuplet saturation. Utilizing this approximation numerical results for three-body force effects in symmetric nuclear matter and pure neutron matter were provided. Interestingly, a moderate repulsion is found increasing with density. The latter effect is going in the direction of the much wanted repulsion expected to solve the hyperonization puzzle in neutron star matter.

A different aspect of hypernuclear physics is considered by the Darmstadt group of Roth and collaborators. In [119] light finite hypernuclei are investigated by no core shell model (NCSM) methods. In that paper, ab initio calculations for p-shell hypernuclei were presented including for the first time hyperon-nucleon-nucleon (YNN) contributions induced by a similarity renormalization group (SRG) transformation of the initial hyperon-nucleon interaction. The transformation including the YNN terms conserves the spectrum of the Hamiltonian while drastically improving model-space convergence of the importance-truncated no-core shell model. In that way a precise extraction of binding and excitation energies was achieved. Results using a hyperon-nucleon interaction at leading order in chiral effective field theory for lower- to mid-p-shell hypernuclei showed a good reproduction of experimental excitation energies but hyperon separation energies are typically overestimated as seen in Fig. 5.19. The induced YNN contributions are strongly repulsive, explained by a decoupling of the Σ hyperons from the hypernuclear system corresponding to a suppression of the $\Lambda - \Sigma$ conversion terms in the Hamiltonian. Thus, a highly

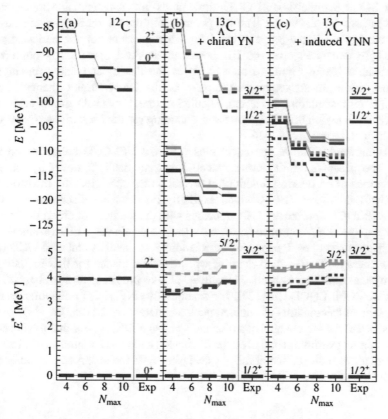

Fig. 5.19 Absolute and excitation energies of $^{13}_{\Lambda}C$. The convergence properties of the calculated energies on the number of harmonic oscillator basis states is displayed, denoted by the number of principal oscillator shells N_{max}. (**a**) Nucleonic parent absolute and excitation energies, (**b**) hypernucleus with bare (dashed line) and SRG-evolved (solid line) YN interaction, (**c**) hypernucleus with added YNN terms for cutoffs 700 MeV/c (solid line) and 600 MeV/c (dotted line). Energies are determined with respect to the corresponding ground states (from Ref. [119])

interesting link to the so-called hyperonization puzzle in neutron star physics is found which provides a basic mechanism for the explanation of strong ΛNN three-baryon forces.

5.4.5 Brief Overview on LQCD Activities

On the QCD-side the lattice groups in Japan (HALQCD) and the Seattle-Barcelona (NPLQCD) collaborations are making strong progress in computing baryon-baryon interactions numerically. The HALQCD method [120] relies on recasting the lattice results into a Schroedinger-type wave equation by which binding and scattering observables of baryonic systems are extracted. Baryon-baryon interactions in three-flavor SU(3) symmetric full QCD simulations are investigated with degenerate quark masses for all flavors. The BB potentials in the orbital S-wave are extracted from the Nambu-Bethe-Salpeter wave functions measured on the lattice. A strong flavor-spin dependence of the BB potentials at short distances is observed, in particular, a strong repulsive core exists in the flavor-octet and spin-singlet 8_s channel, while an attractive core appears in the flavor singlet channel, i.e. the 1 SU(3) representation. In recent calculation, the HALQCD group achieved to approach the region of physical masses, obtaining results for various NN, YN, and YY channels, see e.g. [121–123].

A somewhat different approach is used by the NPLQCD collaboration [124–127]. The effects of a finite lattice spacing is systematically removed by combining calculations of correlation functions at several lattice spacings with the low-energy effective field theory (EFT) which explicitly includes the discretization effects. Thus, NPLQCD combines LQCD methods with the methods of chiral EFT which a particular appealing approach because it allows to match the χ EFT results obtained from hadronic studies. Performing calculations specifically to match LQCD results to low-energy effective field theories will provide a means for first predictions at the physical quark mass limit. This allows also to predict quantities beyond those calculated with LQCD. In [127], for example, the NPLQCD collaboration report the results of calculations of nucleon-nucleon interactions in the $^3S_1 - ^3D_1$ coupled channels and the $^1S_0$ channel at a pion mass $m_\pi = 450\,\text{MeV}$. For that pion mass, the n-p system is overbound and even the di-neutron becomes a bound states. However, extrapolations indicate that at the physical pion mass the observed properties of the two-nucleon systems will be approached.

5.4.6 Infinite Hypermatter

Calculations in infinite matter are simplified because of translational invariance. By that reason, the baryons are in plane wave states and the meson mean-fields become independent of spatial coordinates. Under these conditions the field equations reduce to algebraic equations and many observables can be evaluated in closed form. The total baryon number density becomes

$$\rho_B = \sum_B tr_s \int \frac{d^3k}{(2\pi)^3} n_{sB}(k, k_{F_B}) \quad . \tag{5.75}$$

where the trace is to be evaluated with respect to spin s. In cold spin saturated matter the occupation numbers n_{sB} are independent of s and are given by $n_B = \Theta(k_{F_B}^2 - k^2)$ resulting in

$$\rho_B = \frac{N_s}{3\pi^2} k_{F_B}^3 \tag{5.76}$$

and $N_s = 2$ is the spin multiplicity for a spin-1/2 particle. Frequently we use $\rho_B = \xi_B \rho_B$ where the fractional baryon numbers $\xi_B = \rho_B/\rho_B$ add up to unity. The scalar densities are defined by

$$\rho_B^{(s)} = N_s \int \frac{d^3k}{(2\pi)^3} \frac{M_B^*}{E_B^*(k)} \tag{5.77}$$

where $E_B^*(k) = \sqrt{k^2 + M_B^{*2}}$. The integral is easily evaluated in closed form and one finds $\rho_B^{(s)} = \rho_B f_s(z_B)$ where

$$f_s(z) = \frac{3}{2z^3} \left(z\sqrt{1 + z^2} - \log \left(z + \sqrt{1 + z^2}\right) \right) \quad . \tag{5.78}$$

is a positive transcendental function with $f_s \leq 1$ depending on $z_B = k_{F_B}/M_B^*$. Since M_B^* depends on $\rho_B^{(s)}$ via the scalar fields, the scalar densities are actually defined through a system of coupled algebraic equations which has to be solved iteratively. Thus, already on the mean-field level a theoretically and numerically demanding complex structure has to be handled.

In infinite matter, also the energy-momentum tensor can be evaluated explicitly in mean-field approximation. The energy density in the mean-field sector is

$$\mathcal{E}(\rho_B) = \langle T^{00} \rangle = \sum_B \frac{1}{4} \left[3E_{F_B} \rho_B + m_B^* \rho_B^{(s)} \right]$$

$$+ \frac{1}{2} \left[m_\sigma^2 \Phi_\sigma^2 + m_\delta^2 \Phi_\delta^2 + m_{\sigma'}^2 \Phi_{\sigma'}^2 + m_\omega^2 V_\omega^{02} + m_\rho^2 V_\rho^{02} + m_\phi^2 V_\phi^{02} \right] \tag{5.79}$$

where the sum runs over all baryons and their energies are weighted by the partial vector and scalar densities ρ_B and $\rho^{(s)B}$ respectively. The (classical) field energies of the condensed meson mean-fields are indicated. For completeness the field energy of the SU(3)-singlet scalar and vector mesons σ' and ϕ, respectively, are also included.

An important observable is the binding energy per particle

$$\varepsilon(\rho_{\mathcal{B}}) = \mathcal{E}(\rho_{\mathcal{B}})/\rho_{\mathcal{B}} - \sum_{B} \xi_{B} M_{B} \quad . \tag{5.80}$$

In Fig. 5.20 results of DDRH calculation in the scaling approximation for $\epsilon(\rho_{\mathcal{B}})$ are shown for (p, n, Λ)-matter. A varying fraction of Λ-hyperons is embedded into a background of symmetric (p, n)-matter. Hence, we fix $\xi_{p} = \xi_{n}$ and $\xi_{\Lambda} = 1 - 2\xi_{p}$.

The saturation properties of symmetric pure (p, n)-matter are very satisfactorily described: The saturation point is located within the experimentally allowed region at $\rho_{sat} = 0.166\,\text{fm}^{-3}$ and $\varepsilon(\rho_{sat}) = -15.95\,\text{MeV}$ with an incompressibility $K_{\infty} = 268\,\text{MeV}$ which is at the upper end of the accepted range of values. Adding Λ hyperons the binding energy first increases until a new minimum for 10% Λ-content is reached at $\rho_{min} = 0.21\,\text{fm}^{-3}$ with a binding energy of $\varepsilon(\rho_{min}) = -18\,\text{MeV}$. Increasing either ξ_{Λ} and/or the density, the binding energy approaches eventually zero, as marked by the red line In Fig. 5.20. The minimum, in fact, is located in a rather wide valley, albeit with comparatively steep slopes, thus indicating the possibility of a large variety of bound single and even multiple-Λ hypernuclei. Note, however, that the binding energy per particle considerably weakens at high densities as the Λ-fraction increases: At high values of the density and the Λ fraction finally p, n, Λ matter becomes unbound.

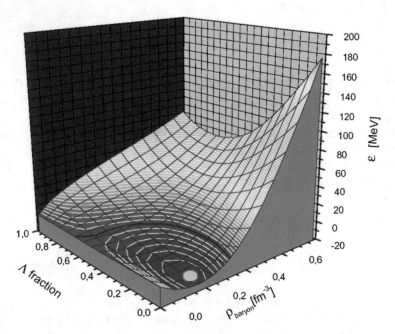

Fig. 5.20 Binding energy per baryon of (n, p, Λ) matter. The Λ fraction is defined as $\xi_{\Lambda} = \rho_{\Lambda}/\rho$ and the background medium is chosen as symmetric (p, n) matter, $\xi_{n} = \xi_{p}$. The absolute minimum is marked by a filled circle. The line $\varepsilon = 0$ is indicated by a red line

5.5   SU(3) Constraints on In-Medium Baryon Interactions

The SU(3) relations among the coupling constants of the octet baryons and the $0^-, 0^+, 1^-$ meson nonets are conventionally used as constraints at the tree-level interactions. In this section we take a different point of view. First of all, the mixing of singlet and octet mesons, to be discussed below, is considered. An interesting observation, closely connected to the mixing, is that in each interaction channel the three fundamental SU(3) constants g_D, g_F, g_S are already fixed by the NN vertices with the isoscalar and isovector octet mesons and the isoscalar singlet meson, under the provision that the octet-singlet mixing angles are known. As shown below, the mixing angles depend only on meson masses. Since the Brueckner-approach retains the meson masses, the relations fixing the mixing angles are conserved by the solutions of the Bethe-Salpeter or Lippmann-Schwinger equation, respectively. Moreover, SU(3) symmetry in general will be conserved, as far as interactions are concerned. The only substantial source of symmetry breaking is due to the use of physical masses from which one might expect SU(3) violating effects of the order of 10%. Although in low-energy baryon interactions the exchange particles are far off their mass shell, the on-shell mixing relations will persist because the BB' T-matrices are symmetry conserving.

5.5.1   Meson Octet-Singlet Mixing

In the quark model, the isoscalar mesons have flavor wave functions

$$|f_1\rangle = \frac{1}{\sqrt{3}} \left(|u\bar{u}\rangle + |d\bar{d}\rangle + |s\bar{s}\rangle \right) \tag{5.81}$$

$$|f_8\rangle = \frac{1}{\sqrt{6}} \left(|u\bar{u}\rangle + |d\bar{d}\rangle - 2|s\bar{s}\rangle \right) \tag{5.82}$$

Since they are degenerate in their spin-flavor quantum numbers the physical states $f_8 \to f$ and $f_1 \to f'$ will be superpositions of the bare states. Taking this into account, the physical mesons are written as

$$|f_m\rangle = \cos\theta_m |f_0\rangle + \sin\theta_m |f_8\rangle \tag{5.83}$$

$$|f'_m\rangle = \sin\theta_m |f_0\rangle - \cos\theta_m |f_8\rangle \tag{5.84}$$

where $f_m \in \{\eta, \omega, \sigma\}$ and $f'_m \in \{\eta', \phi, \sigma_s\}$. *Ideal mixing* is defined if f_m does not contain a $s\bar{s}$ component and f'_m is given as a pure $s\bar{s}$ configuration, requiring $\theta_{ideal} = \pi/2 - \arcsin(\sqrt{(2/3)}) \simeq 35.3°$. *Physical mixing* within a meson nonet, however, is reflected by mass relations of the Gell-Mann Okubo-type. The widely used linear mass relation [65]

$$\tan \theta_m = \frac{4m_K - m_a - 3m_{f'}}{2\sqrt{2}(m_a - m_K)} \tag{5.85}$$

leads to (η, η') mixing with $\theta_P = -24.6°$ and (ω, ϕ)-mixing with $\theta_V = +36.0°$, respectively. For the scalar nonet, the situation is less well understood, mainly because of the unclear structure of those mesons. The 0^{++} multiplets are typically strongly mixed with two- and multi-meson configurations or corresponding multi-$q\bar{q}$-configurations, leading to broad spectral distributions. For the purpose of low-energy baryon-baryon and nuclear structure physics we follow the successful strategy and identify the lowest scalar resonances as the relevant degrees of freedom. Thus, we choose the scalar octet consisting of the isoscalar $\sigma = f_0(500)$, the isovector $a_0(980)$ and the isodoublet $\kappa = K_0^*(800)$ mesons, respectively. While these mesons are observed at least as broad resonances [65], the isoscalar-singlet partner σ' of the σ-meson is essentially unknown. From the SU(3) mixing relation, however, one easily derives the instructive mass relation [65],

$$(m_f + m_{f'})(4m_K - m_a) - 3m_f m_{f'} - 8m_K^2 + 8m_K m_a - 3m_a^2 = 0 \quad , \tag{5.86}$$

serving as a constraint among the physical masses. Solving this equation for the scalar singlet meson we find the mass $m_{f'} = m_{\sigma'} = 936^{+406}_{-88}$ MeV. The large uncertainty range indicates the uncertainties of the choice of the σ- and the $\kappa = K_0^*$-masses, mentioned before. We use the mean mass values $m_\sigma = 475$ MeV and $m_\kappa = 740$ MeV, leading the scalar-singlet mass $m_{\sigma'} = 936$ MeV, which lies close to mass of the $f_0(980)$ state, in good compliance with general expectations [65]. Then, the corresponding scalar mixing angle is $\theta_S = -50.73°$. In Table 5.6 the mixing results are collected.

Meson-mixing affects directly the baryon interactions. The transformed BB-vector nonet coupling constants are displayed in Table 5.7 and corresponding relations hold for the pseudo-scalar $\{\omega, \rho, K^*, \phi\} \rightarrow \{\eta, \pi, K, \eta'\}$ and the scalar nonets, $\{\omega, \rho, K^*, \phi\} \rightarrow \{\sigma, a_0, \kappa, \sigma'\}$.

Table 5.6 Octet-singlet meson mixing used to determine the in-medium vertices

Channel	f	f'	a	$\theta(°)$
Pseudo-scalar	η	η'	π	−25.65
Vector	ω	ϕ	ρ	+36.0
Scalar	σ	σ'	a_0	−50.73

Table 5.7 SU(3) relations
for the ω, ρ and ϕ baryon
coupling constants, relevant
for the mean-field sector of
the theory

Vertex	Coupling constant
$NN\omega$	$g_{NN\omega} = g_S \cos(\theta_v) + \frac{1}{\sqrt{6}}(3g_F - g_D)\sin(\theta_v)$
$NN\phi$	$g_{NN\phi} = g_S \sin(\theta_v) - \frac{1}{\sqrt{6}}(3g_F - g_D)\cos(\theta_v)$
$NN\rho$	$g_{NN\rho} = \sqrt{2}(g_F + g_D)$
$\Lambda\Lambda\omega$	$g_{\Lambda\Lambda\omega} = g_S \cos(\theta_v) - \sqrt{\frac{2}{3}}g_D \sin(\theta_v)$
$\Lambda\Lambda\phi$	$g_{\Lambda\Lambda\phi} = g_S \sin(\theta_v) + \sqrt{\frac{2}{3}}g_D \cos(\theta_v)$
$\Sigma\Sigma\omega$	$g_{\Sigma\Sigma\omega} = g_S \cos(\theta_v) + \sqrt{\frac{2}{3}}g_D \sin(\theta_v)$
$\Sigma\Sigma\phi$	$g_{\Sigma\Sigma\phi} = g_S \sin(\theta_v) - \sqrt{\frac{2}{3}}g_D \cos(\theta_v)$
$\Sigma\Sigma\rho$	$g_{\Sigma\Sigma\rho} = \sqrt{2}g_F$
$\Lambda\Sigma\rho$	$g^\rho_{\Lambda\Sigma} = \sqrt{\frac{2}{3}}g_D$
$\Xi\Xi\omega$	$g_{\Xi\Xi\omega} = g_S \cos(\theta_v) - \frac{1}{\sqrt{6}}(3g_F + g_D)\sin(\theta_v)$
$\Xi\Xi\phi$	$g_{\Xi\Xi\phi} = g_S \sin(\theta_v) + \frac{1}{\sqrt{6}}(3g_F + g_D)\cos(\theta_v)$
$\Xi\Xi\rho$	$g_{\Xi\Xi\rho} = \sqrt{2}(g_F - g_D)$

5.5.2 SU(3) In-medium Vertices

Since we are primarily interested in mean-field dynamics we consider in this section
interactions in the vector and the scalar channels only. From DBHF theory we
have available in-medium isoscalar and isovector NN-vector and NN-scalar vertices
as density dependent functionals $\Gamma_\alpha(\rho)$, see Fig. 5.12. Thus, the SU(3) relation,
Table 5.7, lead to the set of equations for the vector sector

$$\Gamma_\omega(\rho) = g_S^{(v)} \cos(\theta_v) + \frac{1}{\sqrt{6}}\left(3g_F^{(v)} - g_D^{(v)}\right)\sin(\theta_v)$$

$$\Gamma_\phi(\rho) = g_S^{(v)} \sin(\theta_v) - \frac{1}{\sqrt{6}}\left(3g_F^{(v)} - g_D^{(v)}\right)\cos(\theta_v)$$

$$\Gamma_\rho(\rho) = \sqrt{2}(g_D^{(v)} + g_F^{(v)}) \tag{5.87}$$

and accordingly for the scalar sector,

$$\Gamma_\sigma(\rho) = g_S^{(s)}\cos(\theta_s) + \frac{1}{\sqrt{6}}\left(3g_F^{(s)} - g_D^{(s)}\right)\sin(\theta_s)$$

$$\Gamma_{\sigma'}(\rho) = g_S^{(s)}\sin(\theta_s) - \frac{1}{\sqrt{6}}\left(3g_F^{(s)} - g_D^{(s)}\right)\cos(\theta_s)$$

$$\Gamma_\delta(\rho) = \sqrt{2}(g_D^{(s)} + g_F^{(s)}) \tag{5.88}$$

For the present discussion we assume that the NNf'-singlet vertices $\Gamma_{\phi,\sigma'}$ vanish as it would be the case in the quark-model under ideal mixing conditions. Obviously, this constraint is easily relaxed and generalized scenarios with non-vanishing NNf' coupling can be investigated.

The resulting SU(3) vertices are displayed in Figs. 5.21 and 5.22 for the vector and the scalar nonets, respectively. In both the vector and the scalar channel, g_D is found to be negative and with a modulus smaller than g_F, g_S by a factor 5–10 which is in surprisingly good agreement with the general conclusion that g_D should be small. However, here we derive this result from an input of coupling constants which describe perfectly well infinite nuclear matter and nuclear properties. The vertex functionals depend only weakly on the density with variations on the 10% level over the shown density range. The resulting BB vector and scalar vertices are shown in Figs. 5.23 and 5.24. The results may be taken as a confirmation of the widely used scaling hypothesis at least for the Λ-hyperon. Up to saturation density

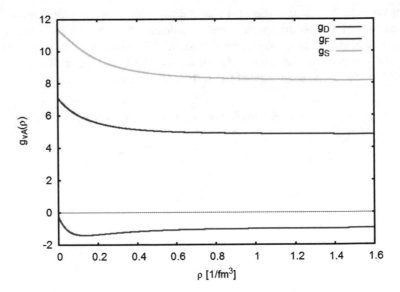

Fig. 5.21 In-medium fundamental SU(3) vector vertices g_{vA} (as indicated) versus the baryon density ρ

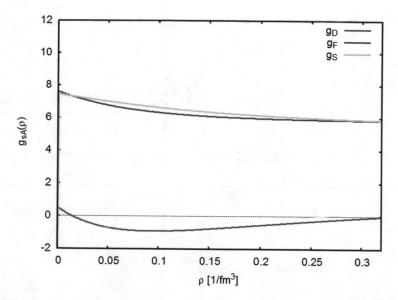

Fig. 5.22 In-medium fundamental SU(3) scalar vertices g_{sA} (as indicated) versus the baryon density ρ

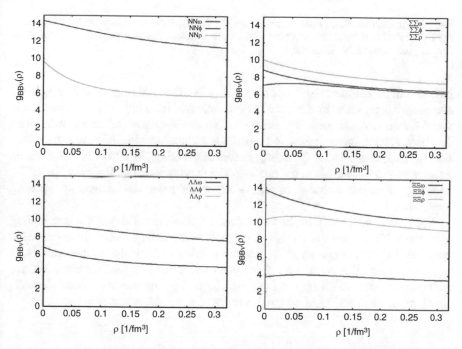

Fig. 5.23 In-medium SU(3) vector vertices. The NN and $\Lambda\Lambda$ vertices are found in the left column, the $\Sigma\Sigma$ and $\Xi\Xi$ vertices are displayed in the right column. Note that the $NN\phi$ and the $\Lambda\Lambda\rho$ coupling constants vanish identically

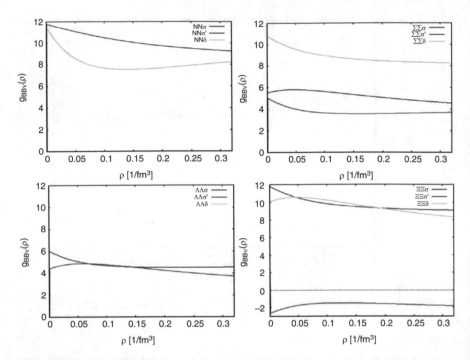

Fig. 5.24 In-medium SU(3) scalar vertices. The NN and $\Lambda\Lambda$ vertices are found in the left column, the $\Sigma\Sigma$ and $\Xi\Xi$ vertices are displayed in the right column. Note that the $NN\sigma'$ and the $\Lambda\Lambda\delta$ coupling constants vanish identically.

indeed almost constant values of about $R_{\omega,\sigma} \sim 0.5\ldots0.6$ are found which are also surprisingly close to the quark model estimate. Roughly the same situation is found for the isoscalar-octet vertices in the Σ- and Ξ-channels: The isoscalar-octet Σ-vertex scaling factors agree to a good approximation with the values found for the Λ. The isoscalar-octet Ξ scaling factors are ranging close to $0.25\cdots0.3$ which again is surprisingly close to the quark-model expectation of $\frac{1}{3}$. However, the $\Xi\Xi\sigma$ vertex involves also a change of sign which would never be obtained by a scaling hypothesis.

The isovector interactions, however, do not follow the naïve quark-model scaling hypothesis. There, one finds scaling constants of the order of unity. In hypermatter with more than a single hyperon, sizable condensed isoscalar-singlet fields will evolve to which the Λ, Σ and Ξ baryons will couple. The Ξ-interactions, for example, are dominated by the isoscalar-singlet and the isovector-octet channels which might shed new light on the dynamics of $S = -2$ hypernuclei.

5.5.3 Mean-Field Self-Energies of Octet Baryons in Infinite Nuclear Matter

An important global test of the SU(3)-constrained approach is the application to nuclear dynamics. For that purpose we consider the mean-field potentials predicted by the present approach. The leading order relativistic baryon potentials in symmetric nuclear matter are shown in Fig. 5.25. Since the isovector self-energy components are vanishing in symmetric matter the members of the nucleon, Sigma and Cascade isospin multiplets are having their respective common mean-field potentials. It is worthwhile to emphasize that without any attempt to fit the value the Λ potential acquires at saturation density a depth of about $-32\,\mathrm{MeV}$. That value is just perfectly in agreement with the afore cited s-wave Lambda separation energy of about $S_\Lambda \simeq 28\,\mathrm{MeV}$. The Σ experiences a slightly attractive potential which, however, is much too weak for the formation of a bound state.

With increasing proton-neutron asymmetry the isovector potentials gain strength and are inducing a splitting of the potentials within the iso-multiplets. The effect is most pronounced in the limiting, albeit hypothetical case of pure neutron matter. The Lambda, however, is not affected because of its isoscalar nature. In Fig. 5.25 the mean-field potentials for the three iso-multiplets are displayed. The proton and

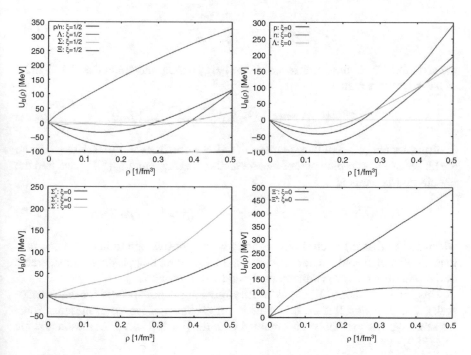

Fig. 5.25 Mean-field potentials of the octet baryons in symmetric nuclear matter (upper right) with proton fraction $\xi = \frac{Z}{A} = \frac{1}{2}$ and in pure neutron matter where $\xi = 0$

neutron potentials are showing the known behavior of a deepening of the proton and a reduction of the neutron potential depth, respectively. The three Sigma hyperons have obtained now mean-fields of a quite different depth: While the Σ^- feels a strongly repulsive interaction the Σ^+ potential has become attractive. The latter may indicate that there might be bound exotic Σ^+ nuclei. With increasing asymmetry the Cascade potentials remain throughout repulsive, indicating that it is unlikely to find exotic bound $\Xi - A$ systems or even double-Lambda hypernuclei with $S = -2$.

5.5.4 SU(3) Symmetry Breaking for Lambda Hyperons

There is an appreciable electro-weak mixing between the ideal isospin-pure Λ and Σ^0 states. Exact SU(3) symmetry of strong interactions predicts $g_{\Lambda\Lambda\pi^0} = 0$. This effect was investigated, in fact, by Dalitz and von Hippel [128] already in the early days of strangeness physics. They derived the effective coupling constant

$$g_{\Lambda\Lambda\pi} = c_b g_{\Lambda\Sigma\pi}, \tag{5.89}$$

with the symmetry breaking coefficient

$$c_b = -2\frac{\langle \Sigma^0|\delta M|\Lambda\rangle}{M_{\Sigma^0} - M_\Lambda}, \tag{5.90}$$

given by the Σ^0-Λ mass difference in the energy denominator and the $\Sigma\Lambda$ element of the octet mass matrix,

$$\langle \Sigma^0|\delta M|\Lambda\rangle = \left[M_{\Sigma^0} - M_{\Sigma^+} + M_p - M_n\right]/\sqrt{3}. \tag{5.91}$$

Substituting the physical baryon masses [65], we obtain $c_b = -0.02699\ldots$. From the nucleon-nucleon-pion part of the interaction Lagrangian, Eq. (5.17), we find the isospin matrix element

$$(\overline{N}\boldsymbol{\tau}N)\cdot\boldsymbol{\pi} = \overline{p}p\pi^0 - \overline{n}n\pi^0 + \sqrt{2}\left(\overline{p}n\pi^+ + \overline{n}p\pi^-\right), \tag{5.92}$$

which shows that the neutral pion couples with opposite sign to neutrons and protons. This implies that the non-zero induced $g_{\Lambda\Lambda\pi^0}$ coupling produces considerable deviations from charge symmetry in Λp and Λn interactions.

This kind of induced SU(3) symmetry breaking is not specific for the pseudo-scalar meson sector. Rather, it is generic for all kinds of isovector mesons. Thus, the same type of mechanism is obtained with the (neutral) ρ vector meson and the $\delta/a_0(980)$ meson:

$$g_{\Lambda\Lambda\rho} = c_b \, g_{\Lambda\Sigma\rho} \quad , \quad g_{\Lambda\Lambda\delta} = c_b \, g_{\Lambda\Sigma\delta} \tag{5.93}$$

and generalizing the above pion-nucleon relation for arbitrary isovector mesons $\mathbf{a} = \rho, \delta \ldots$

$$(\overline{N}\tau N)\cdot\mathbf{a} = \overline{p}pa^0 - \overline{n}na^0 + \sqrt{2}\left(\overline{p}na^+ + \overline{n}pa^-\right), \tag{5.94}$$

confirming the well known fact that isovector mesons couple to the baryonic isovector currents. The symmetry breaking potentials are found as given in terms of the condensed isovector meson fields $\phi_{\delta,\rho}$

$$U_{\Lambda SB}^{(s)} = g_{\Lambda\Lambda\delta}\phi_\delta \tag{5.95}$$

$$U_{\Lambda SB}^{(v)} = g_{\Lambda\Lambda\rho}\phi_\rho \tag{5.96}$$

and SU(3) symmetry breaking resides fully in the coupling constants. In Hartree-approximation, the isovector fields are directly proportional to the differences of vector and scalar proton and neutron densities:

$$\phi_\rho = g_{NN\rho}\frac{1}{m_\rho^2}\left(\rho_p^{(v)} - \rho_n^{(v)}\right) \tag{5.97}$$

$$\phi_\delta = g_{NN\delta}\frac{1}{m_\delta^2}\left(\rho_p^{(s)} - \rho_n^{(s)}\right), \tag{5.98}$$

as anticipated by the isovector vertex structure, Eq. (5.94). The field ϕ_ρ is the static time-like component of the full rho-meson vector field.

The mean-fields of Eq. (5.97) are also defining the SU(3) *symmetry conserving* $\Lambda\Sigma$ mixing potentials

$$U_{\Lambda\Sigma}^{(s)}(\rho) = g_{\Lambda\Sigma\delta}\phi_\delta \tag{5.99}$$

$$U_{\Lambda\Sigma}^{(v)}(\rho) = g_{\Lambda\Sigma\rho}\phi_\rho, \tag{5.100}$$

which allow to express Eqs. (5.95) and (5.96) in a rather intriguing form

$$U_{\Lambda SB}^{(s)}(\rho) = c_b U_{\Lambda\Sigma}^{(s)}(\rho) \tag{5.101}$$

$$U_{\Lambda SB}^{(v)}(\rho) = c_b U_{\Lambda\Sigma}^{(V)}(\rho) \tag{5.102}$$

showing explicitly the intimate relation between the electro-weak SU(3) symmetry violation—contained in the symmetry breaking coefficient c_b—and the SU(3) symmetry conserving $\Lambda\Sigma$ mixing potentials.

In Fig. 5.26 the in-medium $g_{\Lambda\Sigma\rho}$ and $g_{\Lambda\Sigma\delta}$ vertices are shown together with the resulting $\Lambda\Sigma^0$ mixing potentials. At low density, both potentials are of similar strength but the vector potential starts to dominate at higher densities. The SU(3)-symmetry breaking effective scalar and vector potentials are displayed in Fig. 5.27. As an example, results are displayed for asymmetric nuclear matter with $Z/A = 0.4$,

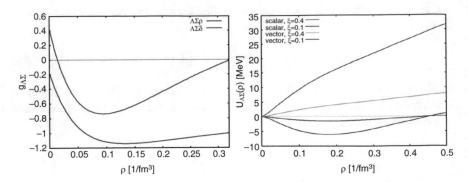

Fig. 5.26 In-medium $\Lambda\Sigma^0$ mixing and SU(3) symmetry breaking. The (symmetry conserving) $\Lambda\Sigma^0$ interaction vertices resulting from the isovector-vector and isovector-scalar interactions are shown in left panel. The $\Lambda\Sigma^0$ mixing scalar (Eq. (5.99), lower curves) and vector (Eq. (5.100), upper corves) potentials are shown in the right panel for asymmetric nuclear matter with $\xi = Z/A = 0.4$ and $\xi = Z/A = 0.1$, respectively

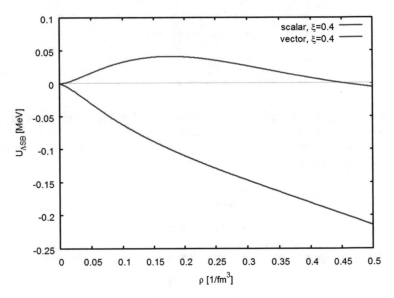

Fig. 5.27 In-medium SU(3) symmetry breaking scalar, Eq. (5.101), and vector potentials, Eq. (5.102), of the Lambda hyperon are shown for asymmetric nuclear matter with $\xi = Z/A = 0.4$

as e.g. found in $^{208}_{\Lambda}Pb$ or also $^{6}_{\Lambda}He$. The symmetry breaking potentials amount, in fact, to contributions of a few hundred keV only. Even in pure neutron matter the magnitudes range well below 1 MeV at saturation density and also at higher densities never exceed a few MeV. However, for high precision spectroscopic investigations the effect must be taken into account.

5.5.5 $\Lambda\Sigma^0$ Mixing in Asymmetric Nuclear Matter

As discussed in the previous section, SU(3) symmetry breaking and SU(3) mixing are intimately connected as stated by Eqs. (5.101) and (5.102), respectively. The SU(3) symmetry conserving $\Lambda\Sigma^0$ mixing by isovector mesons is included, of course, on the level of two-body NY interactions when solving the Bethe-Goldstone equations. This leads to the coupling of $N\Lambda$ and $N\Sigma^0$ channels. A new phenomenon, however, is encountered in asymmetric nuclear matter. The non-vanishing isovector rho- and delta-meson mean-fields are inducing a superposition of Λ and Σ^0 states with respect to the nucleonic core, such that the two hyperon single particle states are becoming mixtures of the free-space hyperon states, i.e.

$$|\tilde{\Lambda}\rangle = \cos\alpha|\Lambda\rangle + \sin\alpha|\Sigma^0\rangle$$
$$|\tilde{\Sigma}^0\rangle = -\sin\alpha|\Lambda\rangle + \cos\alpha|\Sigma^0\rangle. \tag{5.103}$$

The total isospin of the hyperon-nuclear compound system is of course conserved by virtue of the background medium. The coupling is determined by the effective mixing self-energy

$$U_m(\rho) = \gamma^0 U_{ASB}^{(s)}(\rho) + U_{ASB}^{(v)}(\rho) \tag{5.104}$$

and a two-by-two coupled system of equations has to be solved. Without going into the details, here we limit the discussion to two aspects. First, we note that the (density dependent) mixing angle is given in leading order by

$$\tan\alpha(\rho) = \frac{\langle\Lambda|U_m(\rho)|\Sigma^0\rangle}{M_{\Sigma^0-M_\Lambda}} \tag{5.105}$$

where we neglected certain higher order terms from differences of the diagonal Λ and Σ^0 mean-field Hamiltonians.

Secondly, we remark that the diagonal Λ and Σ^0 Hamiltonians should include also weak interaction effects leading to the decay of the octet hyperons. Usually, they are neglected because of their smallness against the strong interaction self-energies. In a mixing situation, however, they should be included. The dispersive interactions may be accounted for by a purely anti-hermitian weak interaction self-energies

$$2\Sigma_{\Lambda,\Sigma}^{(w)} = -i\Gamma_{\Lambda,\Sigma} = -i\frac{1}{\tau_{\Lambda,\Sigma}}, \tag{5.106}$$

where we have denoted the free space lifetimes by $\tau_{\Lambda,\Sigma}$ and the corresponding spectral widths by $\Gamma_{\Lambda,\Sigma}$.

A particular interesting aspect is that the mixing will modify the lifetime of the Λ hyperon. The diagonalization leads to the effective in-medium Λ width

$$\Gamma_\Lambda^*(\rho) = \Gamma_\Lambda \left(1 + \frac{\Gamma_\Sigma}{\Gamma_\Lambda} \tan^2 \alpha(\rho)\right) \tag{5.107}$$

which is density dependent because of the in-medium mixing angle $\alpha(\rho)$. This width is giving rise to a *reduced* in-medium lifetime of the Lambda hyperon

$$\tau_\Lambda^* \simeq \tau_\Lambda \frac{1}{1 + \frac{\tau_\Lambda}{\tau_\Sigma} \tan^2 \alpha}. \tag{5.108}$$

In Fig. 5.28 the lifetime τ_Λ^* is shown together with the mixing angle α. Already small admixtures are changing the Λ lifetime drastically. Assuming that this mechanism is the source for the lifetime reduction observed for the hyper-triton, admixtures of $\alpha \simeq 4 \times 10^{-6}$ rad will be sufficient to produce a lifetime of $\tau_\Lambda^*(^3H) \simeq 180$ ps.

5.6   Theory of Baryon Resonances in Nuclear Matter

5.6.1   Decuplet Baryons as Dynamically Generated, Composite States

Almost all of the decuplet baryons, Fig. 5.1 and Table 5.8, are decaying by strong interactions to octet states [65], thus giving them lifetimes of the order of $t_{\frac{1}{2}} \sim 10^{-23}$ s. An exception is the $S = -3$ Ω^- state with its seminal [sss] valence quark structure. The Ω baryon decays by weak interaction with a probability of $\sim 68\%$ into the ΛK^- channel with $t_{\frac{1}{2}} \sim 0.8 \times 10^{-10}$ s. In Table 5.8 masses, lifetimes, and valence quark configurations of the decuplet baryons are listed.

The large decay widths of the decuplet baryons indicate a strong coupling to the final meson-nucleon decay channels, thus pointing to wave functions with a considerable amount of virtual meson-nucleon admixtures. However, as discussed below, there is theoretical evidence that the amount of mixing varies over the multiplet with the tendency to decrease with increasing mass. At present, QCD-inspired effective models are still highly useful approaches to understand baryons at least until LQCD [129] and functional methods, e.g. [130], will be able to treating decay channels quantitatively. The coupling to meson-baryon configurations has been exploited in a number of theoretical investigations, among others especially by the Valencia group. Aceti and Oset [131, 132] are describing in their chiral unitary formalism the decuplet states and hadronic states above the ground state octets as dynamically generated, composite states in terms of meson-baryon or meson-meson scattering configurations. They apply an extension of the Weinberg compositeness condition on partial waves of $L = 1$ and resonant states to determine the weight of

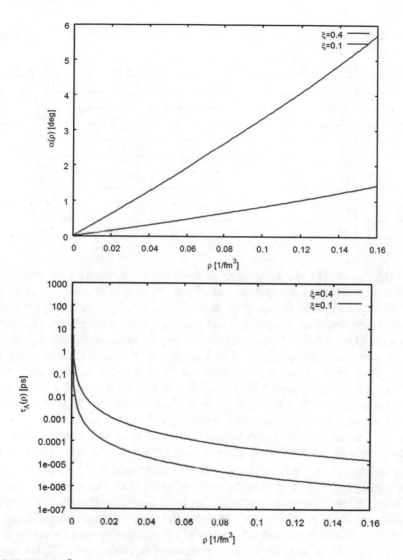

Fig. 5.28 The $\Lambda\Sigma^0$ mixing angle α, Eq. (5.105), and the effective in-medium lifetime τ_Λ^* of the mixed in-medium Λ-like state are shown as functions of the nuclear matter density. Note that the lifetime is given in picoseconds

the meson-baryon component in the $\Delta(1232)$ resonance and the other members of the $J^P = \frac{3}{2}^+$ baryon decuplet.

The calculations predict an appreciable πN fraction in the $\Delta(1232)$ wave function, as large as 60%. At first sight this is a surprising result which, however, looks more acceptable when one recalls that experiments on deep inelastic and Drell-Yan reactions are indicating that already the nucleon contains admixtures of virtual below-threshold pion-like $u\bar{u}N$ and $d\bar{d}N$ components on a level of up to

Table 5.8 Mass, width, lifetime, and valence quark configuration of the $J^P = \frac{3}{2}^+$ decuplet baryons (taken from Ref. [65])

State	Mass (MeV)	Width (MeV)	Lifetime (s)	Configuration
Δ^{++}	1230	120	10^{-23}	[uuu]
Δ^+	1232	120	10^{-23}	[uud]
Δ^0	1234	120	10^{-23}	[udd]
Δ^-	1237	120	10^{-23}	[ddd]
Σ^{*+}	1385	100	10^{-23}	[uus]
Σ^{*0}	1385	100	10^{-23}	[uds]
Σ^{*-}	1385	100	10^{-23}	[dds]
Ξ^{*0}	1530	50	10^{-23}	[uss]
Ξ^{*-}	1530	50	10^{-23}	[dss]
Ω^-	1672	9.9	0.8×10^{-10}	[sss]

30% [133, 134]. The wave functions of the larger mass decuplet baryons contain smaller meson-baryon components, steadily decreasing with mass. Thus, the Σ^*, Ξ^* and especially the Ω^- baryons acquire wave functions in which the meson-baryon components are suppressed and genuine QCD-like configurations start to dominate. Thus, a rather diverse picture is emerging from those studies, indicating the necessity for case-by-case studies, assigning a large pion-nucleon component to the $\Delta(1232)$ but leading to different conclusions about the decuplet baryons with non-vanishing strangeness. These differences have a natural explanation by considering particle thresholds: $S = -1$ baryons should couple preferentially to the $\bar{K}N$ channel but that threshold is much higher than the pion-nucleon one. The $S = -2$ baryons would couple preferentially to $\bar{K}\Lambda$ or $\bar{K}\Sigma$ channels with even higher thresholds and so on. The Aceti-Oset approach was further extended by investigating the formation of resonances by interactions of $\frac{3}{2}^+$ decuplet baryons with pseudo-scalar mesons from the lowest 0^- octet [135] and vector mesons from the lowest 1^- octet [136], respectively, thus investigating even higher resonances.

The coupling to meson-baryon channels will also affect states below the particle emission threshold by virtual admixtures of the meson-baryon continuum. Those effects are found not only for the afore mentioned $\Lambda(1405)$ state [137] but also the $\Lambda(1520)$ [138] resonances. A compelling insight from those and similar studies is that the baryons above the lowest $\frac{1}{2}^+$ octet have much richer structure than expected from a pure quark model with valence quarks only. The same features, by the way, are also found in mesonic systems. The best studied case is probably $\rho(770)$ $J^P = 1^-$ vector meson which is known to be a pronounced $\pi\pi$ p-wave resonance [65]. Also the other members of the 1^- vector meson octet contain strong substructures given by p-wave resonances of mesons from the 0^- pseudo-scalar octet. For example, in [139] the Aceti-Oset approach was used to investigate the $K\pi$-component of the $K^*(800)$ vector meson. Prominent examples are also the scalar mesons. All members of the 0^+ meson octet are dominated by meson-scattering configurations of the 0^- multiplet, as discussed in the previous sections.

Besides spectral studies there is a general interest in meson-baryon interactions as an attempt to generalize the work from NN- and YN-interaction to higher lying multiplets. The chiral SU(3) quark cluster model was used in [140] to derive interactions among decuplet baryons, neglecting, however, the coupling to the decay channels. In the framework of the resonating-group method, the interactions of decuplet baryon-baryon systems with strangeness $S = -1$ and $S = -5$ were investigated within the chiral SU(3) quark model. The effective baryon-baryon interactions deduced from quark-quark interactions and scattering cross sections of the $\Sigma^* \Delta$ and $\Xi^* \Omega$ systems were calculated. The so restricted study led to rather strongly attractive decuplet interaction, producing deeply bound $\Sigma \Delta$ and $\Xi \Omega$ dibaryons with large binding energies exceeding that of the deuteron by at least an order of magnitude. These results resemble the deeply bound $S = -2$ H-dibaryon predicted by Jaffe [141]. Here, we are less ambitious and consider mainly interactions of the Δ baryon and few other resonances in nuclear matter.

5.6.2 The $N^* N^{-1}$ Resonance Nucleon-Hole Model

The Delta resonance is taken here as a representative example but the results can be generalized essentially unchanged also to other resonances N^* after the proper adjustments of vertices and propagators as required by spin, isospin, and parity. The creation of a resonances in a nucleus amounts to transform a nucleon into an excited intrinsic states, Thus, the nucleon is removed from the pre-existing Fermi-sea, leaving the target in a $N^* N^{-1}$ configuration. That state is not an eigenstate of the many-body system but starts to interact with the background medium through residual interactions V_{NN^*}. The appropriate theoretical frame work for that process is given by the polarization propagator formalism [92], also underlying, for example, the approaches in [142–144].

In brief, the Delta-hole approach consists of calculating simultaneously the pion self-energies and effective vertices by the coupling to the ΔN^{-1} excitations of the nuclear medium. These requirements are illustrated in Fig. 5.29 where the diagrams representing the approach are shown. As seen in that figure, the Dyson equations for the propagation of pions and Δ's in nuclear matter have to be solved self-consistently. The ultraviolet divergences of the loop integrals are regularized by using properly defined hadronic vertex functions.

An elegant and transparent formulation of the ΔN^{-1} problem is obtained by the polarization propagator method [92]. We consider first the Green's function of the interacting system. Here, we limit the investigations to the coupling of NN^{-1} and ΔN^{-1} modes. For the non-interacting system the propagator is given by a diagonal matrix of block structure

$$\mathcal{G}^{(0)}(\omega, \mathbf{q}) = \begin{pmatrix} G_{NN^{-1}}^{(0)}(\omega, \mathbf{q}) & 0 \\ 0 & G_{\Delta N^{-1}}^{(0)}(\omega, \mathbf{q}) \end{pmatrix} \tag{5.109}$$

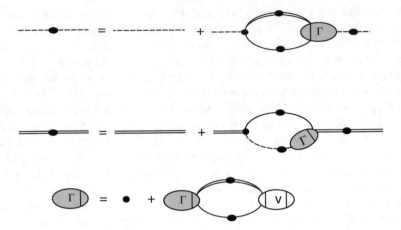

Fig. 5.29 Diagrams entering into the self-consistent description of the dressed pion propagators (upper row), the Δ propagator (second row), and the resulting dressed vertex (third row). Nucleon propagators are given by lines, the Δ propagator is shown as a double-line. Bare vertices are indicated by filled circles, the dressed vertices are denoted by Γ. V is the ΔN^{-1} residual interaction. Note that the Δ-resonance obtains a self-energy due to its decay into intermediate πN configurations

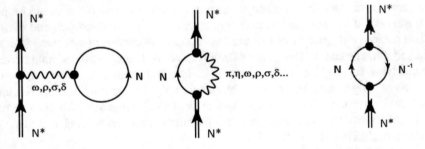

Fig. 5.30 In-medium interactions of a baryon resonance N^* via the static mean-field (left) and the dispersive polarization self-energies (center) indicated here by the decay into intermediate nucleon-meson configurations. Moreover, in nuclear matter the coupling to NN^{-1} excitations contributes a spreading width (right)

where $G^{(0)}_{NN^{-1}}$ and $G^{(0)}_{\Delta N^{-1}}$ describe the unperturbed propagation of two-quasiparticle NN^{-1} and ΔN^{-1} states through the nuclear medium, including, however, their self-energies Σ_{N,N^*} of Fig. 5.30.

These propagators are given by the Lindhard functions [92]

$$\phi_N(k) = i \int \frac{d^4p}{(2\pi)^4} G_N(p) G_N(p+k) , \qquad (5.110)$$

$$\phi_\Delta(\pm k) = i \int \frac{d^4p}{(2\pi)^4} G_N(p) G_\Delta(p \pm k) . \qquad (5.111)$$

Here, p denotes the 4-momentum and $G_N(p)$ and $G_\Delta(p)$ are the nucleon and Δ propagators :

$$G_N(p) = \frac{1}{p^0 - \varepsilon(\mathbf{p}) - \Sigma_N(p) + i0} + 2\pi i n(\mathbf{p}) \delta(p^0 - \varepsilon(\mathbf{p}) - \Sigma_N(p^2)) ,$$

$$(5.112)$$

$$G_\Delta^{\mu\nu}(p) = \frac{1}{p^0 - \varepsilon_\Delta(\mathbf{p}) - \Sigma_\Delta(p^2)} \delta^{\mu\nu} ,$$

$$(5.113)$$

with the single particle (reduced kinetic) energies $\varepsilon_{N,\Delta}$, and $n(\mathbf{p})$ is the nucleon occupation number. We follow the general practice and approximate the Delta-propagator by the leading order term resembling a spin-$\frac{1}{2}$ Green's function, thus leaving away the complexities of a Rarita-Schwinger propagator [145]. This is an acceptable approximation in the low-energy limit used below where the neglected terms will be suppressed, anyway. Since Σ_Δ is generically of non-hermitian character, we can omit the infinitesimal shift into the complex plane.

In the non-relativistic limit of cold infinite matter with nucleons filling up the Fermi sea up to the Fermi momentum k_F, we have $n(\mathbf{p}) = \theta(k_F^2 - \mathbf{p}^2)$. The nucleon propagator Eq. (5.112) consists of the vacuum part and the in-medium part ($\propto n(\mathbf{p})$). The Δ propagator Eq. (5.113) includes the vacuum part only, since we have neglected the presence of Δ excitations in nuclear matter. Both propagators take into account effective mass corrections if present.

After the contour integration over p^0 (c.f. [92]) the nucleon-hole Lindhard function, Eq. (5.110), takes the following form :

$$\phi_N(k) = -\int \frac{d^3p}{(2\pi)^3} \left(\frac{n(\mathbf{p}+\mathbf{k})(1-n(\mathbf{p}))}{\varepsilon^*(\mathbf{p}+\mathbf{k}) - k^0 - \varepsilon^*(\mathbf{p}) + i0} + \frac{n(\mathbf{p})(1-n(\mathbf{p}+\mathbf{k}))}{\varepsilon^*(\mathbf{p}) + k^0 - \varepsilon^*(\mathbf{p}+\mathbf{k}) + i0} \right) .$$

$$(5.114)$$

where for simplicity, we have introduced the quasiparticle energies $\varepsilon_N^*(\mathbf{p}) = \varepsilon_N(\mathbf{p}) + \Sigma_N(p)$. Corresponding expressions are found for the Delta-hole Lindhard function, Eq. (5.111):

$$\phi_\Delta(\pm k) = -\int \frac{d^3p}{(2\pi)^3} \frac{n(\mathbf{p})}{\varepsilon_N^*(\mathbf{p}) - \varepsilon_\Delta^*(\mathbf{p} \pm \mathbf{k}) \pm k^0} .$$

$$(5.115)$$

Replacing the dependence of the self-energies on the integration variable by a conveniently chosen external value, for example by replacing the argument by the pole value, the integration can be performed in closed form in the zero temperature limit. Analytic formulas are found in Ref. [92].

Including the residual NN^{-1} and ΔN^{-1} interactions by

$$\mathcal{V} = \begin{pmatrix} V_{NN} & V_{N\Delta} \\ V_{\Delta N} & V_{\Delta\Delta} \end{pmatrix}$$

$$(5.116)$$

the Green function of the interacting system is given by the Dyson equation for the 4-point function

$$\mathcal{G}(\omega, \mathbf{q}) = \mathcal{G}^{(0)}(\omega, \mathbf{q}) + \mathcal{G}^{(0)}(\omega, \mathbf{q})\mathcal{V}\mathcal{G}(w, \mathbf{q}) \tag{5.117}$$

truncated to the two-quasiparticle sector, i.e. evaluated in Random Phase Approximation (RPA). Actually, the approach discussed below corresponds to a projection, not a truncation, to the 4-point function because the coupling to the hierarchy of higher order propagators is taken into account effectively by induced self-energies and interactions.

The coherent response of the many-body system with ground state $|A\rangle$ to an external perturbation described by an operator $\mathcal{O}_a(\mathbf{q}) \sim e^{i\mathbf{q}\cdot\mathbf{r}}\sigma^{S_a}\tau^{T_a}$ where $a = (S, T)$ denotes spin ($S_a = 0, 1$), isospin ($T_A = 0, 1$) and momentum ($\sim e^{i\mathbf{q}\cdot\mathbf{r}}$) transfer, is described by the polarization propagators of the non-interacting system

$$\Pi_{ab}^{(0)}(\omega, \mathbf{q}) = \langle A|\mathcal{O}_b^\dagger\mathcal{G}(0)\mathcal{O}_a|A\rangle \tag{5.118}$$

and the interacting system

$$\Pi_{ab}(\omega, \mathbf{q}) = \langle A|\mathcal{O}_b^\dagger\mathcal{G}\mathcal{O}_a|A\rangle \tag{5.119}$$

which—by definition—has the same functional structure as the propagator $\mathcal{G}$. For a single resonance the polarization tensor is given by a 2-by-2 tensorial structure

$$\begin{pmatrix} \Pi_{NN} & \Pi_{N\Delta} \\ \Pi_{\Delta N} & \Pi_{\Delta\Delta} \end{pmatrix} = \begin{pmatrix} \Pi_{NN}^{(0)} & \Pi_{N\Delta}^{(0)} \\ \Pi_{\Delta N}^{(0)} & \Pi_{\Delta\Delta}^{(0)} \end{pmatrix} + \begin{pmatrix} \Pi_{NN}^{(0)} & \Pi_{N\Delta}^{(0)} \\ \Pi_{\Delta N}^{(0)} & \Pi_{\Delta\Delta}^{(0)} \end{pmatrix} \begin{pmatrix} V_{NN} & V_{N\Delta} \\ V_{\Delta N} & V_{\Delta\Delta} \end{pmatrix} \begin{pmatrix} \Pi_{NN} & \Pi_{N\Delta} \\ \Pi_{\Delta N} & \Pi_{\Delta\Delta} \end{pmatrix}$$
$$\tag{5.120}$$

where the reference to the transition operators $\mathcal{O}_{a,b}$ is implicit. The diagrammatic structure is shown in Fig. 5.31. The mixing of the NN^{-1} and the ΔN^{-1} configura-

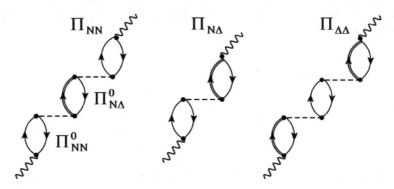

Fig. 5.31 The RPA polarization propagator. The $N^{-1}N \rightarrow N^{-1}N$ (left), the mixed $N^{-1}N \rightarrow N^{-1}\Delta$ and the $N^{-1}\Delta \rightarrow N^{-1}\Delta$ components are displayed. Also the bare particle-hole type propagators are indicated. External fields are shown by wavy lines, the residual interactions are denoted by dashed lines. Only part of the infinite RPA series is shown

tions by the residual interactions is resulting in the mixed polarization tensors $\Pi_{N\Delta}$ and $\Pi_{\Delta N}$, respectively. The polarization tensor contains the full multipole structure as supported by the nuclear system, the interaction $\mathcal{V}$, and the external transition operators $\mathcal{O}_{a,b}$. Thus, a decomposition into irreducible tensor components may be performed. Alternatively, a decomposition into longitudinal and transversal components is frequently invoked. In practice, one often evaluates the tensor elements in infinite nuclear matter and applies the result in local density approximation by mapping the particle densities and Fermi momenta to the corresponding radial-dependent quantities of a finite nucleus, e.g. $\rho_{p,n} \to \rho_{p,n}(\mathbf{r})$, see e.g. [146–148].

Once the polarization propagator is known, observables are easily calculated. Spectral distributions and response function are of particular importance because they are entering directly into cross sections. The response functions are defined by

$$R_{ab}(\omega, \mathbf{q}) = -\frac{1}{\pi} Im\left[\Pi_{ab}(\omega, q)\right] \tag{5.121}$$

The response function techniques are frequently used in lepton induced reactions like $(e, e'p)$ or neutrino-induced pion production $(\nu, \mu\pi)$. In [149, 150] similar methods have been applied in light ion-induced charge exchange reactions up to the Delta-region including also the pion-production channel. A recent application to N^* excitations in heavy ion charge exchange reaction with Sn-projectiles is found in [151, 152].

A widely used choice for $\mathcal{V}$ is a combination of pion-exchange and contact interactions of Landau-Migdal type corresponding to the afore mentioned OPEM approach, see e.g. [142–144]. The diagrammatic structure of the QPEM interactions is shown in Fig. 5.32. Inclusion of other mesons, e.g. the ρ-meson, is easily obtained. The pion exchange part takes care of the long-range interaction component. In non-

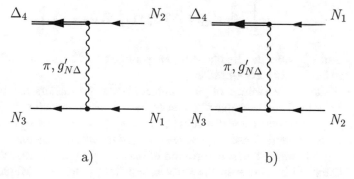

Fig. 5.32 Direct (**a**) and exchange (**b**) diagrams of the OPEM approach for the process $N_1 N_2 \to N_3 \Delta_4$. The wavy line denotes either π (and ρ) exchange or the contact interaction $\propto g'_{N\Delta}$

relativistic reduction (c.f. [153]), but using relativistic kinematics, the πNN and $\pi N\Delta$ interactions, the Lagrangians are:

$$\mathcal{L}_{\pi NN} = \frac{f}{m_\pi}\psi^\dagger \boldsymbol{\sigma}\boldsymbol{\tau}\psi \cdot \nabla\boldsymbol{\pi} \, , \tag{5.122}$$

$$\mathcal{L}_{\pi N\Delta} = \frac{f_\Delta}{m_\pi}\psi_\Delta^\dagger \mathbf{S}\mathbf{T}\psi \cdot \nabla\boldsymbol{\pi} + h.c. \, , \tag{5.123}$$

where ψ, ψ_Δ and $\boldsymbol{\pi}$ are the nucleon, Δ resonance and pion field respectively. The dot-product indicates the contraction of the spin and gradient operators. Typically values for the coupling constants are $f = 1.008$ and $f_\Delta = 2.202$, see e.g. [144, 154]. $\boldsymbol{\sigma}$ and $\boldsymbol{\tau}$ are the spin and isospin Pauli matrices. The $(1/2 \rightarrow 3/2)$ spin and isospin transition operators are given by $\mathbf{S}$ and $\mathbf{T}$, defined according to Ref. [155]. Pion-exchange is seen to be of spin-longitudinal structure. Occasionally, also ρ-meson exchange is treated explicitly, e.g. [146, 150] introducing an explicit spin-transversal interaction component into $\mathcal{V}$ which, however, can be decomposed into spin-spin and spin-longitudinal terms [156].

The short range pieces are subsumed into the contact interactions, defined by the following Lagrangian:

$$\begin{aligned}
\mathcal{L}_{SRC} = &-\frac{f^2}{2m_\pi^2}g'_{NN}(\psi^\dagger\boldsymbol{\sigma}\boldsymbol{\tau}\psi)\cdot(\psi^\dagger\boldsymbol{\sigma}\boldsymbol{\tau}\psi) \\
&-\left[\frac{f_\Delta}{m_\pi^2}g'_{N\Delta}(\psi^\dagger\boldsymbol{\sigma}\boldsymbol{\tau}\psi)\cdot(\psi_\Delta^\dagger\mathbf{S}\mathbf{T}\psi) + h.c.\right] \\
&-\frac{f_\Delta^2}{m_\pi^2}g'_{\Delta\Delta}(\psi_\Delta^\dagger\mathbf{S}\mathbf{T}\psi)\cdot(\psi^\dagger\mathbf{S}^\dagger\mathbf{T}^\dagger\psi_\Delta) \\
&-\left[\frac{f_\Delta^2}{2m_\pi^2}g'_{\Delta\Delta}(\psi_\Delta^\dagger\mathbf{S}\mathbf{T}\psi)\cdot(\psi_\Delta^\dagger\mathbf{S}\mathbf{T}\psi) + h.c.\right]
\end{aligned} \tag{5.124}$$

g'_{NN}, $g'_{N\Delta}$ and $g'_{\Delta\Delta}$ are the Landau-Migdal parameters. The spin-isospin scalar products are indicated as a dot-product.

In the literature the values of the Landau-Migdal parameters are not fixed unambiguously by theory but must be constrained on phenomenological grounds. Within a simple universality assumption $g'_{NN} = g'_{N\Delta} = g'_{\Delta\Delta} \equiv g'_{BW}$, which is the so-called Bäckmann-Weise choice (see Ref. [157] and Refs. therein), one gets $g'_{BW} = 0.7 \pm 0.1$ from the best description of the unnatural parity isovector states in ^{4}He, ^{16}O and ^{40}Ca. However, the same calculations within the Migdal model [158] assumption $g'_{N\Delta} = g'_{\Delta\Delta} = 0$ produce $g'_{NN} = 0.9\ldots1$. The description of the quenching of the Gamow-Teller matrix elements requires, on the other hand, $g'_{\Delta\Delta} = 0.6\ldots0.7$ (assuming $g'_{N\Delta} = g'_{\Delta\Delta}$) [159]. The real part of the pion optical potential in π-atoms implies $g'_{N\Delta} = 0.2$ and $g'_{\Delta\Delta} = 0.8$ [158]. The pion induced two-proton emission is best described with $g'_{N\Delta} = 0.25\ldots0.35$.

The Delta self-energies are a heavily studied subject because they are of particular importance for pion-nucleus interactions. c.f. [156, 160]. The general conclusion is that the modifications of the decay width by the medium can be expressed to a good approximation as a superposition of Pauli-blocking terms from the occupation of the Fermi-sea and an absorption or spreading term due to the coupling to NN^{-1} excitations [161, 162]

$$\Gamma_\Delta(\omega, \rho) = -2Im\Sigma_\Delta(\omega, \rho) \sim \Gamma_{free}(\omega) + \Gamma_{Pauli}(\omega, \rho) + \Gamma_{abs}(\omega, \rho) \qquad (5.125)$$

Since the self-energies are not known over the large energy and momentum regions necessary for theoretical applications, parametrizations are introduced and used for extrapolations. A frequently used parametrization of the in-medium width is

$$\Gamma_\Delta(\omega) = \Gamma_{abs}(\omega)\frac{\rho}{\rho_{sat}} + \Gamma_\Delta^0\left(\frac{q(\omega, m_N^*, m_\pi)}{q(m_\Delta, m_N, m_\pi)}\right)^3 \frac{m_\Delta^*}{\omega} \frac{\beta_0^2 + q^2(m_\Delta^*, m_N^*, m_\pi)}{\beta_0^2 + q^2(\omega, m_N^*, m_\pi)},$$
$$(5.126)$$

where β_0 is an adjustable parameter and $\rho_{sat} = 0.16\,\mathrm{fm}^{-3}$ is the nuclear saturation density. $\omega = \sqrt{s_\Delta}$ is the total relativistic energy of the Delta resonance in the medium as defined in the $\pi + N$ system. The spreading width due to the coupling to NN^{-1} modes is denoted by Γ_{abs} and a simple scaling law is used for the density dependence. The Lorentz-invariant center-of-mass momentum is defined as usual,

$$q^2(\omega, m_1, m_2) = (\omega^2 - (m_1 + m_2)^2)(\omega^2 - (m_1 - m_2)^2)/4\omega^2. \qquad (5.127)$$

The free and effective nucleon and Delta in-medium masses are denoted by $m_{N,\Delta}$ and $m_{N,\Delta}^*$, respectively. Using a (subtracted) dispersion relation the real part can be reconstructed by a Cauchy Principal Value integral

$$Re(\Sigma_\Delta(\omega)) = -\frac{\omega - m_\Delta}{\pi}P\int d\omega' \frac{\Gamma_\Delta(\omega')}{(\omega' - m_\Delta)(\omega' - \omega)}, \qquad (5.128)$$

by which the dispersive self-energy is completely determined.

5.6.3 Δ Mean-Field Dynamics

While the dispersive Delta self-energies have obtained large attention, the static mean-field part is typically neglected or treated rather schematically. Some authors use the universality assumption stating that the Delta mean-field should agree with the nucleon one. In the relativistic Hartree scheme this amounts to use the same scalar and vector coupling constants for nucleons and resonances. Obviously, that assumption comes to an end in charge-asymmetric matter by the fact that the Delta resonance comes in four charge states. A simple but meaningful extension is to introduce also an isovector potential, thus allowing for interactions of the resonance and the background medium through exchange of isovector scalar and vector mesons. In the non-relativistic limit, the Delta Hartree potential becomes a sum of isoscalar and isovector potentials U_0^Δ and U_1^Δ, respectively:

$$U_\Delta^{(H)} = U_0^\Delta + \frac{2}{A}U_1^\Delta \boldsymbol{\tau}^\Delta \cdot \boldsymbol{\tau}^N \tag{5.129}$$

where A is the nucleon number and $\boldsymbol{\tau}^\Delta = 2\mathbf{T}_\Delta$ and $\boldsymbol{\tau}^N$ denote the Delta and nucleon isospin operators, respectively, with the known properties of isospin

$$\tau_3^N|p\rangle = +|p\rangle \quad ; \quad \tau_3^N|n\rangle = -|n\rangle \tag{5.130}$$

and

$$T_3^\Delta|\Delta^{++}\rangle = +\frac{3}{2}|\Delta^{++}\rangle \quad ; T_3^\Delta|\Delta^+\rangle = +\frac{1}{2}|\Delta^+\rangle$$

$$T_3^\Delta|\Delta^0\rangle = -\frac{1}{2}|\Delta^0\rangle \quad ; T_3^\Delta|\Delta^-\rangle = -\frac{3}{2}|\Delta^-\rangle. \tag{5.131}$$

In a nucleus, the resonance is moving in background medium with Z protons and N neutrons and $A = N + Z$. Thus, integrating out the nucleons, our simple model potential becomes

$$U_\Delta^{(H)} \simeq U_0^\Delta - U_1^\Delta \tau_z^\Delta \frac{N-Z}{A}. \tag{5.132}$$

Thus, we find

$$U_{\Delta^{++}}^{(H)} = U_0^\Delta - 3U_1^\Delta \frac{N-Z}{A} \tag{5.133}$$

$$U_{\Delta^+}^{(H)} = U_0^\Delta - U_1^\Delta \frac{N-Z}{A} \tag{5.134}$$

$$U_{\Delta^0}^{(H)} = U_0^\Delta + U_1^\Delta \frac{N-Z}{A} \tag{5.135}$$

$$U_{\Delta^-}^{(H)} = U_0^\Delta + 3U_1^\Delta \frac{N-Z}{A} \tag{5.136}$$

and the universality assumption would mean to use $U_{0,1}^\Delta = U_{0,1}^N$. At the nuclear center, typical values are $U_0^N = -40\ldots-60\,\text{MeV}$ and $U_1^N = +20\ldots+30\,\text{MeV}$, depending on the chosen form factor.[1] These estimates agree quite well with those

---

[1]More meaningful values are in fact the volume integrals per nucleon.

of the involved many-body calculations in [163]. Actually, also spin-orbit potentials will contribute about which nothing is known.

As for nucleons and hyperons the RMF approach is also being used to describe the N^* mean-field dynamics. In Refs. [59–61] and also [62, 63] the Delta resonance was included into the RMF treatment. In [60] Δ dynamics in nuclear matter is described by the mean-field Lagrangian density

$$\mathcal{L}_\Delta = \overline{\psi}_{\Delta\nu} \left[i\gamma_\mu \partial^\mu - (m_\Delta - g_{\sigma\Delta}\sigma) - g_{\omega\Delta}\gamma_\mu\omega^\mu - g_{\rho\Delta}\gamma_\mu I_3 \rho_3^\mu \right] \psi_\Delta^\nu, \qquad (5.137)$$

where ψ_Δ^ν is the Rarita-Schwinger spinor for the for the full set of $\Delta(1232)$-isobars $(\Delta^{++}, \Delta^+, \Delta^0, \Delta^-)$ and $I_3 = \mathrm{diag}(3/2, 1/2, -1/2, -3/2)$ is the matrix containing the isospin charges of the Δs. Assuming SU(6) universality, the same coupling constants as for nucleons may be used. However, contributions form dispersive self-energies will surely spoil SU(6) symmetry and deviations from that rule will occur.

5.6.4 Response Functions in Local Density Approximation

An application of the response function formalism is shown in Fig. 5.33 where the spectra for the Fermi-transition operator $\mathcal{O}_a = \sigma\tau_+$ for $^{58,64,78}Ni$ are shown. The response functions are normalized to the nucleon numbers $A = 58, 64, 78$. As discussed above, the polarization tensor, Eq. (5.120), was evaluated in infinite

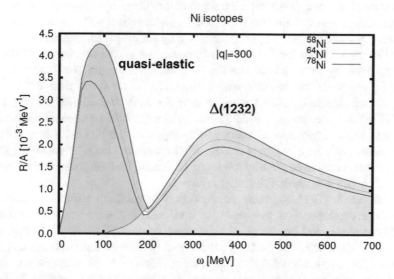

Fig. 5.33 RPA response function per nucleon for the operator $\sigma\tau_+$ in the isotopes $^{58,64,78}Ni$ [164] obtained by an energy density functional as in [94]. Note the apparent shift of the ΔN^{-1} peak to higher energy which is introduced by mixing with the NN^{-1} component due to the residual interactions

matter at a dense mesh of proton and neutron densities, thus leading to $\Pi_{BB'}(\rho_p, \rho_n)$ for $B, B' = N, \Delta$. By mapping the nucleon densities to the radial densities $\rho_{p,n}(r)$ of the Ni-isotopes, we obtain in local density approximation $\Pi_{BB'}(r)$. The response function shown in Fig. 5.33 is obtained finally by integration

$$R(\omega, q) = -\frac{1}{\pi} \frac{1}{A} \int d^3 r \rho_A(\mathbf{r}) Im \left[\Pi_{NN}(\omega, q, \mathbf{r}) + \Pi_{\Delta\Delta}(\omega, q, \mathbf{r}) \right] \qquad (5.138)$$

where $q = |\mathbf{q}|$ and $\rho_A = \rho_p + \rho_n$ is the total nuclear density. In the response functions of Fig. 5.33, the quasi-elastic NN^{-1} and the deep-inelastic ΔN^{-1} are clearly seen. The two components are mixed by the residual interactions, chosen in that calculation according to the (non-relativistic) density functional of Ref. [94]. The configuration mixing induces an upward shift in energy of the Delta-component and a downward shift in energy of the quasi-elastic nucleon component. In this respect, the system behaves in a manner as typical for a coupled two-by-two system.

There is a large body of data available on inclusive (e, e') cross sections [165–168] which are the perfect test case to the response function formalism. Since a detailed discussion of the functional structure of (e, e') cross sections is beyond the scope of the present work, we refer the reader to the ground-breaking monograph of DeForest and Walecka [169] and the more recent review article of Benhar and Sick [170]. The reactions proceed such that the incoming electron couples via virtual photon emission to the charged nuclear currents, involving electric and magnetic interactions. The cross sections are given by the superposition of response functions for operators of spin-longitudinal ($\mathcal{O}_L \sim \boldsymbol{\sigma} \cdot \mathbf{q}$), and spin-transversal ($\mathcal{O}_T \sim \boldsymbol{\sigma} \times \mathbf{q}$) structure, where $\mathbf{q}$ is the momentum transfer. These operators are defining the corresponding response functions (R_{LL}), (R_{TT}), respectively. The cross sections are obtained by weighting the response functions by the proper kinematical factors.

In the high energy limit, the electron scattering waves are approximated sufficiently well by plane waves Results of such a calculation [164] are shown in Fig. 5.34 where the double differential cross section for the inclusive $^{40}Ca(e, e')$ reaction at the electron incident energy $T_{lab} = 500$ MeV at fixed momentum transfer $q = 300$ MeV/c are shown. The data are surprisingly well described by our standard choice of self-energies and interactions although no attempt was made to optimize parameters. The energy gap between the quasi-elastic and the N^* resonance spectral components seems to be slightly too large, indicating that the configuration mixing interaction $V_{N\Delta}$ was chosen slightly too strong.

In Xia et al. [171], the close connection of in-medium pion interactions and Delta-hole excitations on the one side and nuclear charge exchange reactions and photo-absorption on the other side, were considered in detail. In that work it is emphasized that, unlike the conventional picture of level mixing and level repulsion for the pionic and ΔN^{-1} states, the real part of the pion inverse propagator vanishes at only one energy for each momentum because of the width of the Delta-hole excitations. The results of this self-consistent approach has been compared successfully to data on (p, n) charge exchange reactions and photo-absorption on nuclei in the Δ-resonance region. Moreover, the interesting result is found that the

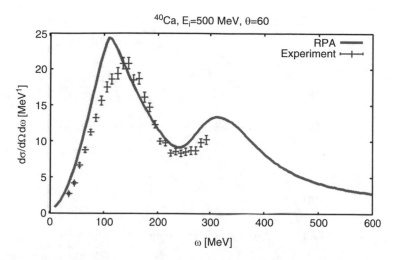

Fig. 5.34 Double differential cross section for the inclusive $^{40}Ca(e, e')$ reaction at $T_{lab} = 500$ MeV [164]. The underlying longitudinal and transversal response functions were obtained by an energy density functional as in [94]. Note the apparent shift of the ΔN^{-1} peak to higher energy which is introduced by mixing with the quasi-elastic NN^{-1} component due to the residual interactions

baryonic vertex form factors obtained for pionic and electromagnetic probes agrees with their interpretation as effective hadronic structure functions.

5.6.5 Resonances in Neutron Stars

Interestingly, neutron stars may be useful systems to study N^* mean-field dynamics [59, 60, 62] and also in [63]. In [59, 60] nucleons, hyperons and Deltas are described within the same RMF approach, used to investigate the composition of neutron star matter. In Fig. 5.35 particle fractions as a function of the baryon density $n_B = \rho_B$ of ρ_{sat} are shown. With the Δ-resonance included the particle mixtures are changed considerably. Furthermore, the onset of the Delta appearance depends on the RMF coupling constant. That effect is illustrated in Fig. 5.35 by varying $x_{\omega\Delta} = \frac{g_{\omega\Delta}}{g_{\omega N}}$.

Moreover, the investigations in [59, 60] lead to the important conclusion that the onset of Δ-isobars is strictly related to the value of the slope parameter L of the density dependence of the symmetry energy. For the accepted range of values of about $40 < L < 120$ MeV [172], the additional Delta degrees of freedom influence the appearance of hyperons and cannot be neglected in the EoS of beta-stable neutron star matter. This correlation of the Δ onset and the symmetry energy slope are indicating also another interesting interrelation between nuclear and sub-nuclear degrees of freedom. These findings are leading immediately to the question to what extent the higher N^* resonances will influence the nuclear and neutron star equations of state.

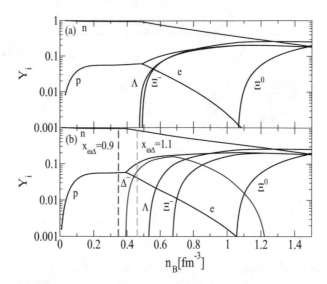

Fig. 5.35 Particle fractions as functions of the baryon density within the SFHo model: only hyperons (panel (**a**)), hyperons and Δs (panel (**b**)) for $x_{\sigma\Delta} = x_{\omega\Delta} = x_{\rho\Delta} = 1$. The red line indicates the fraction of the Δ^- which among the four Δs are the first to appear. The blue and the green vertical lines indicate the onset of the formations of Δ^- for $x_{\omega\Delta} = 0.9$ and $x_{\omega\Delta} = 1.1$, respectively (from Ref. [60])

5.7 Production and Spectroscopy of Baryon Resonances in Nuclear Matter

5.7.1 Resonances as Nuclear Matter Probes

Exploring the spectral properties of resonances and their dynamics in the nuclear medium is the genuine task of nuclear physics. In Fig. 5.36 interactions are indicated which in heavy ion reactions are leading to excitation of nucleon resonances. In fact, already in the 1970s first experiments were performed at the AGS at Brookhaven with proton-induced reaction on a series of targets between Carbon and Uranium [173]. It was recognized soon that charge exchange reactions would be an ideal tool for resonance studies. As early as 1976 high energy (p, n) reactions were used at LAMPF to investigate resonance excitation in heavy target [174, 175]. About the same time also (n, p) reactions were measured at Los Alamos [176, 177]. A few years later, similar experiments utilizing (p, n) charge exchange were started also at Saclay [178]. A broader range of phenomena can be accessed by heavy ion charge exchange reactions with their particular high potential for resonance studies on nuclei by observing the final ions with well defined charge numbers. This implies peripheral reactions corresponding to a gentle perturbation by rearranging of the initial mass and charge distributions by one or a few units. Thus, the colliding ions are left essentially intact and the reaction corresponds to a coherent process in

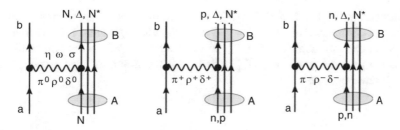

Fig. 5.36 Resonance production by charge-neutral meson exchange (left) and charged meson exchange (right)

which the mass numbers of projectile and target are conserved but the arrangement of protons and neutrons is modified. Experimental groups at SATURNE at Saclay took the first-time chance to initiate dedicated experiments on in-medium resonance physics especially with $(^3He,^3H)$ reactions on heavy targets [179, 180]. In a series of experiments, the excitation of the Delta-resonance was observed. Due to the experimental limitations at that time only fixed target experiments on stable nuclei were possible. A few years later, similar experiments were started at the Synchrophasotron at JINR Dubna, making use of the higher energy at that facility to extend the spectral studies up to the region of the nucleon-pion s-wave and d-wave resonances [181, 182]. While initially those experiments were concentrating on inclusive reactions, a more detailed picture is obviously obtained by observing also the decay of the excited states. Such measurements were indeed performed in the early 1990s with the DIOGENE detector at SATURNE [183, 184], at KEK using the FANCY detector [185]. However, with respect to the first experiments it took another decade or so before those exclusive data were studied theoretically [186]. A couple of years later, corresponding experiments were done at Dubna, taking advantage of the higher energies of up to $p_{lab} = 4.2$ AGeV/c reached at the Synchrophasotron [187–190]. In [191] the measurements were extended to the detection of up to $N^* \rightarrow p\pi^+\pi^-$ three particle decay channels in coincidence allowing to identify also the $N^*(1440)$ Roper resonance and even higher resonances. A spectral distribution is shown in Fig. 5.44.

The generic interaction processes shown in Fig. 5.36 are using a meson exchange picture which describes successfully the dynamics of N^* production in ion-ion reactions. The $\Delta(1232)$ is produced mainly in $NN \rightarrow \Delta N$ reactions and the Delta is subsequently decaying into $N\pi$, thus producing in total a $NN \rightarrow NN\pi$ transition. The pion yield from the Delta source, coming from a p-wave process, competes with direct s-wave pion production, $NN \rightarrow NN\pi$. The intermediate population of higher resonances like $P_{11}(144)$ will lead to $NN \rightarrow NN\pi\pi$ processes. With increasing energy baryons will be excited decaying into channels with higher pion multiplicities. Already the early theoretical studies lead to the conclusion that in heavy-ion collisions at around 1 AGeV up to 30% of the participating nucleons will be excited into resonances. Thus, a kind of short-lived resonance matter is formed before decaying back into nucleons and mesons [192, 193]. That figure is largely confirmed by the work of Ref. [190] on deuteron induced resonance production at the Synchrophasotron at $p_{lab} = 4.2\,\mathrm{AGeV/c}$, although there a somewhat lower resonance excitation rate of $15 \pm 2^{+4}_{-3}\%$ was found which is in the same bulk as the excitation probability of $16 \pm 3^{+4}_{-3}\%$ derived from proton induced reactions at the same beam momentum.

5.7.2 Interaction Effects in Spectral Distribution in Peripheral Reactions

The excitation of the Delta resonance in proton and light ion induced peripheral reactions at intermediate energies was studied theoretically in very detail by Osterfeld and Udagawa and collaborators [149, 150, 194, 195]. The first round of experimental data from 1980s and 1990s were investigated by microscopic theoretical approaches covering inclusive and semi-inclusive reactions. The conclusion from those studies are still valuable and are worth to be recalled. In the Osterfeld-Udagawa approach initial and final state interactions were taken into account by distorted wave methods. A microscopic approach was used to describe the excitation and intrinsic nuclear correlations of $\Delta N^{-}1$ states. This combination of—at that time—very involved theoretical methods was able to explain the observed puzzling shift of the Delta-peak by $\Delta M \sim -70\,\mathrm{MeV}$. In Fig. 5.37 conclusive

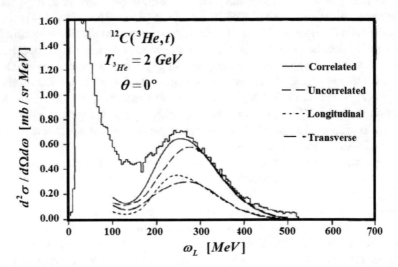

Fig. 5.37 Zero-degree triton spectra for the reaction $^{12}C(^{3}He, t)$ at $T_{lab} = 2\,\mathrm{GeV}$, shown as a function of the excitation energy ω_L. The data are taken from Ref. [179]. The full curve represents the final result including initial and final state interactions, finite size effects, and particle-hole correlations. In addition, the longitudinal (LO) and transverse (TR) partial cross sections are shown (from Ref.[150])

results on that issue are shown. The calculations did not include the quasi-elastic component, produced by single and multiple excitations of NN^{-1} states and knock-out reactions.

In the Delta region the theoretical $^{12}C(He, t)$ cross section matches the experimental data almost perfectly well. Distorted wave effects, i.e. initial and final state interactions of the colliding ions, are of central importance for that kind agreement. They alone provide a shift of about $\Delta M_{DW} \sim -50$ MeV [150]. Finite size and a detailed treatment of particle-hole correlations within the ΔN^{-1} configurations contribute the remaining $\Delta M_c \sim -20$ MeV. The polarization tensor may be decomposed into pion-like longitudinal contributions, the complementary transversal components, representative of vector-meson interactions, and mixed terms, see e.g. [195]. An interesting results, shown in Fig. 5.37, is that the longitudinal (LO) partial cross section appears to be shifted down to a peak values of $\omega_L \sim 240$ MeV, while the transversal partial cross section peaks at $\omega_L \sim 285$ MeV. This is an effect of the $^3He \rightarrow t$ transition form factor which reduces the magnitude of the TR spectrum at high excitation energies because of its exponential falloff at large four-momentum transfer. The shape of the LO spectrum is less strongly affected by this effect. It is remarkable that in contrast to (p, n) reactions the full calculation reproduces the higher energy part of the spectrum so well. This is due to the fact that the high-energy flank of the resonance is practically background-free, since the probability that the excited projectile decays to the triton ground state plus a pion is extremely small. Also a negligible amount of tritons is expected to be contributed from the quasi-free decay of the target. The cross section in the resonance region shows an interesting scaling behaviour: A proportionality following a $(3Z+N)$-law is found where Z and N are the proton and neutron number of the target. This dependence of the cross section reflects that the probability for the $p+p \rightarrow n+\Delta^{++}$ process is three times larger than that for the $p+n \rightarrow n+\Delta^+$ process.

An even more detailed picture emerges from semi-inclusive reactions observing also decay products. For the reactions discussed in [183–185, 196] at incident energies of about 2 AGeV, the $p\pi$ correlations were successfully analyzed and the mass distribution of the $\Delta(1232)$ resonance could be reconstructed. In these peripheral reactions on various targets, the resonance mass was found to be shifted by up to $\Delta M \simeq -70$ MeV towards lower masses compared to those on protons. In reactions on various nuclei at incident energies around 1 AGeV the mass reduction of the $\Delta(1232)$ resonance was traced back to Fermi motion, NN scattering effects, and pion reabsorption in nuclear matter. These findings are in rough agreement with detailed theoretical studies of in-medium properties of the Δ-resonance by the Valencia group [161, 197], considering also the decrease of the Delta-width because of the reduction of the available $N\pi$ decay phase space by Pauli-blocking effects of nucleons in nuclear matter.

5.7.3 Resonances in Central Heavy Ion Collisions

Different aspects of resonance dynamics are probed in central heavy ion collisions. The process responsible for meson production in central heavy-ion collisions at energies of the order of several hundred MeV/nucleon to a few GeV/nucleon is believed to be predominantly driven by the excitation of baryon resonances during the early compression phase of the collision [192, 193, 198–202]. In the later expansion phase these resonances decay into lower mass baryon states and a number of mesons. The influence of the medium is expected to modify mass and width of the resonances by induced self-energies. In high density and heated matter, however, the genuine self-energy effects may be buried behind kinematical effects.

The best studies case is the $\Delta_{33}(1232)$ resonance which, in fact, is a 16-plet formed by four isospin and four spin sub-states. In nucleon-nucleon scattering one observes a centroid mass $M_\Delta \simeq 1232\,\text{MeV}$ and the width $\Gamma_\Delta = 115$–$120\,\text{MeV}$ which is in good agreement with the direct observation in pion-nucleon scattering [65]. Modifications of mass and width of the $\Delta(1232)$ resonances have been observed in central heavy-ion collisions leading to dense and heated hadron matter, e.g. at the BEVALAC by the EOS collaboration [203] and at GSI by the FOPI collaboration [202]. In [203], for example, the mass shift and the width were determined as functions of the centrality, both showing a substantial reduction with decreasing impact parameter. The modification of $\Delta_{33}(1232)$ properties has been interpreted in terms of hadronic density, temperature, and various non-nucleon degrees of freedom in nuclear matter [204–206]. The invariant mass distributions of correlated nucleon and pion pairs provide, in principle, a direct proof for resonance excitation. As discussed in [202], in the early heavy ion collision experiments a major obstacle for the reconstruction of the resonance spectral distribution was the large background of uncorrelated $p\pi$ pairs coming from other sources. Only after their elimination from the data the resonance spectral distributions could be recovered. Results of a first successful resonance reconstruction in central heavy ion collisions are shown in Fig. 5.38.

5.7.4 The Delta Resonance as Pion Source in Heavy Ion Collisions

Transport calculations are describing accurately most of the particle production channels in heavy-ion collisions already in the early days of transport theory [199, 201, 207, 208]. However, surprisingly the pion yield from heavy-ion collisions at SIS energies ($T_{lab} \sim 1\,\text{AGeV}$) could not be reproduced properly by the transport-theoretical description. For a long time the pertinent overprediction of the pion multiplicity [209–216] was a disturbing problem. At the beam energies of a few AGeV the dominating source for pion production is the excitation of the $\Delta(1232)$ resonance in a NN collision $NN \rightarrow N\Delta$ followed by the decay $\Delta \rightarrow N\pi$. In

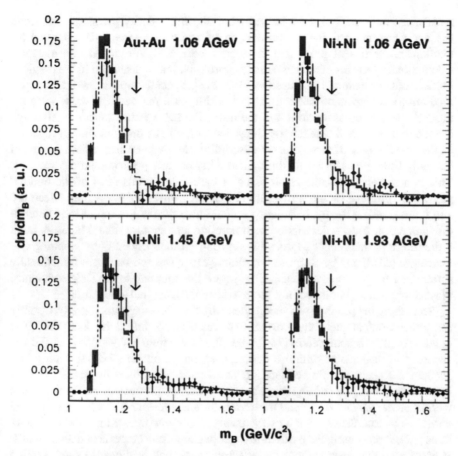

Fig. 5.38 The invariant mass spectrum of baryon resonances excited by the reactions stated in each panel. The filled areas correspond to the analysis of the measured transverse momentum spectra of $\pi^{\pm}$, the full points to the analysis of the measured $p\pi^{\pm}$ pairs. Traces of higher resonances are visible in the high energy tails of the spectral distributions. The arrows point to the maximum of the free $\Delta(1232)$ mass distribution (from Ref. [202])

transport calculations, the pion multiplicity, therefore, depends crucially on the value of the in-medium $NN \rightarrow N\Delta$ cross section. A first attempt to solve that issue was undertaken by Bertsch et al. [142]. In the 1990s Helgesson and Randrup [143, 217] took up that issue anew. In their microscopic $\pi + NN^{-1} + \Delta N^{-1}$ model [143] they considered the excitation of ΔN^{-1} modes in nuclear matter by RPA theory. The coupling to the purely nuclear Gamow-Teller-like NN^{-1} spin-isospin modes and the corresponding pion modes was taken into account. They point out that sufficiently energetic nucleon-nucleon collisions may agitate one or both of the colliding nucleons to a nucleon resonance with especial importance of $\Delta(1232)$, $N^*(1440)$, and $N^*(1535)$. Resonances propagate in their own mean field and may collide with nucleons or other nucleon resonances as well. Moreover, the nucleon resonances may decay by meson emission and these decay processes constitute the main mechanisms for the production of energetic mesons. The derived in-medium properties of pions and Δ isobars were later introduced into transport calculations by means of a local density approximation as discussed in the previous section, but for example also used in [212, 213]. Special emphasis was laid on in-medium pion dispersion relations, the Δ width, pion reabsorption cross sections, the $NN \rightarrow \Delta N$ cross sections and the in-medium Δ spectral function. Although the medium-modified simulations showed strong effects on in-medium properties in the early stages of the transport description the detailed in-medium treatment had only little effect on the final pion and other particle production cross sections. This is a rather reasonable result since in their calculations most of the emitted pions were produced at the surface at low densities where the in-medium effects are still quite small. Actually, in order to account also for the heating of the matter in the interaction zone, a description incorporating temperature should be used. Such a thermo-field theoretical approach was proposed independently by Henning and Umezawa [218], and by Korpa and Malfliet [219]. The approach was intentionally formulated for pion-nucleus scattering, where the coupling to the Delta resonance plays a major role, but it does not seem to have been applied afterwards.

Years later, the problem was reconsidered by the Giessen group. Initially, a purely phenomenological quenching prescription was used for fitting the data [216]. The breakthrough was achieved in [144] when the in-medium $NN \rightarrow N\Delta(1232)$ cross section were calculated within a one pion exchange model (OPEM), taking into account the exchange pion collectivity and vertex corrections by contact nuclear interactions. Also, the (relativistic) effective masses of the nucleon and Δ resonance were considered. The ΔN^{-1} and the corresponding nuclear NN^{-1} modes, discussed above, were calculated again by RPA theory. In infinite matter the Lindhard functions [92], representing the particle-hole propagators, can be evaluated analytically. It was found that even without the effective mass modifications the cross section decreases with the nuclear matter density at high densities already alone by the in-medium Δ width and includes the NN^{-1} Lindhard function (see below) in the calculations. The inclusion of the effective mass modifications for the nucleons and Δ's led to an additional strong reduction of the cross section. Altogether, the total pion multiplicity data [215] measured by the FOPI collaboration on the systems Ca+Ca, Ru+Ru and Au+Au at $T_{lab} = 0.4, 1.0$, and 1.5 AGeV, respectively, could be described by introducing a dropping effective mass with increasing baryon density. The results were found to depend to some extent on the in-medium value of the Δ-spreading width for which the prescription of the Valencia group was used: $\Gamma_{sp} = 80\rho/\rho_0$ MeV [163, 220].

The effect of the medium modifications of the $NN \leftrightarrow N\Delta$ cross sections on the pion multiplicity depends also on the assumption about other channels of the pion production/absorption in NN collisions, most importantly, on the s-wave interaction in the direct channel $NN \leftrightarrow NN\pi$. In [144] it was found that including the effective mass modifications in the $NN \leftrightarrow N\Delta$ channel only, does not reduce pion multiplicity sufficiently, since then more pions are produced in the s-wave channel. An important conclusion for future work is that the in-medium modifications of the higher resonance cross sections do not influence the pion production at 1–2 AGeV collision energy sensitively: other particles like η and ρ mesons are, probably, more sensitive to higher resonance in-medium modifications.

A subtle test for the transport description of pion production is given by $(\pi^{\pm}, \pi^{\mp})$ double charge exchange (DCX) reactions on nuclear targets. In Ref. [221] such reactions were investigated by GiBUU transport calculations. The pionic double charge exchange processes were studied for a series of nuclear targets, including (^{16}O, ^{40}Ca and ^{208}Pb), for pion incident energies $T_{lab} = 120, 150, 180$ MeV covering the Delta-region. As a side aspects, the results could confirm the validity of the so-called parallel ensemble scheme for those reactions in comparison to the more precise but time consuming full ensemble method [221]. GiBUU results for the DCX reaction $\pi^{\pm}O \rightarrow \pi^{\mp}X$ at $T_{lab} = 120, 150$ and 180 MeV are shown Fig. 5.39. A good agreement with data was achieved for the total cross section and also for angular distributions and double differential cross sections. Some strength at backward angles and rather low pion energies below $T_{lab} \approx 30$ MeV is still missed. A striking sensitivity on the thickness of neutron skins was found, indicating that such reactions may of potential advantage for studies of nuclear density profiles.

5.7.5 Perspectives of Resonance Studies by Peripheral Heavy Ion Reactions

The large future potential of resonance physics with heavy ions was demonstrated by recent FRS experiments measuring the excitation of the Delta and higher resonances in peripheral heavy ion charge exchange reactions with stable and exotic secondary beams as heavy as Sn on targets ranging from hydrogen and ^{12}C to ^{208}Pb [12, 152, 223]. With these reactions, exceeding considerably the mass range accessed by of former heavy ion studies, a new territory is explored. A distinct advantage of the FRS and even more so, of the future Super-FRS facility is the high energy of secondary beams, allowing the unique experimental access to sub-nuclear excitations. This allows to perform spectroscopy in the quasi-elastic nucleonic NN^{-1} and the resonance N^*N^{-1} regions at and even beyond the Delta resonance. The elementary excitation mechanisms contributing to peripheral heavy ion charge exchange reactions are shown diagrammatically in Fig. 5.40.

These outstanding experimental conditions open new perspectives for broadening the traditionally strong branch of nuclear structure physics at GSI/FAIR to the new territory of in-medium resonance physics. The most important prerequisites are the high energies and intensities of secondary beams available at the SUPER-FRS. In many cases, inelastic, charge exchange, and breakup or transfer reactions could be done in a similar manner at other laboratories like RIKEN, FRIB, or GANIL, only the combination of SIS18/SIS100 and, in perspective, the Super-FRS provides access beyond the quasi-elastic region allowing to explore sub-nuclear degrees of freedom.

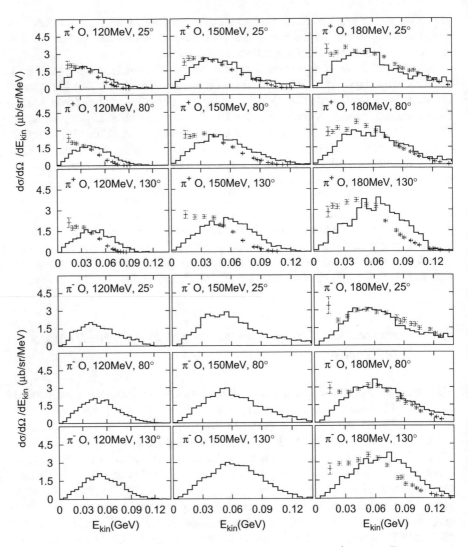

Fig. 5.39 Double differential cross sections for the DCX process $\pi^{\pm}O \rightarrow \pi^{\mp}X$ at $T_{lab} =$ 120, 150 and 180 MeV. The results at different angles are shown as function of the kinetic energies of the produced pions. Data are taken from [222], only statistical errors are shown. The GiBUU results are shown as histograms, where the fluctuations indicate the degree of statistical uncertainty (from Ref.[221])

The experimental conditions at the FRS are providing a stand-alone environment of resonance studies in nuclear matter at large isospin. Reactions at the FRS will focus on peripheral processes. The states of the interacting nuclei will only be changed gently in a well controlled manner. Resonances can be excited in inelastic and charge exchange reactions. In the notation of neutrino physics those reactions are probing neutral current (*NC*) and charge current (*CC*) events and the corre-

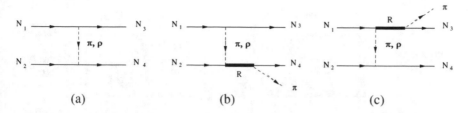

Fig. 5.40 The excitation mechanisms of peripheral heavy ion charge exchange reactions induced by the exchange of charged mesons: quasi-elastic NN^{-1} excitations in target and/or projectile (**a**), RN^{-1} resonance-hole excitations in the target (**b**), and RN^{-1} excitations in the projectile (**c**), where R denotes the Delta $P_{33}(1232)$, the Roper $P_{11}(1440)$ or any other higher nucleon resonance

sponding nuclear response functions. Here, obviously strong interaction vertices are involved but it is worthwhile to point out that the type of nuclear response functions are the same as in the weak interaction processes. By a proper choice of experimental conditions the following reaction scenarios will be accessible in either inelastic or charge exchange reactions with resonance excitation in coherent inclusive reactions or in semi-exclusive coherent reactions with pion detection. In the first type of reaction the energy-momentum distribution of the outgoing beam-like ejectile is observed. Since the charge and mass numbers of that particle are known it must result from a reaction in which it was produced in a bound state. Such reactions primarily record resonances in the target nuclei, folded with the spectrum of bound inelastic or charge exchange excitations, respectively, in the beam-like nucleus. Hence, the reaction is coherent with respect to the beam particles. Results of a recent experiment at the FRS, proving the feasibility of such investigations, are shown in Fig. 5.41. In the spectra, the quasi-elastic and the resonances regions, discussed in Fig. 5.40, are clearly seen and energetically well resolved.

The second scenario is different by the detection of pions emitted by the highly excited intermediate nuclei. By tagging on the pions from the beam-like nuclei one obtains direct information on the spectral distributions of the pion sources, i.e. the nucleon resonances. This scenario, illustrated in Fig. 5.42, will provide access to resonance studies in nuclei with exotic charge-to-mass ratios. Obviously, also pions from the target nuclei can be observed which corresponds to similar early experiments at SATURNE [183, 184], the Dubna Synchrophasotron [191] and, at slightly lower energies, at the FANCY detector at KEK [185]. In the Dubna experiments single and double pion channels have been measured. The gain in spectroscopic information is already visible in the $p\pi$ singles spectra, Fig. 5.43, and even more so in the $p\pi\pi$ spectra in Fig. 5.44. In coincidence experiments measuring the decay particles of in-medium resonances are obviously complementary by establishing a connection of meson production on the free nucleon and on nuclei.

Heavy ions and pions are strongly absorbed particles. Therefore, resonances will be excited mainly at the nuclear surface. Also pions from grazing reactions will carry signals mainly from the nuclear periphery. However, the high energies allow resonance excitation also in deeper density layers of the involved nuclei. In order

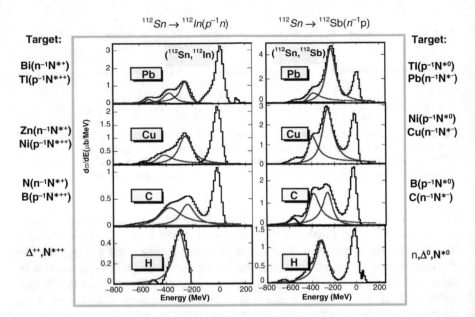

Fig. 5.41 Results of recent heavy ion charge exchange reactions at the FRS at $T_{lab} = 1$ AGeV. Both $p \rightarrow n$ and $n \rightarrow p$ projectile branches were measured on the indicated targets. The excitations reached in the target are shown on the left and right of the data panels. Spectral distributions obtained from a peak-fitting procedure are also shown exposing the Delta resonance and higher resonance-like structures like the $P_{11}(1440)$ Roper resonance (data taken from Ref. [152])

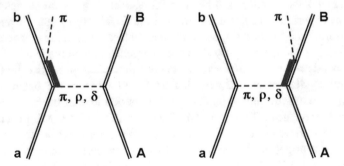

Fig. 5.42 Exclusive resonance production processes by observation of the decay products

to overcome those limitations at least on the decay side it might be worthwhile to consider as a complementary branch to record also dilepton signals.

The scientific perspectives of resonance physics at a high-energy nuclear facility like the Super-FRS at FAIR is tremendous. Pion emission will serve as indicator for resonance excitation and record the resonance properties by their spectral distribution. In the past, theoretically as well as experimentally the Delta resonance has obtained the largest attention. The work, however, was almost exclusively focused to nuclei close to stability, i.e. in symmetric nuclear matter. On the theoretical side, the

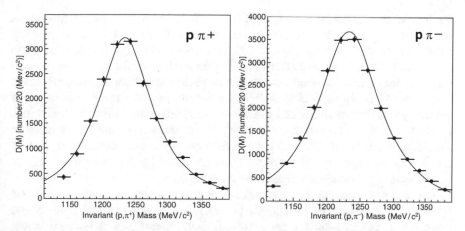

Fig. 5.43 Proton-single pion yields measured at the Synchrophasotron in the reaction $^{12}C +^{12} C$ at $p_{lab} = 4.2$ GeV/c per nucleon (from Ref. [182])

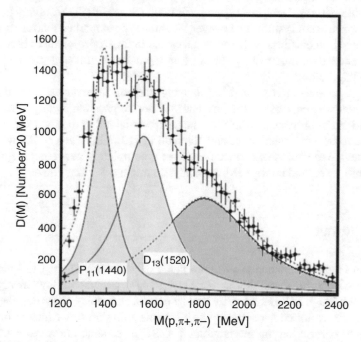

Fig. 5.44 Observation of higher resonances in $C+C$ collisions at the Synchrophasotron by $N^* \rightarrow p\pi^+\pi^-$ two-pion decay spectroscopy at the beam momentum $p_{lab} = 4.2$ GeV/c. Expected N^* states are indicated for the two lower structures. The data are from Ref. [191]

main reason for that self-imposed constraint is our lack of knowledge on resonance dynamics in nuclei far off stability, although in principle theory is well aware of the complexities and changes of resonance properties in nuclear matter. Despite the multitude of published work, until today we do know surprisingly little about the isospin dependence of resonance self-energies. There is an intimate interplay between in-medium meson physics and resonance self-energies. Since the width and the mass location of resonances is closely determined by the coupling to meson-nucleon decay channels modifications in those sectors affect immediately resonance properties. At the Super-FRS such dependencies can be studied over wide ranges of neutron-proton asymmetries and densities of the background medium. Since such effects are likely to be assigned selectively to the various channels, a variation of the charge content will allow to explore different aspects of resonance dynamics, e.g. distinguishing charge states of the Delta resonance by the different in-medium interactions of positively and negatively charged pions.

Last but not least, resonance physics at fragment separators will also add new figures to the astrophysical studies. In supernova explosions and neutron star mergers high energy neutrinos are generated. Their interaction with matter proceeds through quasi-elastic and, to a large extent, through resonance excitation. The assumed neutrino reheating of the shock wave relies on the knowledge of neutrino-nuclear interactions. Neutrino experiments themselves lack the resolving power for detailed spectroscopic studies. However, the same type of nuclear matrix elements is encountered in inelastic and charge exchange excitations of resonances in secondary beam experiments thus testing the nuclear input to neutral and charged current neutrino interactions.

A large potential is foreseen for studying nucleon resonances in exotic nuclei which never was possible in the past and will not be possible by any other facility worldwide in the foreseeable future. The results obtained until now from the FRS-experiment are very interesting and stimulating [12, 152]. Super-FRS will be an unique device to access resonance physics in a completely new context giving the opportunity to extend nuclear structure physics into a new direction.

5.8 Summary

Strangeness and resonance physics are fields of particular interest for our under-standing of baryon dynamics in the very general context of low-energy flavor physics. Although SU(3) symmetry is not a perfect symmetry, the group-theoretical relations are exploited successfully as a scheme bringing order into to multitude of possible baryon-baryon interactions. The SU(3) scheme allows to connect the various interaction vertices of octet baryons and meson multiplets, thus reducing the number of free parameters significantly by relating coupling constants to a few elementary parameters. Since single hyperons and resonances, immersed into a nuclear medium, are not subject of the Pauli exclusion principle, their implementation into nuclei is revealing new aspects of nuclear dynamics. In that sense, hyperons and resonances may serve as probes for the nuclear many-body system. At first, hyperons, their interactions in free-space and nuclear matter was discussed.

For the description of hypernuclei density functional theory was introduced. Except for the lightest nuclei, the DFT approach is applicable practically over the whole nuclear chart and beyond to nuclear matter and neutrons star matter. The DFT discussion was following closely the content of the Giessen DDRH theory. As an appropriate method to describe the density dependence of dressed meson-baryon vertices in field-theoretical approach, nucleon-meson vertices were introduced which are given as Lorentz-invariant functionals on the matter fields. The theory was evaluated in the relativistic mean-field limit. The Lambda separation energies of the known $S = -1$ single-Λ hypernuclei are described satisfactory well, however, with the caveat that the experimental uncertainties lead to a spread in the derived parameters of about 20%. In hypermatter the minimum of the binding energy per particle was found to be shifted to larger density ($\rho_{hyp} \sim 0.21\,\text{fm}^{-3}$) and stronger binding ($\varepsilon(\rho_{hyp}) \sim -18\,\text{MeV}$) by adding Λ hyperons. The minimum is reached for a Λ-content of about 10% as shown in Sect. 5.4.6. The DFT results are found to be in good agreement with other theoretical calculations. Overall, on the theory side convergence seems to be achieved for single Λ-hypernuclei. However, open issues remain about the nature and mass dependence of the crucial Lambda spin-orbit strength. The existence of bound Σ hypernuclei is still undecided although the latest theoretical results are in clear favor of a weak or repulsive Σ potential. $S = -2$ double-Lambda hypernuclei would be an important—if not only—source of information on $\Lambda\Lambda$ interactions. Until now, the results rely essentially on a single case, the famous $^{6}_{\Lambda\Lambda}He$ Nagara event observed years ago in an emulsion experiment at KEK. Future research on the production and spectroscopy of those systems— as planned e.g. for the $\overline{\text{PANDA}}$ experiment at FAIR—is of crucial importance for the fields of hyperon and hypernuclear physics. The lack of detailed knowledge on interactions is also part of the problem of our persisting ignorance on the notorious hyperon puzzle in neutron stars.

A topic of own interest—which was not discussed here—is the theory of hypernuclear production reactions. For a detailed review we refer to Ref. [18] where the description of hypernuclear production reactions is discussed. In that article a broad range of production scenarios, ranging from proton and antiproton induced reactions to the collision of massive ions, is studied theoretically. Strangeness and resonance production is driven by the excitation of a sequence of intermediate N^* states. For the production of hypernuclei, fragmentation reactions are playing the key role. The hyperon production rates, especially for baryons with strangeness $|S| > 1$, depend crucially on the accumulation of strangeness in a sequence of reactions by intermediate resonance excitation, involving also Σ^* and Ξ^* states from the $\frac{3}{2}^+$ decuplet. As a concluding remark on strangeness and hypernuclear production, we emphasize that in-medium reactions induced by heavy-ion beams represent an excellent tool to study in detail the strangeness sector of the hadronic equation of state. The knowledge of the in-medium interactions is essential for a deeper understanding of baryon-baryon interactions in nuclear media over a large range of densities and isospin. It is crucial also for nuclear astrophysics by giving access to the high density region of the EoS of hypermatter, at least for a certain amount of hyperon fractions. In addition, reaction studies on bound superstrange hypermatter offer unprecedented opportunities to explore the hitherto unobserved regions of exotic bound hypernuclear systems.

Resonance physics is obviously of utmost importance for nuclear strangeness physics because they are the initial source for hyperon production. But the physics of N^* states in the nuclear medium is an important field of research by its own. From the side of hadron spectroscopy, there is large interest to use the nuclear medium as a probe for the intrinsic structure of resonances. As pointed out repeatedly, the intrinsic configurations of resonances are mixtures of meson-baryon and 3-quark components, the latter surrounded by a polarization cloud of virtual mesons and $q\bar{q}$ states. The various components are expected to react differently on the polarizing forces of nuclear matter thus offering a more differentiating access to N^* spectroscopy. From the nuclear physics side, resonances are ideal probes for various aspects of nuclear dynamics which are not so easy to access by nucleons alone. They are emphasizing certain excitation modes as the spin-isospin response of nuclear matter. For that purpose, studies of the Δ resonance are the perfect tool. Resonances are also thought to play a key role in many-body forces among nucleons, implying that nucleons in nuclear matter are in fact part of their time in (virtually) excited states. To the best of our knowledge stable nuclear matter exists only because three-body (or multi-body) forces, contributing the correct amount of repulsion already around the saturation point and increasingly so at higher densities. Peripheral heavy ion reactions are the method of choice to produce excited nucleons under controlled conditions in cold nuclear matter below and close to the saturation density. It was pointed, that in central collisions the density of N^* states will increase for a short time to values comparable to the density in the center of a nucleus. The existing data confirm that in peripheral reactions the excitation probability is sizable.

Physically, the production of a resonance in a nucleus corresponds to the creation of N^*N^{-1} particle-hole configuration. The description of such configurations was discussed for the case of $N^* = \Delta$. Extensions to higher resonances are possible and in fact necessary for the description of the nuclear response already observed at high excitation energies. In future experiments a particular role will be played by the decay spectroscopy which is a demanding task for nuclear theory. In any case, the nuclear structure and reaction theory are asked to extend their tool box considerably for a quantitative description of hyperons and resonances in nuclear matter. In this respect, neutrons star physics is a step ahead: As discussed, there is a strong need to investigate also resonances in neutron star matter. In beta-equilibrated matter, resonances will appear at the same densities as hyperons. Thus, in addition to the *hyperon puzzle* there is also a *resonance puzzle* in neutron stars.

Overall, in-medium hadronic reactions offer ample possibilities of studying sub-nuclear degrees of freedom. By using beams of the heaviest possible nuclei at beam energies well above the strangeness production thresholds, one can probe definitely superstrange and resonance matter at baryon densities far beyond saturation, e.g. coming eventually close to the conditions in the deeper layers of a neutron star. Theoretically, such a task is of course possible and the experimental feasibility will come in reach at the Compressed-Baryonic-Experiment (CBM) at FAIR which is devoted specifically to investigations of baryonic matter. The LHC experiments are covering already a sector of much higher energy density but their primary layout is for physics at much smaller scales.

On the side of hadron and nuclear theory, LQCD and QCD-oriented effective field theories may bring substantial progress in the not so far future, supporting a new understanding of sub-nuclear degrees of freedom and in-medium baryon physics in a unified manner. We have mentioned their achievements and merits on a few places. However, both LQCD and EFT approaches, would deserve a much deeper discussion as could be done here. In conjunction with appropriate many-body methods, such as provided by density functional theory, Green's function Monte Carlo techniques, and modern shell model approaches are apt to redirect nuclear and hypernuclear physics into the direction of ab initio descriptions. The explicit treatment of resonances will be a new demanding step for nuclear theory (and experiment!) adding to the breadth of the field.

Acknowledgements Many members and guests of the Giessen group have been contributing to the work summarized in this article. Contributions especially by C. Keil and A. Fedoseew, S. Bender, Th. Gaitanos (now at U. Thessaloniki), R. Shyam (Saha Institute, Kolkatta), and V. Shklyar are gratefully acknowledged. Supported by DFG, contract Le439/9 and SFB/TR16, project B7, BMBF, contract 05P12RGFTE, GSI Darmstadt, and Helmholtz International Center for FAIR.

References

1. G.D. Rochester, C.C. Butler, Nature **160**, 855 (1947). https://doi.org/10.1038/160855a0
2. M. Danysz, J. Pniewski, Phil. Mag. **44**, 348 (1953)
3. M. Agnello et al., Nucl. Phys. A **881**, 269 (2012). https://doi.org/10.1016/j.nuclphysa.2012.02.015
4. T.R. Saito et al., Nucl. Phys. A **954**, 199 (2016). https://doi.org/10.1016/j.nuclphysa.2016.05.011
5. C. Rappold et al., Phys. Lett. B **728**, 543 (2014). https://doi.org/10.1016/j.physletb.2013.12.037
6. N. Shah, Y.G. Ma, J.H. Chen, S. Zhang, Phys. Lett. B **754**, 6 (2016). https://doi.org/10.1016/j.physletb.2016.01.005
7. N. Shah, Y.G. Ma, J.H. Chen, S. Zhang, Production of multistrange hadrons, light nuclei and hypertriton in central Au+Au collisions at $\sqrt{s_{NN}}$ = 11.5 and 200 GeV. Phys. Lett. **B754**, 6–10 (2016). https://doi.org/10.1016/j.physletb.2016.01.005
8. T.P. Cheng, L.F. Li, *Gauge Theory of Elementary Particle Physics* (Clarendon (Oxford Science Publications), Oxford, 1984), 536 pp.
9. J.I. Friedman, H.W. Kendall, Ann. Rev. Nucl. Part. Sci. **22**, 203 (1972). https://doi.org/10.1146/annurev.ns.22.120172.001223
10. A. Gal, E.V. Hungerford, D.J. Millener, Rev. Mod. Phys. **88**(3), 035004 (2016). https://doi.org/10.1103/RevModPhys.88.035004
11. I. Bednarek, P. Haensel, J.L. Zdunik, M. Bejger, R. Manka, Astron. Astrophys. **543**, A157 (2012). https://doi.org/10.1051/0004-6361/201118560
12. J. Benlliure et al., JPS Conf. Proc. **6**, 020039 (2015). https://doi.org/10.7566/JPSCP.6.020039
13. O. Hashimoto, H. Tamura, Prog. Part. Nucl. Phys. **57**, 564 (2006). https://doi.org/10.1016/j.ppnp.2005.07.001

14. A. Gal, O. Hashimoto, J. Pochodzalla, Nucl. Phys. A **881**, 1 (2012)
15. A. Feliciello, T. Nagae, Rep. Prog. Phys. **78**(9), 096301 (2015). https://doi.org/10.1088/0034-4885/78/9/096301
16. A. Gal, J. Pochodzalla (eds.), Nucl. Phys. A **954** (2016)
17. D. Blaschke et al. (eds.), Eur. Phys. J. A **52** (2016)
18. H. Lenske, M. Dhar, T. Gaitanos, X. Cao, Baryons and baryon resonances in nuclear matter. Prog. Part. Nucl. Phys. **98**, 119–206 (2018). https://doi.org/10.1016/j.ppnp.2017.09.001
19. T.A. Rijken, V.G.J. Stoks, Y. Yamamoto, Phys. Rev. C **59**, 21 (1999). https://doi.org/10.1103/PhysRevC.59.21
20. T.A. Rijken, Phys. Rev. C **73**, 044007 (2006). https://doi.org/10.1103/PhysRevC.73.044007
21. T.A. Rijken, Y. Yamamoto, Phys. Rev. C **73**, 044008 (2006). https://doi.org/10.1103/PhysRevC.73.044008
22. T.A. Rijken, M.M. Nagels, Y. Yamamoto, Prog. Theor. Phys. Suppl. **185**, 14 (2010). https://doi.org/10.1143/PTPS.185.14
23. Y. Yamamoto, T. Furumoto, N. Yasutake, T.A. Rijken, Phys. Rev. C **90**, 045805 (2014). https://doi.org/10.1103/PhysRevC.90.045805
24. T.A. Rijken, H.J. Schulze, Eur. Phys. J. A **52**(2), 21 (2016). https://doi.org/10.1140/epja/i2016-16021-6
25. J. Haidenbauer, U.G. Meissner, A. Nogga, H. Polinder, Lect. Notes Phys. **724**, 113 (2007). https://doi.org/10.1007/978-3-540-72039-3_4
26. B. Holzenkamp, K. Holinde, J. Speth, Nucl. Phys. A **500**, 485 (1989). https://doi.org/10.1016/0375-9474(89)90223-6
27. A. Reuber, K. Holinde, J. Speth, Nucl. Phys. A **570**, 543 (1994). https://doi.org/10.1016/0375-9474(94)90073-6
28. M. Dhar, H. Lenske, in *Proceedings of the 12th International Conference on Hypernuclear and Strange Particle Physics (HYP2015)* (2017). https://doi.org/10.7566/HYP2015; http://journals.jps.jp/doi/abs/10.7566/HYP2015
29. M. Dhar, Dissertation JLU Giessen (2016)
30. Y. Fujiwara, M. Kohno, Y. Suzuki, Mod. Phys. Lett. A **24**, 1031 (2009). https://doi.org/10.1142/S0217732309000528
31. M. Kohno, Y. Fujiwara, Phys. Rev. C **79**, 054318 (2009). https://doi.org/10.1103/PhysRevC.79.054318
32. E. Epelbaum, H.W. Hammer, U.G. Meissner, Rev. Mod. Phys. **81**, 1773 (2009). https://doi.org/10.1103/RevModPhys.81.1773
33. E. Hiyama, Nucl. Phys. A **914**, 130 (2013). https://doi.org/10.1016/j.nuclphysa.2013.05.011
34. J. Carlson, S. Gandolfi, F. Pederiva, S.C. Pieper, R. Schiavilla, K.E. Schmidt, R.B. Wiringa, Rev. Mod. Phys. **87**, 1067 (2015). https://doi.org/10.1103/RevModPhys.87.1067
35. D.J. Millener, Nucl. Phys. A **881**, 298 (2012). https://doi.org/10.1016/j.nuclphysa.2012.01.019
36. D.J. Millener, Nucl. Phys. A **914**, 109 (2013). https://doi.org/10.1016/j.nuclphysa.2013.01.023
37. M. Stoitsov, H. Nam, W. Nazarewicz, A. Bulgac, G. Hagen, M. Kortelainen, J.C. Pei, K.J. Roche, N. Schunck, I. Thompson, J.P. Vary, S.M. Wild, UNEDF: advanced scientific computing transforms the low-energy nuclear many-body problem. J. Phys. Conf. Ser. **402**, 12033 (2012). https://doi.org/10.1088/1742-6596/402/1/012033
38. H. Nam et al., J. Phys. Conf. Ser. **402**, 012033 (2012). https://doi.org/10.1088/1742-6596/402/1/012033
39. S. Bogner et al., Comput. Phys. Commun. **184**, 2235 (2013). https://doi.org/10.1016/j.cpc.2013.05.020
40. C.M. Keil, F. Hofmann, H. Lenske, Phys. Rev. C **61**, 064309 (2000). https://doi.org/10.1103/PhysRevC.61.064309
41. H. Lenske, C. Fuchs, Phys. Lett. B **345**, 355 (1995). https://doi.org/10.1016/0370-2693(94)01664-X

42. C. Fuchs, H. Lenske, H.H. Wolter, Phys. Rev. C **52**, 3043 (1995). https://doi.org/10.1103/PhysRevC.52.3043
43. F. Hofmann, C.M. Keil, H. Lenske, Phys. Rev. C **64**, 034314 (2001). https://doi.org/10.1103/PhysRevC.64.034314
44. C. Keil, H. Lenske, Phys. Rev. C **66**, 054307 (2002). https://doi.org/10.1103/PhysRevC.66.054307
45. H. Lenske, Lect. Notes Phys. **641**, 147 (2004). https://doi.org/10.1007/978-3-540-39911-7_5.
46. A. Fedoseew, H. Lenske, Phys. Rev. C **91**(3), 034307 (2015). https://doi.org/10.1103/PhysRevC.91.034307
47. S. Typel, H.H. Wolter, Nucl. Phys. A **656**, 331 (1999). https://doi.org/10.1016/S0375-9474(99)00310-3
48. D. Vretenar, A.V. Afanasjev, G.A. Lalazissis, P. Ring, Phys. Rep. **409**, 101 (2005). https://doi.org/10.1016/j.physrep.2004.10.001
49. H. Liang, J. Meng, S.G. Zhou, Phys. Rep. **570**, 1 (2015). https://doi.org/10.1016/j.physrep.2014.12.005
50. T. Nikšić, N. Paar, P.G. Reinhard, D. Vretenar, J. Phys. G **42**(3), 034008 (2015). https://doi.org/10.1088/0954-3899/42/3/034008
51. H.J. Schulze, M. Baldo, U. Lombardo, J. Cugnon, A. Lejeune, Phys. Rev. C **57**, 704 (1998). https://doi.org/10.1103/PhysRevC.57.704
52. I. Vidana, A. Polls, A. Ramos, M. Hjorth-Jensen, Nucl. Phys. A **644**, 201 (1998). https://doi.org/10.1016/S0375-9474(98)00599-5
53. D.E. Lanskoy, Y. Yamamoto, Phys. Rev. C **55**, 2330 (1997). https://doi.org/10.1103/PhysRevC.55.2330
54. H.J. Schulze, E. Hiyama, Phys. Rev. C **90**(4), 047301 (2014). https://doi.org/10.1103/PhysRevC.90.047301
55. S. Petschauer, J. Haidenbauer, N. Kaiser, U.G. Meissner, W. Weise, Nucl. Phys. A **957**, 347 (2017). https://doi.org/10.1016/j.nuclphysa.2016.09.010
56. N.K. Glendenning, S.A. Moszkowski, Phys. Rev. Lett. **67**, 2414 (1991). https://doi.org/10.1103/PhysRevLett.67.2414
57. N.K. Glendenning, D. Von-Eiff, M. Haft, H. Lenske, M.K. Weigel, Phys. Rev. C **48**, 889 (1993). https://doi.org/10.1103/PhysRevC.48.889
58. E.N.E. van Dalen, G. Colucci, A. Sedrakian, Phys. Lett. B **734**, 383 (2014). https://doi.org/10.1016/j.physletb.2014.06.002
59. A. Drago, A. Lavagna, G. Pagliara, D. Pigato, Phys. Rev. C **90**(6), 065809 (2014). https://doi.org/10.1103/PhysRevC.90.065809
60. A. Drago, A. Lavagna, G. Pagliara, D. Pigato, Eur. Phys. J. A **52**(2), 40 (2016). https://doi.org/10.1140/epja/i2016-16040-3
61. A. Drago, G. Pagliara, Eur. Phys. J. A **52**(2), 41 (2016). https://doi.org/10.1140/epja/i2016-16041-2
62. K.A. Maslov, E.E. Kolomeitsev, D.N. Voskresensky, J. Phys. Conf. Ser. **798**(1), 012070 (2017). https://doi.org/10.1088/1742-6596/798/1/012070
63. H. Lenske, M. Dhar, N. Tsoneva, J. Wilhelm, EPJ Web Conf. **107**, 10001 (2016). https://doi.org/10.1051/epjconf/201610710001
64. S. Schramm, V. Dexheimer, R. Negreiros, Eur. Phys. J. A **52**(1), 14 (2016). https://doi.org/10.1140/epja/i2016-16014-5
65. C. Patrignani et al., Chin. Phys. C **40**(10), 100001 (2016). https://doi.org/10.1088/1674-1137/40/10/100001
66. J.D. Bjorken, S. Drell, *Relativistic Quantum Mechanics* (McGraw-Hill Book Company, New York, 1964), 300 pp.
67. J.J. de Swart, Rev. Mod. Phys. **35**, 916 (1963) [Erratum: Rev. Mod. Phys. **37**, 326 (1965)]. https://doi.org/10.1103/RevModPhys.35.916.
68. R. Machleidt, K. Holinde, C. Elster, Phys. Rep. **149**, 1 (1987). https://doi.org/10.1016/S0370-1573(87)80002-9
69. R. Machleidt, Adv. Nucl. Phys. **19**, 189 (1989)

70. R. Blankenbecler, R. Sugar, Phys. Rev. **142**, 1051 (1966). https://doi.org/10.1103/PhysRev. 142.1051
71. J. Haidenbauer, U.G. Meissner, S. Petschauer, Nucl. Phys. A **954**, 273 (2016). https://doi.org/ 10.1016/j.nuclphysa.2016.01.006
72. W. Briscoe et al. (2016). http://gwdac.phys.gwu.edu/
73. K. Miyagawa, H. Kamada, W. Gloeckle, Nucl. Phys. A **614**, 535 (1997). https://doi.org/10. 1016/S0375-9474(96)00479-4
74. F. de Jong, H. Lenske, Phys. Rev. C **57**, 3099 (1998). https://doi.org/10.1103/PhysRevC.57. 3099
75. H. Feshbach, *Theoretical Nuclear Physics: Nuclear Reactions* (Wiley, New York, 1992), 960 p
76. J. Haidenbauer, U.G. Meissner, Nucl. Phys. A **881**, 44 (2012). https://doi.org/10.1016/j. nuclphysa.2012.01.021
77. N.K. Glendenning, *Compact Stars: Nuclear Physics, Particle Physics, and General Relativity* (Springer, New York, 1997)
78. K. Langanke, S.E. Koonin, J.A. Maruhn (eds.), *Computational Nuclear Physics. Vol. 2: Nuclear Reactions* (Springer, New York, 1993)
79. Y. Lim, C.H. Hyun, K. Kwak, C.H. Lee, Int. J. Mod. Phys. E **24**, 1550100 (2015)
80. H.P. Duerr, Phys. Rev. **103**, 469 (1956). https://doi.org/10.1103/PhysRev.103.469
81. J.D. Walecka, Ann. Phys. **83**, 491 (1974). https://doi.org/10.1016/0003-4916(74)90208-5
82. J. Boguta, A.R. Bodmer, Nucl. Phys. A **292**, 413 (1977). https://doi.org/10.1016/0375-9474(77)90626-1
83. B.D. Serot, J.D. Walecka, Int. J. Mod. Phys. E **6**, 515 (1997). https://doi.org/10.1142/ S0218301397000299
84. K. Maslov, E.E. Kolomeitsev, D.N. Voskresensky, Phys. Lett. B **748**, 369 (2015)
85. A. Ohnishi, K. Tsubakihara, K. Sumiyoshi, C. Ishizuka, S. Yamada, H. Suzuki, Nucl. Phys. A **835**, 374 (2010). https://doi.org/10.1016/j.nuclphysa.2010.01.222
86. S. Weissenborn, D. Chatterjee, J. Schaffner-Bielich, Phys. Rev. C **85**(6), 065802 (2012) [Erratum: Phys. Rev. C **90**(1), 019904 (2014)]. https://doi.org/10.1103/PhysRevC.85.065802; https://doi.org/10.1103/PhysRevC.90.019904
87. F. Hofmann, C.M. Keil, H. Lenske, Phys. Rev. C **64**, 025804 (2001). https://doi.org/10.1103/ PhysRevC.64.025804
88. C.J. Horowitz, B.D. Serot, Phys. Lett. B **137**, 287 (1984). https://doi.org/10.1016/0370-2693(84)91717-9
89. C.J. Horowitz, B.D. Serot, Phys. Lett. B **140**, 181 (1984). https://doi.org/10.1016/0370-2693(84)90916-X
90. C.J. Horowitz, B.D. Serot, Nucl. Phys. A **464**, 613 (1987) [Erratum: Nucl. Phys. A **473**, 760 (1987)]. https://doi.org/10.1016/0375-9474(87)90370-8; https://doi.org/10.1016/0375-9474(87)90281-8
91. A. Migdal, Sov. Phys. JETP **7**, 996 (1958)
92. A. Fetter, J. Walecka, *Quantum Theory of Many-Particle Systems* (McGraw-Hill, New York, 1971)
93. G. Bertsch, J. Borysowicz, H. McManus, W.G. Love, Nucl. Phys. A **284**, 399 (1977). https:// doi.org/10.1016/0375-9474(77)90392-X
94. F. Hofmann, H. Lenske, Phys. Rev. C **57**, 2281 (1998). https://doi.org/10.1103/PhysRevC.57. 2281
95. J. Meng, P. Ring, P. Zhao, Int. Rev. Nucl. Phys. **10**, 21 (2016). https://doi.org/10.1142/ 9789814733267_0002
96. P. Ring, Int. Rev. Nucl. Phys. **10**, 1 (2016). https://doi.org/10.1142/9789814733267_0001
97. J.W. Negele, Rev. Mod. Phys. **54**, 913 (1982). https://doi.org/10.1103/RevModPhys.54.913
98. H. Lenske, C. Xu, M. Dhar, T. Gaitanos, R. Shyam, in *Proceedings of the 12th International Conference on Hypernuclear and Strange Particle Physics (HYP2015)* (2017). https://doi.org/ 10.7566/HYP2015; http://journals.jps.jp/doi/abs/10.7566/HYP2015

99. H. Hotchi et al., Phys. Rev. C **64**, 044302 (2001). https://doi.org/10.1103/PhysRevC.64.044302
100. M. May et al., Phys. Rev. Lett. **78**, 4343 (1997). https://doi.org/10.1103/PhysRevLett.78.4343
101. T. Hasegawa et al., Phys. Rev. C **53**, 1210 (1996). https://doi.org/10.1103/PhysRevC.53.1210
102. S. Ajimura et al., Nucl. Phys. A **585**, 173C (1995). https://doi.org/10.1016/0375-9474(94)00562-2
103. P.H. Pile et al., Phys. Rev. Lett. **66**, 2585 (1991). https://doi.org/10.1103/PhysRevLett.66.2585
104. J. Mares, B.K. Jennings, Phys. Rev. C **49**, 2472 (1994). https://doi.org/10.1103/PhysRevC.49.2472
105. M. Danysz et al., Phys. Rev. Lett. **11**, 29 (1963). https://doi.org/10.1103/PhysRevLett.11.29
106. H. Takahashi et al., Phys. Rev. Lett. **87**, 212502 (2001). https://doi.org/10.1103/PhysRevLett.87.212502
107. J.K. Ahn et al., Phys. Rev. C **88**(1), 014003 (2013). https://doi.org/10.1103/PhysRevC.88.014003
108. Y. Fujiwara, Y. Suzuki, C. Nakamoto, Prog. Part. Nucl. Phys. **58**, 439 (2007). https://doi.org/10.1016/j.ppnp.2006.08.001
109. I.N. Filikhin, A. Gal, Phys. Rev. C **65**, 041001 (2002). https://doi.org/10.1103/PhysRevC.65.041001
110. I.R. Afnan, B.F. Gibson, Phys. Rev. C **67**, 017001 (2003). https://doi.org/10.1103/PhysRevC.67.017001
111. I. Vidana, A. Ramos, A. Polls, Phys. Rev. C **70**, 024306 (2004). https://doi.org/10.1103/PhysRevC.70.024306
112. T. Yamada, Phys. Rev. C **69**, 044301 (2004). https://doi.org/10.1103/PhysRevC.69.044301
113. Q.N. Usmani, A.R. Bodmer, B. Sharma, Phys. Rev. C **70**, 061001 (2004). https://doi.org/10.1103/PhysRevC.70.061001
114. L. Adamczyk et al., Phys. Rev. Lett. **114**(2), 022301 (2015). https://doi.org/10.1103/PhysRevLett.114.022301
115. K. Morita, T. Furumoto, A. Ohnishi, Phys. Rev. C **91**(2), 024916 (2015). https://doi.org/10.1103/PhysRevC.91.024916
116. P. Finelli, N. Kaiser, D. Vretenar, W. Weise, Nucl. Phys. A **831**, 163 (2009). https://doi.org/10.1016/j.nuclphysa.2009.10.083
117. S. Petschauer, N. Kaiser, J. Haidenbauer, U.G. Meissner, W. Weise, Phys. Rev. C **93**(1), 014001 (2016). https://doi.org/10.1103/PhysRevC.93.014001
118. S. Petschauer, J. Haidenbauer, N. Kaiser, U.G. Meissner, W. Weise, Eur. Phys. J. A **52**(1), 15 (2016). https://doi.org/10.1140/epja/i2016-16015-4
119. R. Wirth, R. Roth, Phys. Rev. Lett. **117**, 182501 (2016). https://doi.org/10.1103/PhysRevLett.117.182501
120. T. Inoue, N. Ishii, S. Aoki, T. Doi, T. Hatsuda, Y. Ikeda, K. Murano, H. Nemura, K. Sasaki, Prog. Theor. Phys. **124**, 591 (2010). https://doi.org/10.1143/PTP.124.591
121. T. Doi et al., PoS **LATTICE2016**, 110 (2017)
122. H. Nemura et al., PoS **LATTICE2016**, 101 (2017)
123. N. Ishii et al., PoS **LATTICE2016**, 127 (2017)
124. S.R. Beane, M.J. Savage, Phys. Lett. B **535**, 177 (2002). https://doi.org/10.1016/S0370-2693(02)01762-8
125. S.R. Beane, W. Detmold, K. Orginos, M.J. Savage, Prog. Part. Nucl. Phys. **66**, 1 (2011). https://doi.org/10.1016/j.ppnp.2010.08.002
126. S.R. Beane, E. Chang, S.D. Cohen, W. Detmold, H.W. Lin, T.C. Luu, K. Orginos, A. Parreno, M.J. Savage, A. Walker-Loud, Phys. Rev. Lett. **109**, 172001 (2012). https://doi.org/10.1103/PhysRevLett.109.172001
127. K. Orginos, A. Parreno, M.J. Savage, S.R. Beane, E. Chang, W. Detmold, Phys. Rev. D **92**(11), 114512 (2015). https://doi.org/10.1103/PhysRevD.92.114512
128. R.H. Dalitz, F. Von Hippel, Phys. Lett. **10**, 153 (1964). https://doi.org/10.1016/0031-9163(64)90617-1

129. Z. Fodor, C. Hoelbling, Rev. Mod. Phys. **84**, 449 (2012). https://doi.org/10.1103/RevModPhys.84.449
130. H. Sanchis-Alepuz, C.S. Fischer, Phys. Rev. D **90**(9), 096001 (2014). https://doi.org/10.1103/PhysRevD.90.096001
131. F. Aceti, E. Oset, Phys. Rev. D **86**, 014012 (2012). https://doi.org/10.1103/PhysRevD.86.014012
132. F. Aceti, L.R. Dai, L.S. Geng, E. Oset, Y. Zhang, Eur. Phys. J. A **50**, 57 (2014). https://doi.org/10.1140/epja/i2014-14057-2
133. W.C. Chang, J.C. Peng, Phys. Rev. Lett. **106**, 252002 (2011). https://doi.org/10.1103/PhysRevLett.106.252002
134. W.C. Chang, J.C. Peng, Phys. Lett. B **704**, 197 (2011). https://doi.org/10.1016/j.physletb.2011.08.077
135. S. Sarkar, E. Oset, M.J. Vicente Vacas, Nucl. Phys. A **750**, 294 (2005) [Erratum: Nucl. Phys. A **780**, 90 (2006)]. https://doi.org/10.1016/j.nuclphysa.2005.01.006; https://doi.org/10.1016/j.nuclphysa.2006.09.019
136. S. Sarkar, B.X. Sun, E. Oset, M.J. Vicente Vacas, Eur. Phys. J. A **44**, 431 (2010). https://doi.org/10.1140/epja/i2010-10956-4
137. Y. Kamiya, K. Miyahara, S. Ohnishi, Y. Ikeda, T. Hyodo, E. Oset, W. Weise, Nucl. Phys. A **954**, 41 (2016). https://doi.org/10.1016/j.nuclphysa.2016.04.013
138. F. Aceti, E. Oset, L. Roca, Phys. Rev. C **90**(2), 025208 (2014). https://doi.org/10.1103/PhysRevC.90.025208
139. C.W. Xiao, F. Aceti, M. Bayar, Eur. Phys. J. A **49**, 22 (2013). https://doi.org/10.1140/epja/i2013-13022-y
140. Q.B. Li, P.N. Shen, A. Faessler, Phys. Rev. C **65**, 045206 (2002). https://doi.org/10.1103/PhysRevC.65.045206
141. R.L. Jaffe, Phys. Rev. Lett. **38**, 195 (1977) [Erratum: Phys. Rev. Lett. **38**, 617 (1977)]. https://doi.org/10.1103/PhysRevLett.38.195
142. G.F. Bertsch, B.A. Li, G.E. Brown, V. Koch, Nucl. Phys. A **490**, 745 (1988). https://doi.org/10.1016/0375-9474(88)90024-3
143. J. Helgesson, J. Randrup, Ann. Phys. **244**, 12 (1995). https://doi.org/10.1006/aphy.1995.1106
144. A.B. Larionov, U. Mosel, Nucl. Phys. A **728**, 135 (2003). https://doi.org/10.1016/j.nuclphysa.2003.08.005
145. V. Shklyar, H. Lenske, Phys. Rev. C **80**, 058201 (2009). https://doi.org/10.1103/PhysRevC.80.058201
146. M. Martini, M. Ericson, G. Chanfray, J. Marteau, Phys. Rev. C **80**, 065501 (2009). https://doi.org/10.1103/PhysRevC.80.065501
147. M. Martini, M. Ericson, G. Chanfray, J. Marteau, Phys. Rev. C **81**, 045502 (2010). https://doi.org/10.1103/PhysRevC.81.045502
148. J. Nieves, I. Ruiz Simo, M.J. Vicente Vacas, Phys. Rev. C **83**, 045501 (2011). https://doi.org/10.1103/PhysRevC.83.045501
149. F. Osterfeld, Rev. Mod. Phys. **64**, 491 (1992). https://doi.org/10.1103/RevModPhys.64.491
150. T. Udagawa, P. Oltmanns, F. Osterfeld, S.W. Hong, Phys. Rev. C **49**, 3162 (1994). https://doi.org/10.1103/PhysRevC.49.3162
151. I. Vidana, J. Benlliure, H. Geissel, H. Lenske, C. Scheidenberger, J. Vargas, EPJ Web Conf. **107**, 10003 (2016). https://doi.org/10.1051/epjconf/201610710003
152. J. Benlliure et al., Nuovo Cim. C **39**(6), 401 (2016). https://doi.org/10.1393/ncc/i2016-16401-0
153. T.E.O. Ericson, W. Weise, *Pions and Nuclei*, vol. 74 (Clarendon Press, Oxford, 1988). http://www-spires.fnal.gov/spires/find/books/www?cl=QC793.5.M42E75::1988
154. V. Dmitriev, O. Sushkov, C. Gaarde, Nucl. Phys. A **459**, 503 (1986). https://doi.org/10.1016/0375-9474(86)90158-2
155. G.E. Brown, W. Weise, Phys. Rep. **22**, 279 (1975). https://doi.org/10.1016/0370-1573(75)90026-5

156. E. Oset, H. Toki, W. Weise, Phys. Rep. **83**, 281 (1982). https://doi.org/10.1016/0370-1573(82)90123-5
157. J. Meyer-Ter-Vehn, Phys. Rep. **74**(4), 323 (1981). https://doi.org/http://dx.doi.org/10.1016/0370-1573(81)90151-4; http://www.sciencedirect.com/science/article/pii/0370157381901514
158. A.B. Migdal, E.E. Saperstein, M.A. Troitsky, D.N. Voskresensky, Phys. Rep. **192**, 179 (1990). https://doi.org/10.1016/0370-1573(90)90132-L
159. E. Oset, M. Rho, Phys. Rev. Lett. **42**, 47 (1979). https://doi.org/10.1103/PhysRevLett.42.47
160. E. Oset, W. Weise, Nucl. Phys. A **319**, 477 (1979). https://doi.org/10.1016/0375-9474(79)90527-X
161. C. Garcia-Recio, E. Oset, L.L. Salcedo, D. Strottman, M.J. Lopez, Nucl. Phys. A **526**, 685 (1991). https://doi.org/10.1016/0375-9474(91)90438-C
162. K. Wehrberger, R. Wittman, Nucl. Phys. A **513**, 603 (1990). https://doi.org/10.1016/0375-9474(90)90400-G
163. E. Oset, L.L. Salcedo, Nucl. Phys. A **468**, 631 (1987). https://doi.org/10.1016/0375-9474(87)90185-0
164. A. Fedoseew, JLU Giessen (2017)
165. R.M. Sealock et al., Phys. Rev. Lett. **62**, 1350 (1989). https://doi.org/10.1103/PhysRevLett.62.1350
166. P. Barreau et al., Nucl. Phys. A **402**, 515 (1983). https://doi.org/10.1016/0375-9474(83)90217-8
167. J.S. O'Connell et al., Phys. Rev. Lett. **53**, 1627 (1984). https://doi.org/10.1103/PhysRevLett.53.1627
168. P.K.A. de Witt Huberts, J. Phys. G **16**, 507 (1990). https://doi.org/10.1088/0954-3899/16/4/004
169. T. De Forest Jr., J.D. Walecka, Adv. Phys. **15**, 1 (1966). https://doi.org/10.1080/00018736600101254
170. O. Benhar, D. day, I. Sick, Rev. Mod. Phys. **80**, 189 (2008). https://doi.org/10.1103/RevModPhys.80.189
171. L.h. Xia, P.J. Siemens, M. Soyeur, Nucl. Phys. A **578**, 493 (1994). https://doi.org/10.1016/0375-9474(94)90757-9
172. J. Piekarewicz, Int. J. Mod. Phys. E **24**(09), 1541003 (2015). https://doi.org/10.1142/S0218301315410037
173. R.M. Edelstein, E.J. Makuchowski, C.M. Meltzer, E.L. Miller, J.S. Russ, B. Gobbi, J.L. Rosen, H.A. Scott, S.L. Shapiro, L. Strawczynski, Phys. Rev. Lett. **38**, 185 (1977). https://doi.org/10.1103/PhysRevLett.38.185
174. C.G. Cassapakis et al., Phys. Lett. B **63**, 35 (1976). https://doi.org/10.1016/0370-2693(76)90462-7
175. D.A. Lind, Can. J. Phys. **65**, 637 (1987). https://doi.org/10.1139/p87-090
176. B.E. Bonner, J.E. Simmons, C.R. Newsom, P.J. Riley, G. Glass, J.C. Hiebert, M. Jain, L.C. Northcliffe, Phys. Rev. C **18**, 1418 (1978). https://doi.org/10.1103/PhysRevC.18.1418
177. B.E. Bonner, J.E. Simmons, C.L. Hollas, C.R. Newsom, P.J. Riley, G. Glass, M. Jain, Phys. Rev. Lett. **41**, 1200 (1978). https://doi.org/10.1103/PhysRevLett.41.1200
178. T. Hennino et al., Phys. Rev. Lett. **48**, 997 (1982). https://doi.org/10.1103/PhysRevLett.48.997
179. D. Contardo et al., Phys. Lett. B **168**, 331 (1986). https://doi.org/10.1016/0370-2693(86)91639-4
180. D. Bachelier et al., Phys. Lett. B **172**, 23 (1986). https://doi.org/10.1016/0370-2693(86)90209-1
181. D. Krpic, J. Puzovic, S. Drndarevic, R. Maneska, S. Backovic, J. Bogdanowicz, A.P. Cheplakov, S.Y. Sivoklokov, V.G. Grishin, Phys. Rev. C **46**, 2501 (1992). https://doi.org/10.1103/PhysRevC.46.2501
182. D. Krpic, G. Skoro, I. Picuric, S. Backovic, S. Drndarevic, Phys. Rev. C **65**, 034909 (2002). https://doi.org/10.1103/PhysRevC.65.034909

183. T. Hennino et al., Phys. Lett. B **283**, 42 (1992). https://doi.org/10.1016/0370-2693(92)91423-7

184. T. Hennino et al., Phys. Lett. B **303**, 236 (1993). https://doi.org/10.1016/0370-2693(93)91426-N

185. J. Chiba et al., Phys. Rev. Lett. **67**, 1982 (1991). https://doi.org/10.1103/PhysRevLett.67.1982

186. S. Das, Phys. Rev. C **66**, 014604 (2002). https://doi.org/10.1103/PhysRevC.66.014604

187. K.K. Olimov, S.L. Lutpullaev, K. Olimov, K.G. Gulamov, J.K. Olimov, Phys. Rev. C **75**, 067901 (2007). https://doi.org/10.1103/PhysRevC.75.067901

188. K.K. Olimov, Phys. Atom. Nucl. **71**, 93 (2008) [Yad. Fiz. **71**, 94 (2008)]. https://doi.org/10.1007/s11450-008-1010-2.

189. L. Simic, M.V. Milosavljevic, I. Mendas, D. Krpic, D.S. Popovic, Phys. Rev. C **80**, 017901 (2009). https://doi.org/10.1103/PhysRevC.80.017901

190. K.K. Olimov, M.Q. Haseeb, I. Khan, A.K. Olimov, V.V. Glagolev, Phys. Rev. C **85**, 014907 (2012). https://doi.org/10.1103/PhysRevC.85.014907

191. D. Krpic, S. Drndarevic, J. Ilic, G. Skoro, I. Picuric, S. Backovic, Eur. Phys. J. A **20**, 351 (2004). https://doi.org/10.1140/epja/i2003-10175-2

192. V. Metag, Prog. Part. Nucl. Phys. **30**, 75 (1993). https://doi.org/10.1016/0146-6410(93)90007-3

193. V. Metag, Nucl. Phys. A **553**, 283C (1993). https://doi.org/10.1016/0375-9474(93)90629-C

194. T. Udagawa, S.W. Hong, F. Osterfeld, Phys. Lett. B **245**, 1 (1990). https://doi.org/10.1016/0370-2693(90)90154-X.

195. B. Koerfgen, F. Osterfeld, T. Udagawa, Phys. Rev. C **50**, 1637 (1994). https://doi.org/10.1103/PhysRevC.50.1637

196. M. Trzaska et al., Z. Phys. A **340**, 325 (1991). https://doi.org/10.1007/BF01294681

197. E. Oset, L.L. Salcedo, D. Strottman, Phys. Lett. B **165**, 13 (1985). https://doi.org/10.1016/0370-2693(85)90681-1

198. J. Aichelin, Phys. Rep. **202**, 233 (1991). https://doi.org/10.1016/0370-1573(91)90094-3

199. W. Cassing, V. Metag, U. Mosel, K. Niita, Phys. Rep. **188**, 363 (1990). https://doi.org/10.1016/0370-1573(90)90164-W

200. H. Stoecker, W. Greiner, Phys. Rep. **137**, 277 (1986). https://doi.org/10.1016/0370-1573(86)90131-6

201. U. Mosel, Ann. Rev. Nucl. Part. Sci. **41**, 29 (1991). https://doi.org/10.1146/annurev.ns.41.120191.000333

202. M. Eskef et al., Eur. Phys. J. A **3**, 335 (1998). https://doi.org/10.1007/s100500050188

203. E.L. Hjort et al., Phys. Rev. Lett. **79**, 4345 (1997). https://doi.org/10.1103/PhysRevLett.79.4345

204. G.E. Brown, M. Rho, Phys. Rev. Lett. **66**, 2720 (1991). https://doi.org/10.1103/PhysRevLett.66.2720

205. R. Rapp, Nucl. Phys. A **725**, 254 (2003). https://doi.org/10.1016/S0375-9474(03)01581-1

206. E.V. Shuryak, G.E. Brown, Nucl. Phys. A **717**, 322 (2003). https://doi.org/10.1016/S0375-9474(03)00672-9

207. G. Bertsch, S. Das Gupta, Phys. Rep. **160**, 189 (1988)

208. W. Ehehalt, W. Cassing, A. Engel, U. Mosel, G. Wolf, Phys. Rev. C **47**, R2467 (1993). https://doi.org/10.1103/PhysRevC.47.R2467

209. S.A. Bass, C. Hartnack, H. Stoecker, W. Greiner, Phys. Rev. C **51**, 3343 (1995). https://doi.org/10.1103/PhysRevC.51.3343

210. S. Teis, W. Cassing, M. Effenberger, A. Hombach, U. Mosel, G. Wolf, Z. Phys. A **356**, 421 (1997). https://doi.org/10.1007/BF02769248,10.1007/s002180050198

211. S. Teis, W. Cassing, M. Effenberger, A. Hombach, U. Mosel, G. Wolf, Z. Phys. A **359**, 297 (1997). https://doi.org/10.1007/s002180050405

212. J. Helgesson, J. Randrup, Phys. Lett. B **411**, 1 (1997). https://doi.org/10.1016/S0370-2693(97)00965-9

213. J. Helgesson, J. Randrup, Phys. Lett. B **439**, 243 (1998). https://doi.org/10.1016/S0370-2693(98)01109-5

214. H. Weber, E.L. Bratkovskaya, W. Cassing, H. Stoecker, Phys. Rev. C **67**, 014904 (2003). https://doi.org/10.1103/PhysRevC.67.014904
215. D. Pelte et al., Z. Phys. A **357**, 215 (1997). https://doi.org/10.1007/s002180050236
216. A.B. Larionov, W. Cassing, S. Leupold, U. Mosel, Nucl. Phys. A **696**, 747 (2001). https://doi.org/10.1016/S0375-9474(01)01216-7
217. J. Helgesson, J. Randrup, Ann. Phys. **274**, 1 (1999). https://doi.org/10.1006/aphy.1999.5906
218. P.A. Henning, H. Umezawa, Nucl. Phys. A **571**, 617 (1994). https://doi.org/10.1016/0375-9474(94)90713-7
219. C.L. Korpa, R. Malfliet, Phys. Rev. C **52**, 2756 (1995). https://doi.org/10.1103/PhysRevC.52.2756
220. M. Hirata, J.H. Koch, E.J. Moniz, F. Lenz, Ann. Phys. **120**, 205 (1979). https://doi.org/10.1016/0003-4916(79)90287-2
221. O. Buss, L. Alvarez-Ruso, A.B. Larionov, U. Mosel, Phys. Rev. C **74**, 044610 (2006). https://doi.org/10.1103/PhysRevC.74.044610
222. S.A. Wood, J.L. Matthews, E.R. Kinney, P.A.M. Gram, G.A. Rebka, D.A. Roberts, Phys. Rev. C **46**, 1903 (1992). https://doi.org/10.1103/PhysRevC.46.1903
223. J. Vargas, J. Benlliure, M. Caamano, Nucl. Instrum. Methods A **707**, 16 (2013). https://doi.org/10.1016/j.nima.2012.12.087

Chapter 6
Particle Acceleration Driven by High-Power, Short Pulse Lasers

Peter G. Thirolf

Abstract The availability of high-power, short-pulse laser systems has created over the last two decades a novel branch of accelerator physics. Laser-driven electron and ion acceleration based on the enormous acceleration fields that can be realized over very short acceleration lengths during ultra-short pulse periods has gained enormous interest and huge efforts are conducted worldwide to realize optimized acceleration schemes and beam properties, drawing on the rapid development of the underlying laser technology. The present chapter reviews this development, starting with a primer on the basic features of relativistic laser-plasma interaction. Subsequently, laser-driven acceleration mechanisms for electrons and ions are introduced. Finally, an example for the application of laser-driven ion bunches for nuclear astrophysics is given, based on the unprecedented high density of laser-accelerated ion bunches.

6.1 Introduction

Huge particle accelerators have been the workhorses of research in particle physics for more than eight decades. Through high-energy collisions of accelerated particles, the fundamental building-blocks and forces of nature have been revealed. Most prominent, the Large Hadron Collider (LHC) at CERN in Geneva, achieved to find the Higgs boson, a particle associated with the mechanism through which all other known particles are thought to acquire their masses. But the size and cost of such machines—for the LHC, a 27-km circumference and several billion euros—are raising skepticism if a next-generation elementary particle physics accelerator could realistically be built based on conventional accelerator technology. Therefore serious efforts are ongoing to develop new and more compact accelerator technologies, able to target the ultimate energy frontier as well as providing new properties for a wide range of applications.

P. G. Thirolf (✉)
Ludwig-Maximilians-Universität München, Garching, Germany
e-mail: Peter.Thirolf@Physik.Uni-Muenchen.de

© Springer International Publishing AG, part of Springer Nature 2018
C. Scheidenberger, M. Pfützner (eds.), *The Euroschool on Exotic Beams - Vol. 5*,
Lecture Notes in Physics 948, https://doi.org/10.1007/978-3-319-74878-8_6

The availability of high-power, short-pulse laser systems has created over the last two decades a novel branch of accelerator physics. Laser-driven electron and ion acceleration has gained enormous interest and huge efforts are conducted worldwide to realize optimized acceleration schemes and beam properties [1, 2], drawing on the rapid development of the underlying laser technology. The world map of intense laser laboratories, compiled by the International Committee on Ultra-High Intensity Lasers (ICUIL, http://www.icuil.org/activities/laser-labs.html) and shown in Fig. 6.1, illustrates a cumulative laser peak power (operational or planned for 2017) of about 132 PW, worth an investment of more than 4 billion US dollars and representing the effort of more than 1500 laser scientists and engineers (not counting the large nuclear-fusion laser projects in the US (National Ignition Facility NIF [3] and France (Laser Megajoule [4]).

Nowadays available focused laser intensities well beyond 10^{20} W/cm^2 have enabled, e.g., the generation of multi-GeV electron beams in laser wake field acceleration (LWFA) experiments over a few cm in typically 10^{17-18} e/cm^3 dense plasmas [5, 6].

At present, the achieved laser peak power is in the petawatt regime (PW, 10^{15} W) [7, 8], which is one million times higher than a gigawatt electric power plant, although the lasers deliver this power in an ultra-short pulse (typically few tens of fs, 10^{-15} s) only. Multi-PW and exawatt (EW, 10^{18} W) lasers are now under construction or planned to be built [9, 10]. The rapid progress of laser technology since the invention of the laser by Maiman in 1960 [11] can be seen in Fig. 6.2, where the development of laser intensity and laser power (left axes) as well as laser-accelerated ion energies (right axis) are indicated as a function of time since 1960 (adapted from [8]).

The steep rise went along with a continuous shortening of the laser pulse duration from μs to fs for highest focal intensities. The key enabling technology that allowed overcoming the stagnation visible in the development of the laser intensity in the 1970s to mid 1980s is chirped pulse amplification (CPA) [12].

Here, as shown in Fig. 6.3, the potential damage of intense laser pulses to the gain medium was circumvented by stretching out an ultrashort laser pulse in time prior to introducing it to the gain medium, using a pair of gratings that are arranged such that the low-frequency component of the laser pulse travels a shorter path than the high-frequency component. After going through the grating pair, the laser pulse becomes positively chirped, that is, the high-frequency component lags behind the low-frequency component and exhibits a longer pulse duration than the original by a factor of 10^3–10^5. Then the stretched pulse is safely introduced to the gain medium and amplified by a factor 10^6 or more. Finally, the amplified laser pulse is recompressed back to the original pulse width through the reversal process of stretching, achieving orders of magnitude higher peak power than laser systems could generate before the invention of CPA.

Also indicated in Fig. 6.2 by dashed lines are the relativistic limits (for a laser wavelength of 1 μm) for laser-accelerated electrons (10^{18} W/cm^2) and protons (5×10^{24} W/cm^2), characterized by their normalized vector potential a = eA/mc^2 with A being the electromagnetic vector potential of the laser wave and m denoting

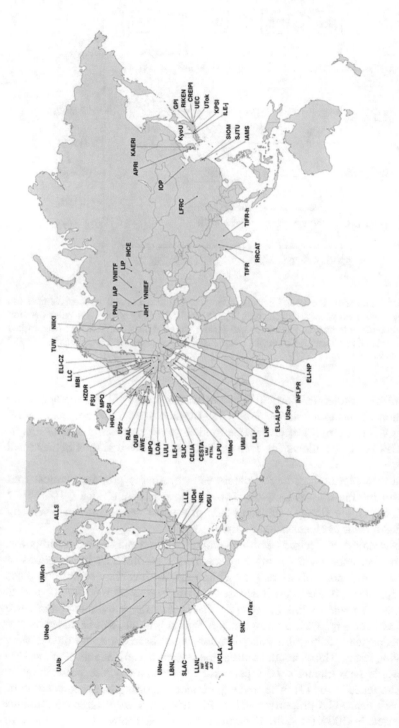

Fig. 6.1 World map of ultra-high intensity ($>10^{19}$ W/cm^2) laser facilities, compiled by the International Committee on Ultra-High Intensity Lasers (ICUIL, http://www.icuil.org/activities/laser-labs.html)

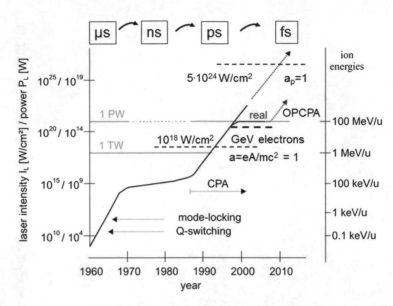

Fig. 6.2 Development of the laser intensity, laser power (left axes) and laser-accelerated ion energies (right axis) vs. years (adapted from [8]). Drastic shortening of the laser pulse duration from microseconds to femtoseconds was crucial for the achieved progress. Indicated with dashed lines are the relativistic limits for electrons (normalized vector potential a = eA/mc^2 = 1) and protons (a_p = 1)

the mass of the accelerated particle (e.g. electron or proton). While the optimistic extrapolation of the trend started with CPA (dashed arrow) did not materialize, an alternative laser amplification scheme (optical parametric chirped pulse amplification, OPCPA [13]) allows for even shorter laser pulses and thus higher focal intensities.

Laser-driven plasma-based accelerators were originally proposed almost four decades ago by Tajima and Dawson [14]. In a low-density plasma (10^{-5}–10^{-4} of solid density), the slowly oscillating field of a wake wave with phase velocity approaching c leads to electron acceleration.

The observation of electrons and ions emitted in laser-plasma interactions was first reported in experiments using high-intensity laser pulses with durations of a few ns to some hundreds of ps [15]. Following the invention of chirped pulse amplification, laser pulses with intensities well in excess of 10^{18} W/cm^2 and fs duration were realized. A link between the driver intensity and the resulting pulse duration has been suggested by Mourou and Tajima [16] and can be stated as: "In order to decrease the achievable pulse duration, we must first increase the intensity of the driving laser." This is not the same as the reverse trivial statement "to increase the achievable peak intensity of a pulse for a given energy, the pulse duration must be shortened" [16, 17]. The first experiments reporting laser acceleration of protons with beam-like properties and multi-MeV energies in laser experiments were reported in 2000 [18–20]. Experiments over the following 16 years have

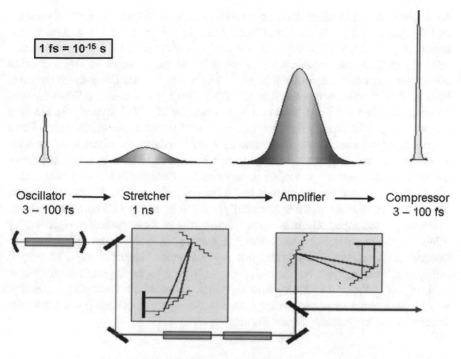

Fig. 6.3 Principle of the chirped pulse amplification scheme (for details see text)

demonstrated, over a wide range of laser and target parameters, the generation of multi-MeV proton and ion beams with unique properties, such as ultrashort burst emission, high brilliance, and low emittance, which have in turn stimulated ideas for a wide range of innovative applications.

6.2   Relativistic Laser-Plasma Interaction

When an ultra-high peak power laser pulse interacts with a target, a plasma is inevitably created due to heating and breakdown by either a relatively long preceding light or, if such pre-pulses are suppressed, due to the optical field ionization [21, 22] by the main pulse. The plasma consists of charged particles, i.e. electrons and ions. The particles with the largest charge-to-mass ratio, the electrons, are accelerated first by the laser. The motion of electrons in this regime is governed by the dimensionless laser amplitude a_0 (the normalized vector potential)

$$a_0 = \frac{eE_0}{m_e c \omega} = \sqrt{\frac{I_0}{I_l}}, \quad I_l = 1.37 \, x10^{18} \frac{W}{cm^2} (\mu m/\lambda)^2, \tag{6.1}$$

where e and m_e are the electron charge and mass, c is the velocity of light in vacuum, and I_0, E_0, λ and ω are the laser peak intensity, electric field, wavelength and angular frequency, respectively. The value of I_l is given here for linear polarization of the laser light, for circular polarization it is two times larger. In terms of other commonly used units, $a_0 = 0.85 \times 10^{-9}\sqrt{I}\lambda$, where I is the intensity of the laser light in W/cm^2 and λ is the wavelength of the laser light in microns. When $a_0 \approx 1$, which is satisfied for 1 μm light at a laser intensity of $\sim 10^{18}$ W/cm^2, the electron mass m_e begins to change significantly compared to the electron rest mass. For a typical laser wavelength of 1 μm and $I_0 \gg 10^{18}$ W/cm^2, the dimensionless laser amplitude is much larger than unity. In this case, a significant number of plasma electrons is accelerated up to relativistic velocities during a few cycles of the laser electric field. Such a plasma is called a "relativistic plasma" [8, 23, 24], which has several unique features, e.g. it is essentially collisionless [25]. In contrast to vacuum, in a plasma and especially in a dense plasma, the particle motion is significantly affected by the collective fields, and the laser pulse itself is modified by non-linearities. Nowadays, based on the previously described laser technique of chirped pulse amplification, laser pulses with focused intensities far beyond the relativistic limit of 10^{18} W/cm^2 can be generated, which generate macroscopic amounts of relativistic plasma in dense matter with novel properties concerning laser-matter interaction. A key quantity is the plasma frequency

$$\omega_p^{rel} = \sqrt{\frac{4\pi e^2 n_e}{m \langle \gamma \rangle}}, \tag{6.2}$$

which is determined by the electron density n_e and the electron mass m. The increase of mass by the relativistic factor $\langle \gamma \rangle$, averaged locally over many electrons, reduces the plasma frequency with far-reaching consequences for light propagation in the plasma. The most prominent ones are: (a) laser-induced transparency, (b) relativistic self-focusing, (c) profile steepening at the pulse front. They are illustrated in Fig. 6.4.

Qualitatively, these effects can be understood in terms of the dispersion relation

$$\omega^2 = \left(\omega_p^{rel}\right)^2 + c^2 k^2, \tag{6.3}$$

which controls the propagation of laser light in a plasma with frequency ω, wave vector k and vacuum velocity c. It leads to the plasma index of refraction

$$n_R = \sqrt{1 - \left(\omega_p^{rel}\big/\omega\right)^2} = \sqrt{1 - \left(\frac{n_e}{n_c}\right)\big/\langle \gamma \rangle}, \tag{6.4}$$

where n_c is the critical density, at which a plasma becomes non-transparent in the non-relativistic regime $\langle \gamma \rangle = 1$ (with $\omega^{rel} = \omega_p/\langle \gamma \rangle$). At relativistic laser intensities, however, the $\langle \gamma \rangle$- factor depends on the local intensity I, approximately $\langle \gamma \rangle \propto I^{1/2}$, and one enters a regime of relativistic non-linear optics as well as relativistic

Fig. 6.4 Effects of relativistic non-linear optics: (**a**) induced transparency. A plasma with $\omega_p > \omega_{Las}$ becomes transparent for high intensities producing $\langle\gamma\rangle \gg 1$. (**b**) Phase fronts in the central region of the laser pulse with high intensity and enhanced index of refraction n_R move with lower phase velocity $v_p = c/n_R$ than in the wings, where the intensity is smaller. The plasma then acts like a positive lens and relativistic self-focusing occurs. (**c**) Central regions of the laser pulse with peak intensities move with a larger group velocity $v_g = cn_R$ than the front region, where the intensity is lower, this leads to profile steepening

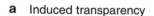

a   Induced transparency

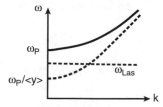

b   relativistic self-focussing: $V_P = c / n_R$

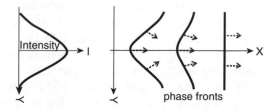

c   profile steepening: $V_g = c \, n_R$

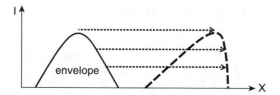

plasma physics. Electrons oscillate at relativistic velocities in laser fields that exceed 10^{11} V/cm, which results in relativistic mass changes exceeding the electron rest mass. At this point, the magnetic field of the electromagnetic wave also becomes important. Electrons behave in such fields as if the light wave was rectified. If the plasma frequency exceeds the frequency of the laser light ($\omega_p > \omega_{Las}$), the plasma becomes transparent for high intensities producing $\langle\gamma\rangle \gg 1$ (induced transparency illustrated in panel a) of Fig. 6.4). Phase fronts in the central region of the laser pulse with high intensity and an enhanced index of refraction n_R will propagate with a lower phase velocity $v_p = c/n_R$ than in the wings, where the intensity is smaller. Therefore, the plasma then acts like a positive lens and relativistic self-focusing occurs (Fig. 6.4, panel b). Moreover, central regions of the laser pulse with peak intensities move with a larger group velocity $v_g = cn_R$ than the front region, where the intensity is lower, this leads to a steepening of the pulse profile (panel c). Of particular importance is the effect of relativistic self-focusing, which means that the quasistatic magnetic field may become strong enough to pinch the relativistic electrons and that the path of light follows the electron deflection. As a consequence, the pattern of current and light filaments coalesces into a single narrow channel containing a significant part of the initial laser power. Relativistic self-focusing occurs, when the laser power exceeds a critical power, given by:

$$P_c = 17(\omega_{las}/\omega_p)^2 \text{ GW},$$

with ω_{las} denoting the laser frequency, while ω_p represents the plasma frequency. On the other hand, photo-ionization can defocus the light and thus increase the self-focusing threshold, by increasing the on-axis density and the refractive index. When this focusing effect just balances the defocusing due to diffraction, the laser pulse can be self-guided, or propagate over a long distance with high intensity. The effect of self-focusing can be seen in Fig. 6.5, where the result of a 3D particle-in-cell (PIC) simulation of a relativistic laser pulse (10^{19} W/cm^2) creating a plasma (with electron density $n_e/n_c = 0.6$) is displayed [26]. Particle-in-Cell (PIC) simulations are the standard computational tool to study the behaviour of laser-plasma interactions over macroscopic volumes [27]. The laser plasma-interaction is solved at the fundamental level of Maxwell's equations and the equation of motion for relativistic particles moving in the electromagnetic fields, which are averaged over cells. In the example shown in Fig. 6.5, the incident laser beam with enough intensity to create a relativistic plasma first propagates through an unstable filamentory stage (blue section) and then collapses into a single channel with a width of (1–2) λ (laser wavelength).

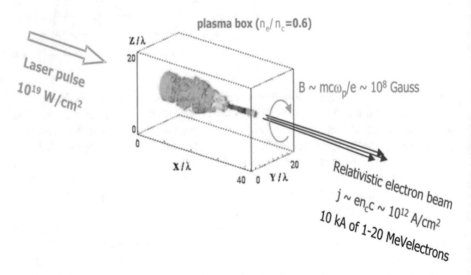

Fig. 6.5 Result of a 3D particle-in-cell simulation illustrating the effect of relativistic self-focusing of an incident relativistic laser beam via laser-plasma interaction. The initially filamented light path follows the path of the resulting relativistic electron beam, accelerated with high current (ca. 10 kA) and accompanied by a high quasistatic magnetic field (ca. 100 MG). Figure taken from [26]. (Reprinted with permission from A. Pukhov and J. Meyer-ter-Vehn, Phys. Rev. Lett. 76, 3975 (1996). Copyright 1996 by the American Physical Society)

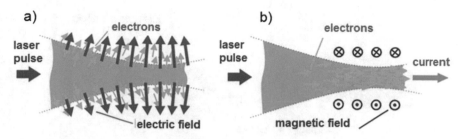

Fig. 6.6 (**a**) Illustration of the macroscopic charge separation induced by the light pressure of an intense laser pulse creating a relativistic plasma and pushing aside the free electrons, resulting in a quasi-static electric field of up to TV/m. (**b**) A strong electron current (kA) is dragged by the laser pulse, accompanied by a strong quasi-static magnetic field (ca. 10^4 T)

The new regime of relativistic laser-plasma interaction became accessible, when focused laser intensities in excess of 10^{18} W/cm^2 could be realized with laser pulses of fs duration. Here, the quiver velocity of electrons in the electromagnetic field of the laser approaches the speed of light. Moreover, the $v \times B$-term of the Lorentz force starts to dominate and pushes the electrons into the direction of the laser propagation, creating high electron pulse currents (typically several kA) with induced quasi-static magnetic fields (in the order of 10^8 Gauss), as illustrated by Fig. 6.6.

The ability to generate such relativistic electrons was a mandatory prerequisite for the acceleration of ions to high energies.

6.3   Laser-Driven Electron Acceleration

Using the interaction of ultra-intense lasers with relativistic plasmas is an elegant and efficient way to generate electron beams. Moreover, laser-accelerated electrons will always accompany the generation of laser-accelerated protons and heavier ions. Therefore in this section an overview about the methods and status of laser-driven electron acceleration will be given. For a review of laser-driven plasma-based electron acceleration see Ref. [28]. This acceleration scheme was originally proposed by Tajima and Dawson almost 40 years ago [14]. Laser-plasma accelerator experiments prior to 2004 have demonstrated acceleration gradients >100 GV/m, accelerated electrons with energies >100 MeV and accelerated bunch charges >1 nC [29–35]. However, the quality of the accelerated electron bunches did not meet the expectations. Typically, the accelerated bunch was characterized by an exponentially decreasing energy distribution, with most of the electrons at low energies (<10 MeV) and a long tail extending to higher energies (>100 MeV). This changed drastically, when in 2004 a new milestone in laser plasma acceleration was reached, with three groups simultaneously reporting on the production of high-quality, quasi-monoenergetic electron beams, characterized by significant charge

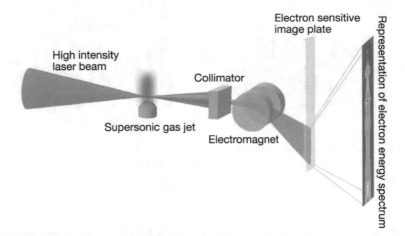

Fig. 6.7 Experimental setup for quasi-monoenergetic laser acceleration of electrons (for details see text). Figure taken from [37]. (Reprinted by permission from MacMillan Publishers Ltd.: Nature 431, 535 (2004), copyright 2004)

($\geq$100 pC) at high average energy (100 MeV), with small energy spread (ca. a few percent) and low divergence (few milliradians) [36–38]. A typical setup for these experiments is shown in Fig. 6.7 [37]. The experiment used the high-power titanium: sapphire laser system at the Rutherford Appleton Laboratory (Astra). The laser pulses ($\lambda = 800$ nm, $\tau = 40$ fs with energy approximately 0.5 J on target) were focused with an f/16.7 off-axis parabolic mirror (f: ratio between focal length and mirror diameter) onto the edge of a 2-mm-long supersonic jet of helium gas to produce peak intensities up to 2.5×10^{18} W cm^{-2}. The electron density (n_e) of the plasma was observed to vary linearly with backing pressure within the range $n_e = 3 \times 10^{18}$ cm^{-3}–5×10^{19} cm^{-3}. Electron spectra were measured using an on-axis magnetic spectrometer. Other diagnostics that was used included transverse imaging of the interaction, and radiochromic film stacks to measure the divergence and total number of accelerated electrons.

The electron acceleration mechanism in a relativistic plasma is based on the extremely high electric field gradients (up to 1 TV/m) achievable with multi-terawatt lasers as discussed before. One of the appealing applications for these lasers is laser wake field acceleration (LWFA) of charged particles in plasma. When a laser pulse propagates through an underdense plasma, it excites a running plasma wave oscillating with the plasma frequency $\omega_p{}^2 = 4\pi e^2 n_e/m$, where e, m, and n_e denote charge, mass, and density of electrons, respectively. The wave trails the laserpulse with phase velocity set by the laser pulse group velocity

$$v_{ph}^{wake} = v_g = c\left(1 - \omega_p^2/\omega_0^2\right)^{1/2},$$

where ω_0 is the laser frequency. The electric field of this plasma wave is longitudinal, i.e. it points in the propagation direction. A relativistic electron can ride this

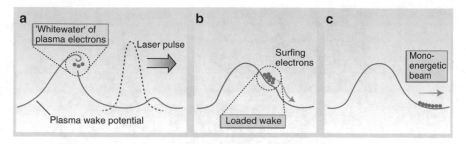

Fig. 6.8 (**a**) In a plasma excited by a laser pulse, the wake potential rises until it steepens and breaks. Electrons from the plasma are caught in the "whitewater" and surf the wave. (**b**) The load of the electrons deforms the wake, stopping further trapping of electrons from the plasma. (**c**) As the electrons surf to the bottom of the wake potential, they each arrive bearing a similar amount of energy. Figure taken from [39]. (Reprinted by permission from MacMillan Publishers Ltd.: Nature 431, 515 (2004), copyright 2004)

plasma wave, staying in phase with this longitudinal electric field and be accelerated to high energies. This scheme is illustrated in Fig. 6.8.

Although this acceleration mechanism was known since the seminal work of Tajima and Dawson [14], the energy spread of the beams accelerated hereby amounted initially to about as much as 100%. This wide range of energies occurred, because the particles were trapped from the background plasma—in much the same way that whitewater gets trapped and accelerated in an ocean wave—rather than injected into a single location near the peak of the wave (as is done in a conventional RF-based accelerator). But injection is difficult in a wake field accelerator, because the wavelength of the plasma wave is tiny—typically 10,000 times shorter than the usual 10-cm wavelengths of the radio-frequency fields in conventional accelerators. Successfully injecting tightly packed bunches of particles near the plasma-wave peak turned out to be the main challenge. The solution was inspired by the theoretical finding of the so-called "bubble-regime" [40] at sufficiently high laser intensities, where self-trapping of plasma electrons occurs and peaked spectra of accelerated electrons are found. Bubble acceleration differs from traditional wake-field acceleration in that one drives the plasma far beyond the wave-breaking limit, such that a single wake rather than a regular plasma wave train is formed. In this regime, large amounts of plasma electrons (in the order of nano-Coulombs) self-trap in the wake and are accelerated without interacting directly with the laser field. The bubble length is approximately one plasma wavelength λ_p, and an important condition to generate electron spectra peaked in energy is that the laser pulse extends only over half of the plasma wavelength, thus favoring short laser pulses of the order of 20–50 fs. For longer pulses, the accelerating electrons interact directly with the laser pulse, and one obtains broad thermal-like spectral shapes, see e.g. [33]. Still the early peaked electron spectra [36–38, 41–44] were surprising, considering that the corresponding experiments were performed with laser pulses longer than $\lambda_p/2$. The reason why the bubble regime could be accessed nevertheless is that nonlinear laser plasma interaction tends to shape the incident laser pulses until they fit into the

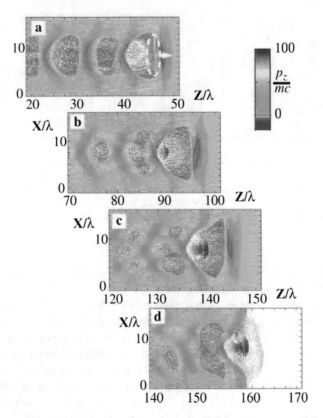

Fig. 6.9 Formation of the "bubble" regime of laser electron acceleration: Wake field behind a 20-mJ, 6.6-fs laser pulse propagating in a plasma layer from left to right. Cuts along the propagation axis show the evolution of the electron-density of the plasma wave (green-blue) and high-energy pulse (orange-red) at times (**a**) $t = 50\lambda/c$, (**b**) $t = 100\lambda/c$, (**c**) $t = 150\lambda/c$, $t = 170\lambda/c$. The color coding corresponds to the longitudinal electron momentum. The dashed ellipse in frame (**a**) indicates the position of the laser pulse. Figure taken from [40]. (Reprinted from A. Pukhov and J. Meyer-ter-Vehn, Appl. Phys. B 74, 355 (2002) with permission of Springer)

first wake and can produce mono-energetic pulses. Apparently, the bubble regime has the properties of an attractor. This was concluded from simulations presented in [38, 43, 45]. Figure 6.9 (taken from [40]) displays the formation of the acceleration "bubble" behind the laser pulse.

The most favorable regime for laser-driven electron acceleration occurs for laser pulses shorter than λ_p, as described above and for relativistic intensities high enough to break the plasma wave after the first oscillation. The electric field required for wave breaking can be estimated from the expression $E_{wb}/E_0 = \sqrt{(2(\gamma_p) - 1)}$, with $\gamma_p = (1 - v_g^2/c^2)^{-1/2} = \omega_0/\omega_p$, with v_g as the plasma wave group velocity and ω_0 as the laser frequency, which holds for low-temperature planar plasma waves [46]. In the present relativistic regime, however, the wave fronts are curved and first break near the wave axis and for lower values than in the plane-wave limit. Wave breaking turns out to be of central importance, because it leads to abundant self-trapping of electrons in the potential of the wave bucket, which are then accelerated in large numbers. This results in high conversion efficiency of laser energy into relativistic beam energy. Figure 6.9 shows four movie frames, following the front of the plasma wave as it propagates in the z direction [40]. The laser pulse is indicated only in the first frame by the dashed white ellipsis behind the wave front. The frames show electron density in cuts along the laser-beam propagation axis. Each dot corresponds to one electron and is colored according to its momentum p_z/mc. This way of plotting exhibits both the structure of the plasma wave and the self-trapped electrons accelerated to high energies in the first wave bucket. Figure 6.9a shows the wave when it has propagated 45 laser wavelengths into the layer. Like a snow-plough, the laser pulse pushes the front layer of compressed few-MeV electrons (dark green) and leaves a region with low electron density behind. It still contains a population of low-momentum electrons (in blue), which stream from right to left and feed the wave structure trailing the pulse. The green wave crests are curved and start to break at their vertex near the propagation axis. The curvature reflects the 3D structure of a plasma wave with transverse size of order λ_p. In the prototypical 3D simulations of [40], wave breaking occurred already for $E_{max}/E_{wb} \approx 0.3$ and strong electron depletion was observed only in the first wave bucket. When comparing with non-broken wake fields, the striking new feature in Fig. 6.9b–d is the stem of relativistic electrons growing out of the base. The electron energy rises towards the top of the stem, with high-momentum electrons (in red), which have been trapped first and accelerated over the full propagation distance, exhibiting the highest energies. The cavity with the accelerating electron stem is a rather stable structure, hardly affected by the drop in laser intensity, which decreases steadily due to energy transfer to electrons. The efficiency of electron acceleration in the cavity amounts to a rather high value of about 15% of the initial laser energy that can be transferred to the electron bunch. In Fig. 6.9d, the cavity arrives at the rear boundary of the plasma layer and bursts, releasing the relativistic electron load into vacuum.

Following these seminal findings, the parameter space for optimized laser-driven electron acceleration has been systematically explored in a plethora of 2D and 3D PIC simulations. An overview can be found in the review article [28] and its references.

The theory of bubble formation, its stability, and its scaling properties has been developed by Gordienko and Pukhov [47, 48]. A limiting property of the achievable electron acceleration is the dephasing length L_d, defined as the length the electron must travel before it phase slips by one half of the plasma wave (i.e. the length after which an electron to be accelerated by a plasma wake wave outruns the wave and cannot gain further momentum. A way to extend the dephasing length could be to decrease the electron density in the plasma generated from (hydrogen) gas-jet targets below 10^{19} cm^{-3} and to create a plasma channel to guide the relativistic laser pulse—which, however, turned out to be impractical due to inefficient laser heating at low densities [49]. To overcome the limitations of gas jets,

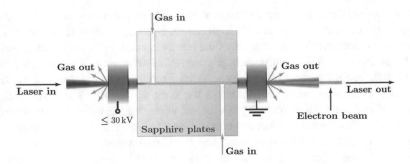

Fig. 6.10 Schematical layout of a laser-driven electron acceleration setup based on an extended plasma channel serving as plasma waveguide and created by a discharge capillary of a few tens of mm length

a gas-filled capillary discharge waveguide (ca. 30 mm long, diameter ca. 200–300 μm, machined from sapphire blocks) was used to produce centimeter-scale lower-density plasma channels via ionization in the hydrogen gas from a discharge induced at the capillary ends [50, 51]. PIC simulations in 2D and 3D showed that the initial profile of the laser pulse injected into the channel produces a wake with an amplitude that is too low to produce self-trapping. Over the first few millimeters of propagation, the plasma wake feedbacks on the laser pulse, leading to self-modulation and self-steepening, which further increases the wake amplitude. A blow-out or cavitated wake is eventually produced of sufficient amplitude so as to allow self-trapping. Trapping continues until there is sufficient trapped charge to "beam load" the wake, i.e. electrons acting back on the wake field potential by reducing its amplitude and terminating the self-trapping process. Over the next 1 cm of propagation, the bunch accelerates as the laser energy depletes. Laser depletion occurs after approximately a dephasing length, resulting in the production of a narrow energy spread electron bunch with an energy near 1 GeV [28]. Injection and acceleration of electrons is found to depend sensitively on the delay between the onset of the discharge current and the arrival of the laser pulse [52]. This technique is schematically displayed in Fig. 6.10.

Various numerical studies predicted the access to multi-GeV electron beams with laser powers in the range of 0.1–1 PW [47, 53–57], where higher-energy gain scan be achieved for a given plasma laser power by the use of channel guiding [56, 57]. This technology finally allowed to generate high-energy electron beams at the GeV level in several laboratories [58–60]. With the availability of petawatt class lasers, electron beams were produced in non-preformed plasmas with energies up to 2 GeV using a 7 cm long gas cell and 150 J laser energy in 50 fs long laser pulses [61], and using a dual gas jet system of 1.4 cm, beams with energy tails up to 3 GeV were observed [5]. Significantly lower laser pulse energies of 16 J, coupling (ca. 40 fs) laser pulses with high mode quality to preformed plasma channel waveguides produced by a 9 cm-long capillary discharge (electron density ca. 7×10^{17} cm^{-3}),

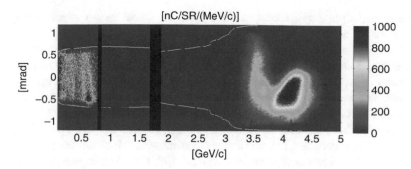

Fig. 6.11 Energy spectrum of a 4.2 GeV electron beam, generated by focusing laser pulses with energies of 16 J and a duration of 40 fs (i.e. 400 TW) into a 9 cm-long discharge capillary (500 μm diameter) plasma waveguide (plasma electron density: 7×10^{17} cm^{-3}). The electron energy was measured using a broadband magnetic spectrometer. The white lines show the angular acceptance of the spectrometer. The two black vertical stripes are areas not covered by the phosphor screen. Figure taken from [6]. (Reprinted with permission from W. Leemans et al., Phys. Rev. Lett. 113, 245002 (2014). Copyright 2014 by the American Physical Society)

were sufficient to generate the present world record of laser accelerated electron beam energy of ca. 4.2 GeV (see Fig. 6.11) [6].

Future experiments will employ techniques to both provide better guiding of the laser pulse and triggered injection [62] for acceleration at lower densities (with longer dephasing and pump depletion lengths), as well as improved reproducibility. Simulations indicate that this will allow for the generation of electron beam energies at the 10 GeV level using 40 J, 100 fs laser pulses [63].

6.4   Laser-Driven Ion Acceleration

The first proposal of plasma-based ion acceleration dates back already to the 1950s [64]. The concept was tested utilizing an induced electric field driven by an electron beam injected into a plasma [65]. The main motivation of these plasma expansion studies was laser-plasma interaction and the energetic or fast ion energy loss for laser-induced fusion. The accelerated proton or ion energy was far below 1 MeV, even though large, building-sized laser installations were employed. The community had to wait another decade for ultra-short pulse and ultra-high-intensity lasers and corresponding experiments, which are described below. Starting in the 1990s, ion acceleration in a relativistic plasma was demonstrated with ultra-short pulse lasers based on the chirped pulse amplification technique, which can provide not only picosecond or femtosecond laser pulse duration, but simultaneously ultra-high peak power of terawatt to petawatt levels. Ion acceleration experiments were based on theoretical and computational studies [66–68]. Laser-driven ion acceleration attracted considerable attention, when an intense proton beam with a maximum energy of 55 MeV was observed in 2000 [20, 69]. Since these early proof-of-principle experiments, various laser ion acceleration experiments were performed to establish knowledge on the fundamental underlying processes and their mechanisms [19, 70–97], as well as on applications of the proton and ion beams [98]. Starting from the year 2000, several groups demonstrated low transverse emittance, tens of MeV proton beams with a conversion efficiency of up to several percent. The laser-accelerated particle beams have a duration of the order of a few picoseconds at the source, an ultra-high peak current and a broad energy spectrum, which make them suitable for many, including several unique, applications [1], ranging from a compact ion source for conventional, RF-based accelerators [99–101], (isochoric) heating of warm dense matter [90, 102], to compact proton sources for particle cancer therapy [103, 104]. Laser-accelerated proton and ion beams may become compelling alternatives to electrons used for fast ignition in laser-driven fusion [105, 106]. Alaser-driven proton beam can be used for the production of radioactive materials for medical applications [83, 107]. For a detailed discussion of all relevant aspects the reader is referred to the excellent review article [1].

6.4.1 Mechanisms of Laser-Driven Ion Acceleration

In contrast to laser-driven electron acceleration, where an intense driver laser pulse is focused into a (hydrogen) gas jet, laser-driven ion acceleration mainly uses solid targets, typically foils in the micrometer thickness range. In fact, the development of an optimized targetry is an indispensable part of the development of laser-driven ion beams. Still, the laser pulse interacts mostly with the electrons of the target material and ions are affected by the resulting plasma fields. To maximize laser energy absorption and to prevent potentially harmful back reflections from the target surface to the laser chain, p-polarization at oblique incidence is mostly chosen (p-polarized light is understood to have an electric field direction parallel to the plane of incidence on a device). When discussing laser-driven ion acceleration mechanisms, one should be aware that in the real scenario many different physical phenomena will be involved at each interaction stage, starting from ionization, initial pre-plasma formation, coupling of the main pulse energy to the electrons via absorption, plasma evolution driven by the laser and collective plasma fields, and finally ion propagation during and after the acceleration process [1]. Based on eventually limited diagnostic capabilities, it may be difficult to determine adominant acceleration mechanism and this explains also the multitude of different mechanisms that can be found in the literature of the last decade. Here we will focus on two prototypical mechanisms.

6.4.1.1 Target Normal Sheath Acceleration (TNSA)

In the intensity regime of relevance (as a guideline, $I\lambda^2 > 10^{18}$ W/cm^2), the laser pulse can efficiently transfer energy into relativistic electrons, mainly through ponderomotive processes (e.g., the relativistic $j \times B$ heating mechanism, arising from the oscillating component of the ponderomotive force [108]). The average energy of the electrons is typically of MeV order, e.g., their collisional range is much larger than the foil thickness. This means that the laser pulse irradiating the target transfers its energy into hot electrons, which propagate through the target and form a strong electric charge-separation field at the rear side of the target, whereas the foil remains opaque to the laser light throughout the interaction. The charge-separation field ionizes atoms from surface layers, and the resulting ions start to expand into the vacuum behind the target, following the electrons [109, 110]. While a limited number of energetic electrons will effectively leave the target, most of the hot electrons will be held back within the target volume by the space charge, and will form a sheath extending by approximately a Debye length $\lambda_D = (\varepsilon_0 T_e k_B / n_e e^2)^{1/2}$ from the initially unperturbed rear surface (with T_e and n_e denoting the hot electron temperature and density, respectively). Consequently, a quasi-static charge separation field builds up at the target rear side, reaching values on the order of TV/m. This field is maintained by an equilibrium of electrons being pulled back into the target and new electrons arriving from the front surface plasma. According to the model developed in [100], the initial accelerating field will be given by

$$E_{sheath,0} = \sqrt{\frac{2}{e_N} \frac{T_e}{e\lambda_D}} = \left(\frac{8\pi}{e_N} n_e T_e\right)^{1/2}, \tag{6.5}$$

where e_N is Euler's number (i.e. the base of the natural logarithm). Ions are accelerated in the quasi-static sheath field, which is orientated normal to the surface, giving the process its name "target normal sheath acceleration" (TNSA). Figure 6.12 schematically depicts the main mechanisms for laser-driven ion acceleration from overdense, opaque plasmas created from μm-scale foils [111]. Besides the described TNSA mechanism, also "hole-boring RPA" is included, which will be discussed in the following section.

6.4.1.2 Radiation Pressure Acceleration (RPA)

In the proposal of a new nuclear reaction scenario that will be introduced in the following section, it is envisaged to exploit a different laser ion acceleration mechanism, called Radiation Pressure Acceleration (RPA), which is expected to become dominant at higher laser intensities (ca. $\geq 10^{23}$ W/cm^2) than realized in present facilities, but accessible for the next generation of multi-PW-class laser facilities planned or already under construction. As will be outlined, RPA promises several considerable advantages compared to the currently used TNSA scheme. It was first proposed theoretically [112–117]. Special emphasis has been given to RPA with circularly polarized laser pulses, as this suppresses fast electron generation and leads to the interaction dominated by the radiation pressure [112, 113]. When a high-intensity laser pulse interacts with a thin foil (in RPA typical target thicknesses

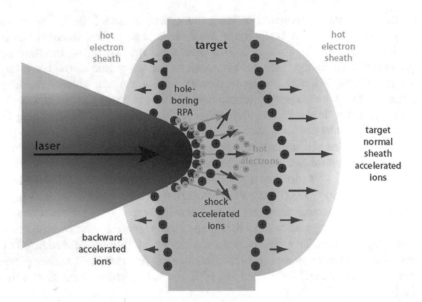

Fig. 6.12 Schematical illustration of the main mechanisms for ion acceleration from an overdense, opaque target. The highly intense laser pulse incident from the left gives rise to rapid ionization and the generation of hot electrons which are able to traverse the foil. A sheath field of transverse dimension much larger than the focal spot size is set up at both surfaces, accelerating ions along the target normal (TNSA). Additionally, the laser beam piles up a layer of electrons at its front, while it bores a hole into the target plasma, thus pulling ions into the substrate (hole-boring radiation pressure acceleration (RPA)). This flow of ions can launch a shock wave, which reflects ions to twice the shock speed while crossing the target. In case of a radially varying intensity distribution irradiating a planar surface, the shock front will not be planar and the shock-accelerated ion component will exhibit a large angle of divergence [111]

range in the nm regime compared to μm in TNSA), it can ideally push forward all electrons due to the radiation pressure. The ions respond slowly, and a large charge-separation field builds up and efficiently accelerates the main body of the irradiated target area, i.e. the ions. If this charge-separation field is strong enough to accelerate ions quickly to relativistic velocity, the distance between the electrons and ions remains relatively small, and instabilities do not have time to develop [114]. Even if at the early stage of the acceleration process the foil is partly transparent to the laser due to relativistic effects, at a later time, when the foil velocity approaches c, it becomes highly reflective due to the laser frequency downshift in the co-moving frame. Further, due to the double Doppler effect, the frequency of the reflected light becomes $\omega_r \approx \omega_0/4\gamma^2$, where $\gamma \gg 1$ is the relativistic gamma-factor of the accelerated foil, and the reflected light energy significantly decreases; almost all of the laser pulse energy is transferred to the foil. This makes the RPA acceleration scheme much more efficient compared to TNSA, with the achievable ion energy scaling linearly with the laser intensity: $E_{ion} \propto I_{laser}$, whereas in TNSA the maximum ion energy correlates with $E_{ion} \propto \sqrt{I_{laser}}$. At the final acceleration phase, the ions

moving with nearly the same velocity as electrons take most of this laser pulse
energy due to their much larger mass. The radiation pressure, which is a relativistic
invariant, is given by [1]

$$P_{rad} = R' \left(\frac{E_0'}{2\pi} \right)^2 = R' \frac{c-V}{c+V} \frac{E_0^2}{2\pi} = R' \frac{c-V}{c+V} \frac{2I_0}{c}, \tag{6.6}$$

Here E_0' and R' are the laser field and target foil reflectivity in the boosted reference
frame moving with the foil. I_0 is the laser intensity, circular polarization is assumed.
R' is identical to the reflectivity in terms of the photon number, absorption is
assumed to be negligible, so $R' + T' = 1$, with T' as the transmission. The foil
velocity is denoted by $V = dx/dt = pc/(p^2 + m_i^2 c^2)^{1/2}$, where p is the individual
ion momentum. In order to achieve the maximum ion acceleration, all electrons need
to be expelled, hence the laser electric field should be of the order of the maximum
charge-separation field, which is near the onset of the relativistic transparency of
the target foil [118]. Figure 6.13 (left) illustrates the described scheme of radiation
pressure laser ion acceleration, the right panel displays the ion density isosurface for
$n = 8n_{cr}$ (n_{cr}: critical electron density: cut-off frequency where the light frequency
in a plasma equals the plasma frequency. If the critical density is exceeded, the
plasma is called over-dense) from a simulation study of a laser with peak intensity
$I = 1.37 \times 10^{23}$ W/cm^2 (and a pulse energy of 10 kJ) interacting with a fully ionized
plasma ($n_e = 5.5 \times 10^{22}$/cm^3).

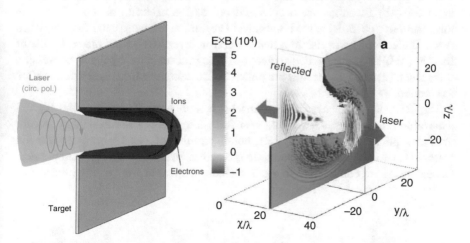

Fig. 6.13 Left: Schematical illustration of the radiation pressure laser ion acceleration mechanism
(RPA): a (preferably circularly polarized) laser pulses hit a thin foil target, expelling electrons
and consecutively accelerating ions in the emerging charge-separation field. Ion bunches are
accelerated at near-solid-state density. Right: Ion density for $n = 8n_{cr}$ (n_{cr}: critical electron density,
see text) and $I = 1.37 \times 10^{23}$ W/cm^2 with partially reflected and transmitted laser beam. Right
part of Figure reprinted from [114]. (Reprinted with permission from T. Esirkepov et al., Phys.
Rev. Lett. 92, 175003 (2004). Copyright 2004by the American Physical Society)

If the laser field amplitude is too high, most of the laser pulse is transmitted before the foil reaches the relativistic velocity, if, on the other hand, the foil is "too thick", the acceleration will remain small. For protons, it can be shown that optimum acceleration conditions can be achieved for a value of the dimensionless laser amplitude $a \approx \pi\, n_e I/n_{cr}\lambda_0$ (with electron density n_e, critical density n_{cr} and laser wavelength and intensity λ_0 and I) of $a \approx 300$, corresponding to a laser intensity of $I_0 \approx 1.2 \cdot 10^{23}\, W/cm^2\, (\mu m/\lambda_0)^2$, achievable with laser systems e.g. presently being set up for the ELI project (https://www.eli-beams.eu/en/research/laser-technology/petawatt-lasers/) [10, 119].

It has been shown that RPA operates in two (main) modes. In the first one, called "hole-boring" [120] or sometimes "collisionless shock acceleration" [113, 121, 122], the laser pulses interact with targets thick enough to end up in an overdense, opaque plasma, allowing the laser to drive target material ahead of it as a piston, but without interacting with the target rear surface [112]. The first tentative experimental observation of RPA in the "hole-boring" regime was reported in experiments led by the Munich group [123, 124]. The second, and most simple model of RPA is the one of a "perfect" (i.e., totally reflecting) plane mirror boosted by a light wave at perpendicular incidence [125], which is also known as the "light sail" regime. In contrast to the hole-boring version of RPA, in the light-sail regime only a layer of the foil at its rear side is accelerated, since the target is sufficiently thin for the laser pulse to punch through the foil and accelerate a slab of the plasma as a single object.

Figure 6.14 illustrates the light sail mechanism: (a) a (circularly polarized) laser pulse is focused on an ultra-thin foil (ca. 1–10 nm), thereby fully ionizing the target foil. (b) Electrons are accelerated out of the backside of the foil via the ponderomotive potential of the laser pulse. The attractive electrostatic force between ions and electrons gives rise to a dense electron layer just behind the back side of the foil. (c) Like a sail, the dense electron layer is accelerated by the light pressure of the laser. (d) The heavier ions are pulled by the dense electron layer and therefore accelerated.

The RPA laser ion acceleration mechanism in general provides the highest achievable efficiency for the conversion from laser energy to ion energy and for circularly polarized laser light RPA holds promise of quasi-monoenergetic ion beams. Due to the circular polarization, electron heating is strongly suppressed. The electrons are compressed to a dense electron sheet in front of the laser pulse,

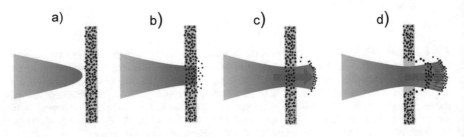

Fig. 6.14 Illustration of the "light sail" laser ion acceleration mechanism (for details see text)

which then via the Coulomb field accelerates the ions. This mechanism requires very thin targets and ultra-high contrast laser pulses to avoid the pre-heating and expansion of the target before the interaction with the main laser pulse. The RPA mechanism allows to produce ion bunches with solid-state density (10^{22}–10^{23}/cm^3), which thus are 10^{14} times more dense than ion bunches from classical accelerators. Correspondingly, the areal densities of these bunches are $\sim 10^7$ times larger. It is important to note that these ion bunches are accelerated as neutral ensembles together with the accompanying electrons and thus do not Coulomb explode.

When aiming at optimized laser pulse properties for RPA-based laser ion acceleration, the temporal pulse intensity contrast is a critical parameter. Disintegration (e.g. plasma formation) of the target has to be delayed as much as possible until the laser pulse maximum reaches the target foil, requiring a high (10–12 orders of magnitude) temporal intensity contrast within picoseconds from the pulse maximum to avoid premature target expansion. The maximum achievable ion energy is correlated with the laser pulse energy available in the focal spot on target, as can be seen from Fig. 6.15. As a useful, yet slightly optimistic rule of thumb for considering the low energy laser systems of Joule pulse energy level, Macchi et al. found that "under the right/clean conditions, protons can gain 10 MeV of kinetic energy per 1 J of laser energy that one manages to concentrate in the laser focal spot" [126]. This

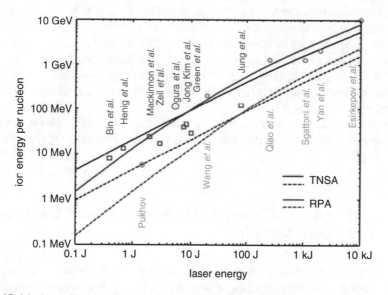

Fig. 6.15 Maximum laser-accelerated ion energies predicted by models for the TNSA and RPA acceleration mechanisms. Some selected experimental results are included as blue squares (Bin et al. [131], Henig et al. [123, 124], Mackinnon et al. [74], Zeil et al. [132], Ogura et al. [133], Jong Kim et al. [134], Green et al. [135], Jung et al. [136]) and theoretical results obtained from PIC simulations are marked by green circles (Pukhov [109], Wang et al. [137], Quiao et al. [138], Sgattoni et al. [139], Yan et al. [140], Esirkepov et al. [114]). Figure taken from Schreiber et al. [128]. (Reprinted with permission from J. Schreiber, F. Bell and Z. Najmudin, High Power Laser Science and Engineering 2, e41 (2014). Copyright 2014 by Cambridge University Press)

condition would be found in the upmost part of the grey area in Fig. 6.15 (and even slightly exceed it) [127, 128]. Besides a maximum proton energy of about 85 MeV (and particle numbers of about 10^9 in an energy bin of 1 MeV around this maximum) [129], also carbon ion beams up to 60 MeV/u were reported so far [130]. A further favorable characteristics of laser-driven ion bunches is their excellent longitudinal emittance, due to the short (fs-scale) time structure of the pulse even in view of the typically broad (typically 20–100%) energy width. Recently, also proton beams accelerated from nanometer thick diamond-like carbon (DLC) foils, irradiated by an intense laser with high contrast were observed with extremely small divergence (half angle) of $2°$, showing one order of magnitude reduction of the divergence angle in comparison to the results from micrometer thick targets [131]. It was demonstrated that this reduction arises from a steep longitudinal electron density gradient and an exponentially decaying transverse profile at the rear side of the ultrathin foils.

In summary, the significant features of laser-driven ion beams compared with conventional ion accelerators are that the effective source size of ion emission is extremely small, typically of the order of 10 μm. Another important feature is the ultra-short duration at the source of the ion bunch, which is of the order of picoseconds. The acceleration gradients are of the order of MeV/μm, compared to about MeV/m provided by conventional radio-frequency (RF) wave based accelerators. Moreover, laser-driven ion bunches exhibit a drastically increased density compared to conventionally accelerated ion beams, as will be discussed in more detail in the following chapter. Although the laser-driven technique inherently provides significant advantages from the point of view of many unique applications, one still has numerous issues to overcome, such as increasing the particle energy, spectral and angular control of the beam, conversion efficiency from laser energy into the ion beam, possible activation issues, as well as stability of the acceleration parameters.

Further practical considerations on the application of laser-driven ion sources in the framework of an integrated laser-driven accelerator system (ILDIAS) can be found in [127].

6.5 Application of Ultra-Dense Laser-Accelerated Ion Beams for Nuclear Astrophysics

This section exemplifies, based on the novel nuclear "fission-fusion" reaction mechanism, how the unique properties of laser-driven ions compared to conventionally accelerated particle beams can be exploited to access research fields that may otherwise keep out of reach for direct experimental observation. The presentation follows the description given in the technical design report of the ELI-Nuclear Physics facility in Magurele (Bucharest/Romania) [141] that is presently under construction.

6.5.1 The Quest for the Waiting Point at N = 126 of the Astrophysical r-Process Nucleosynthesis

Elements like platinum, gold, thorium and uranium are produced via the rapid neutron capture process (r-process) at astrophysical sites like merging neutron star binaries or (core collapse) supernova type II explosions. We aim at improving our understanding of these nuclear processes by measuring the properties of heavy nuclei on (or near) the r-process path. While the lower-mass path of the r-process is well explored, the nuclei around the $N = 126$ waiting point critically determine this nucleosynthesis mechanism. At present, basically nothing is known about these nuclei. Figure 6.16 shows the chart of nuclides marked with different nucleosynthesis pathways for the production of heavy elements in the Universe: the thermonuclear fusion processes in stars producing elements up to iron (orange arrow), the slow neutron capture process (s-process) along the valley of stability leading to about half of the heavier nuclei (red arrow) and the rapid neutron capture process (r-process). The astrophysical site of the r-process nucleosynthesis is still under debate: it may be cataclysmic core collapse supernovae (II) explosions with neutrino winds [142–145] or mergers of neutron-star binaries [146–148]. For the

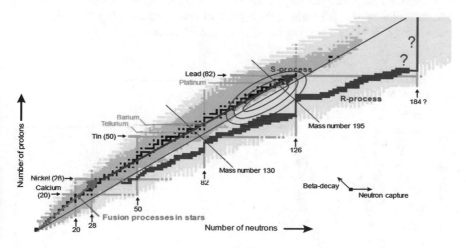

Fig. 6.16 Chart of the nuclides indicating various pathways for astrophysical nucleosynthesis: thermonuclear fusion reactions in stars (orange vector), s-process path (red vector) and the r-process generating heavy nuclei in the Universe (red pathway). The nuclei marked in black indicate stable nuclei. For the green nuclei some nuclear properties are known, while the yellow, yet unexplored regions extend to the neutron and proton drip lines as well as into the region of superheavy elements (SHE) at the upper end of the chart. The blue line connects nuclei with the same neutron/proton ratio as for (almost) stable actinide nuclei. On this line the maximum yield of nuclei produced via fission-fusion (without neutron evaporation) will be located. The elliptical contour lines correspond to the expected maximum laser-driven fission-fusion cross sections decreased to 50, 10 and 0.1%, respectively, for primary ^{232}Th beams. Figure taken from [149]. (Reprinted from D. Habs et al., Appl. Phys. B 103, 471 (2011) with permission of Springer)

heavier elements beyond barium, the isotopic abundances are always very similar (called universality) and the process seems to be very robust. Perhaps also the recycling of fission fragments from the end of the r-process strengthens this stability. Presently, it seems more likely that a merger of neutron star binaries is the source for the heavier r-process branch, while core collapsing supernova explosions contribute to the lighter elements below barium.

The modern nuclear equations of state, neutrino interactions and recent supernova explosion simulations [143] lead to detailed discussions of the waiting point $N = 126$. Here measured nuclear properties along the $N = 126$ waiting point may help to clarify the sites of the r-process.

A detailed knowledge of nuclear lifetimes and binding energies in the region of the $N = 126$ waiting point will narrow down the possible astrophysical sites. If, e.g., no shell quenching could be found in this mass range, the large dip existing for this case in front of the third abundance peak (as visible in the solar elemental abundance pattern shown in Fig. 6.17 with special emphasis to the third abundance peak near A 190–200) would have to be filled up by other processes like neutrino wind interactions. Considering the still rather large difficulties to identify convincing astrophysical sites for the third peak of the r-process with sufficiently occurrence rates, measurements of the nuclear properties around the $N = 126$ waiting point will represent an important step forward in solving the difficult and yet confusing site selection of the third abundance peak of the r-process. The key bottleneck nuclei of the $N = 126$ waiting point around $Z \sim 70$ are about 15 neutrons away

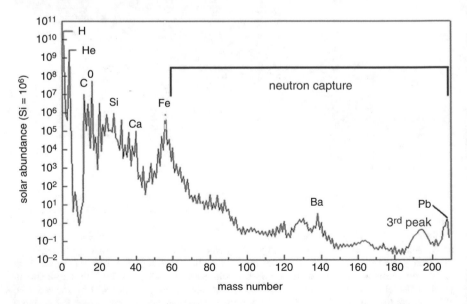

Fig. 6.17 Elemental abundance pattern for the solar system, indicating the region of heavy elements formed by neutron capture in the r-process, particularly emphasizing the third peak at A 190–200, relevant for the presented fission-fusion mechanism

from presently known nuclei (see Fig. 6.16), with a typical drop of the production cross section for classical radioactive beam production schemes of about a factor of 10–20 for each additional neutron towards more neutron-rich isotopes. Thus presently nothing is known about these nuclei and even next-generation large-scale "conventional" radioactive beam facilities like FAIR [150], SPIRAL II [151] or FRIB (http://www.frib.msu.edu/aboutmsu-frib-proposal) will hardly be able to grant experimental access to the most important isotopes on the r-process path. The third peak in the abundance curve of r-process nuclei is due to the $N = 126$ waiting point as visible in Fig. 6.16. These nuclei are expected to have rather long half-lives of a few 100 ms. This waiting point represents the bottleneck for the nucleosynthesis of heavy elements up to the actinides. From the view point of astrophysics, it is the last region, where the r-process path gets close to the valley of stability and thus can be studied with the new isotopic production scheme discussed below. While the waiting point nuclei at $N = 50$ and $N = 82$ have been studied rather extensively [152–155], nothing is known experimentally about the nuclear properties of waiting point nuclei at the $N = 126$ magic number. Nuclear properties to be studied here are nuclear masses, lifetimes, beta-delayed neutron emission probabilities P_n and the underlying nuclear structure. If we improve our experimental understanding of this final bottleneck to the actinides at $N = 126$, many new visions open up: (1) as predicted by many mass formulas (e.g. [156]), there is a branch of the r-process leading to extremely long-lived super-heavy elements beyond $Z = 110$ with lifetimes of about 10^9 years. If these predictions could be made more accurately, a search for these super-heavy elements in nature would become more promising. (2) At present the prediction for the formation of uranium and thorium elements in the r-process is rather difficult, because there are no nearby magic numbers and those nuclei are formed during a fast passage of the nuclidic area between shells. Such predictions could be improved, if the bottleneck of actinide formation would be more reliably known. (3) Also the question could be clarified if fission fragments are recycled in many r-process loops or if only a small fraction is reprocessed. This description of our present understanding of the r-process underlines the importance of the present project for nuclear physics and, particularly, for astrophysics.

6.5.2 The Novel Laser-Driven "Fission-Fusion" Nuclear Reaction Mechanism

The basic concept of the fission-fusion reaction scenario draws on the ultra-high density of laser accelerated ion bunches, quantitatively outlined in Ref. [149]. Choosing fissile isotopes as target material for a first target foil accelerated by an intense driver laser pulse will enable the interaction of a dense beam of fission fragments with a second target foil, also consisting of fissile isotopes. So finally in a second step of the reaction process, fusion between (neutron-rich) beam-like and target-like (light) fission products will become possible, generating extremely neutron-rich ion species.

For our discussion, we choose ^{232}Th (the only component of chemically pure Th) as fissile target material, primarily because of its long half-life of 1.4×10^{10} years, which avoids extensive radioprotection precautions during handling and operation. Moreover, metallic thorium targets are rather stable in a typical laser vacuum of 10^{-6} mbar, whereas, e.g., metallic ^{238}U targets would quickly oxidize. Nevertheless, in a later stage it may become advantageous to use also heavier actinide species in order to allow for the production of even more exotic fusion products. In general, the fission process of the two heavy thorium nuclei from beam and target will be preceded by the deep inelastic transfer of neutrons between the inducing and the fissioning nuclei. Here the magic neutron number in the super deformed fissile nucleus with $N = 146$ [157, 158] may drive the process towards more neutron-rich fissioning nuclei, because the second potential minimum acts like a doorway state towards fission. Since in the subsequent fission process the heavy fission fragments keep their A and N values [159], these additional neutrons will show up in the light fission fragments and assist to reach more neutron-rich nuclei.

Figure 6.18 shows a sketch of the proposed fission-fusion reaction scenario. The accelerated thorium ions will be fissioned in the CH_2 layer of the reaction target, whereas the accelerated carbon ions and deuterons from the production target generate thorium fragments in the thick thorium layer of the reaction target. This scenario is more efficient than the one where fission would be induced by the thorium ions only. In view of the available energy in the accelerating driver laser pulse, the optimized production target should have a thickness of about 0.5 μm for the thorium as well as for the CD_2 layers. The thorium layer of the reaction target

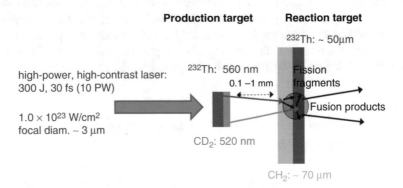

Fig. 6.18 Sketch of the target arrangement envisaged for the fission-fusion reaction process based on laser ion acceleration, consisting of a production and a reaction target from a fissile material (here ^{232}Th), each of them covered by a layer of low-Z materials (CD_2 and CH_2, respectively). The thickness of the CH_2 layer as well as the second thorium reaction target have to be limited to 70 μm and 50 μm, respectively, in order to enable fission of beam and target nuclei. This will allow for fusion between their light fragments, as well as enable the fusion products to leave the second thorium reaction target [141]

would have a thickness of about 50 μm. Using a distance of 2.8 Å between atoms in solid layers of CH_2, the accelerated light ion bunch (1.4×10^{11} ions) corresponds to 1860 atomic layers in case of a 520 nm thick CD_2 target.

In order to allow for an optimized fission of the accelerated Th beam, the thicker Th layer of the reaction target, which is positioned behind the production target, is covered by about 70 μm of polyethylene. This layer serves a twofold purpose: Primarily it is used to induce fission of the impinging Th ion beam, generating the beam-like fission fragments. Here polyethylene is advantageous compared to a pure carbon layer because of the increased number of atoms able to induce fission on the impinging Th ions. In addition, the thickness of this CH_2 layer has been chosen such that the produced fission fragments will be decelerated to a kinetic energy which is suitable for optimized fusion with the target-like fission fragments generated by the light accelerated ions in the Th layer of the reaction target, minimizing the amount of evaporated neutrons. For practical reasons, we propose to place the reaction target about 0.1 mm behind the production target, as indicated in Fig. 6.18. After each laser shot, a new double-target has to be rotated into position.

In general, the fission process proceeds asymmetric [159]. The heavy fission fragment for ^{232}Th is centred at $A = 139.5$ (approximately at $Z = 54.5$ and $N = 84$) close to the magic numbers $Z = 50$ and $N = 82$. Accordingly, the light fission fragment mass is adjusted to the mass of the fixed heavy fission fragment, thus resulting for ^{232}Th in $A_L = 91$ with $Z_L \approx 37.5$. The width (FWHM) of the light fission fragment peak is typically $\Delta A_L = 14$ mass units, the 1/10 maximum width about 22 mass units [159].

So far we have considered the fission process of beam-like Th nuclei in the CH_2 layer of the reaction target. Similar arguments can be invoked for the deuteron- (and carbon-) induced generation of (target-like) fission products in the subsequent thicker thorium layer of the reaction target, where deuteron- and carbon-induced fission will occur in the ^{232}Th layer of the reaction target. Based on the chosen target layer thicknesses (see Fig. 6.18) and a typical conversion efficiency from laser pulse energy to ion energy of about 10% (estimated from the 1D RPA-model in the hole-boring mode as outlined in (http://www.frib.msu.edu/aboutmsu-frib-proposal)), we end up with 2.8×10^{11} laser-accelerated deuterons (plus 1.4×10^{11} carbon ions) impinging on the second target per laser pulse. In a simplistic approach, based on the solid-state density of the accelerated ion bunches, they can be considered as 1860 consecutive atomic layers, we conclude a corresponding fission probability in the Th layer of the reaction target of about 2.3×10^{-5}, corresponding to 3.2×10^6 target-like fission fragments per laser pulse. A thickness of the thorium layer of the reaction target of about 50 μm could be exploited, where the kinetic proton energy would be above the Coulomb barrier to induce fission over the full target depth. In a second step of the fission-fusion scenario, we consider the fusion between the light fission fragments of beam and target to a compound nucleus with a central value of $A \sim 182$ and $Z \sim 75$. Again we employ geometrical arguments for an order-of-magnitude estimate of the corresponding fusion cross section. For a typical light fission fragment with $A = 90$, the nuclear radius can be estimated as 5.4 fm.

Considering a thickness of 50 μm for the Th layer of the reaction target that will be converted to fission fragments, equivalent to 1.6×10^5 atomic layers, this results in a fusion probability of about 1.8×10^{-4}. Very neutron-rich nuclei still have comparably small production cross sections, because weakly bound neutrons ($S_n \sim 3$ MeV) will be evaporated easily. The optimum range of beam energies for fusion reactions resulting in neutron-rich fusion products amounts to about 2.8 MeV/u according to PACE4 [160, 161] calculations. So, e.g., the fusion of two neutron-rich $^{98}_{35}$Br fission products with a kinetic energy of the beam-like fragment of 275 MeV leads with excitation energy of about 60 MeV to a fusion cross section of 13 mb for $^{189}_{70}$Yb$_{119}$, which is already eight neutrons away from the last presently known Yb isotope. One should note that the well-known hindrance of fusion for nearly symmetric systems (break-down of fusion) only sets in for projectile and target masses heavier than about 100 amu [162, 163]. Thus for the fusion of light fission fragments, we expect an unhindered fusion evaporation process. A detailed discussion of the achievable fission-fusion reaction yield is given in Ref. [164]. In addition to the scenario discussed above, the exceptionally high ion bunch density may lead to collective effects that do not occur with conventional ion beams: when sending the energetic, solid-state density ion bunch into a solid carbon or thorium target, the plasma wavelength ($\lambda_p \approx 5$ nm, driven by the ion bunch with a phase velocity corresponding to the thorium ion velocity) is much smaller than the ion bunch length (≈ 560 nm) and collective acceleration and deceleration effects cancel. As will be discussed in the next paragraph, only the binary collisions remain and contribute to the stopping power. In this case the first layers of the impinging ion bunch will attract the electrons from the target and like a snow plough will take up the decelerating electron momenta. Hence the predominant part of the ion bunch is screened from electrons and we expect a drastic reduction of the stopping power. The electron density n_e will be strongly reduced in the channel defined by the laser-accelerated ions, because many electrons are expelled by the ion bunch and the laser pulse. This effect requires detailed experimental investigations planned for the near future, aiming at verifying the perspective to use a significantly thicker reaction target, which in turn would significantly boost the achievable fusion yield.

Figure 6.19 displays a closer view into the region of nuclides around the N = 126 waiting point of the r-process, where nuclei on the r-process path are indicated by the green colour, with dark green highlighting the key bottleneck r-process isotopes [165] at N = 126 between Z = 66 (Dy) and Z = 70 (Yb). One should note that, e.g., for Yb the presently last known isotope is 15 neutrons away from the r-process path at N = 126. The isotopes in light blue mark those nuclides, where recently beta-half-lives could be measured following projectile fragmentation and in-flight separation at GSI [166]. Again the elliptical contour lines indicate the range of nuclei accessible with our new fission-fusion scenario on a level of 50%, 10% and 10^{-3} of the maximum fusion cross section between two neutron-rich light fission fragments in the energy range of about 2.8 MeV/u, respectively.

Besides the fusion of two light fission fragments, other reactions may happen. The fusion of a light fission fragment and a heavy fission fragment would lead back to the original Th nuclei, with large fission probabilities, thus we can neglect

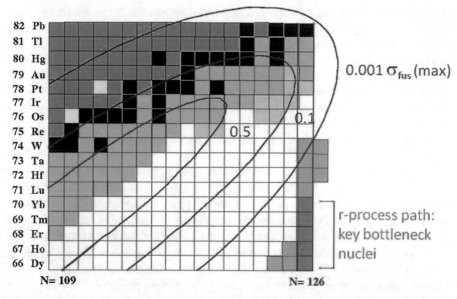

Fig. 6.19 Chart of nuclides around the N = 126 waiting point of the r-process path. The blue ellipses denote the expected range of isotopes accessible via the novel fission-fusion process. The indicated lines represent 0.5, 0.1 and 0.001 of the maximum fusion cross section after neutron evaporation. In green the N = 126 nuclides relevant for the r-process are marked, with the dark green colour indicating the key bottleneck nuclei for the astrophysical r-process. Figure taken from [149]. (Reprinted from D. Habs et al., Appl. Phys. B **103**, 471 (2011) with permission of Springer)

these fusion cross sections. The fusion of two heavy fission fragments would lead to nuclei with A   278, again nuclei with very high fission probability. Hence we have also neglected these rare fusion cross sections, although they may be of interest on their own. However, the multitude of reaction channels will require conclusive experimental precautions for a separation of the fusion reaction products of interest in the diagnostics and identification stage of the experimental setup.

6.5.3   Stopping Power of Very Dense Ion Bunches

In nuclear physics, the Bethe-Bloch formula [167] is used to calculate the atomic stopping of energetic individual electrons [168] by ionization and atomic excitation. For relativistic electrons, the other important energy loss is bremsstrahlung. The radiation loss is dominant for high energy electrons, e.g., $E \geq 100$ MeV and $Z = 10$. If, however (see below), the atomic stopping becomes orders of magnitude larger by collective effects, the radiation loss can be neglected. For laser acceleration, the electron and ion bunch densities reach solid state densities, which are about 14–15 orders of magnitude larger compared to beams from classical accelerators. Here collective effects become important. One can decompose the Bethe-Bloch equation according to [169] into a first contribution describing binary collisions and a second term describing long range collective contributions. Reference [170] discusses the mechanism of collective deceleration of a dense particle bunch in a thin plasma, where the particle bunch fits into part of the plasma oscillation and is decelerated 10^5–10^6 stronger than predicted by the classical Bethe-Bloch equation [167] due to the strong collective wake field. For ion deceleration targets are envisaged with suitably low density. These new laws of deceleration and stopping of charged particles have to be established to use them later in experiments in an optimum way.

In the following, the opposite effect with a strongly reduced atomic stopping power that occurs when sending an energetic, solid state density ion bunch into a solid target, will be discussed. For this target the plasma wavelength ($\lambda_p \approx 1$ nm) is much smaller than the ion bunch length (≈ 500 nm) and collective acceleration and deceleration effects cancel each other. Only the binary collisions are important.

Hence, one may consider the dense ion bunch as consisting (in a simplistic view) of 300 layers with Angstrom distances. Here the first layers of the bunch will attract the electrons from the target and—like a snow plough—will take up the decelerating electron momenta. The predominant part of the ion bunch is screened from electrons and we expect a significant (here assumed as $\approx 10^2$ fold) reduction in stopping power. The electron density n_e is strongly reduced in the channel because many electrons are driven out by the ion bunch and the laser. Again, all these effects have to be studied in detail. It is expected that the resulting very dense ion bunches should have a time evolution and the reaction products are emitted at different times and angles. Therefore, for the characterization of the dense bunches and their time evolution, the detection system needs to capture the reaction products, emitted at different times (analogous to time of flight measurements), and measure their angular distributions. Of course, the temporal evolutions, which can be followed, vary greatly depending on the temporal resolution of the diagnosis system. In a preliminary phase, it is expected that electrons and ions are emitted due to the Coulomb explosion of a part of the initially formed bunch (pre-bunch emission). Then, the remaining bunch will have a slower temporal evolution, which can be followed as a function of its time of flight in free space. The experimental study of deceleration of dense, high speed bunches of electrons and ions will require:

- Bunch characterization in free space: its components, their energies and the ion charge states, their angular distribution and temporal evolution; due to the large number of particles, the detection solid angles must be small (of the order of 10^{-7} sr or less).
- Tracking the changes introduced by bunches passing through different materials (solid or gas) and their deceleration study. Studies will be carried out depending on laser power and target type and thickness and for deceleration—depending on material type and its thickness.

The same detection system could be used for both diagnosis in free space and diagnosis after passing through a material. A rapid characterization may be done with a Thomson parabola ion spectrometer, and an electron magnetic spectrometer, implying measurements of the emissions at different times and possibly their angular distribution; in relevant case. A more complete analysis will require a diagnosis system working in real-time, using magnetic spectrometers and detection systems with high granularity or with position sensitive readout in the focal plane (e.g., stacks of ΔE-E detectors, with ionization chambers and Si or scintillation detectors). Even if the laser pulse frequency is small, the nuclear electronics can be triggered in the usual way.

6.6 Conclusion

In conclusion, laser-driven particle beams with their unique properties, even though differing in many regards from conventionally accelerated ion bunches, will open new perspectives for a large variety of specific applications in physics, radiation chemistry, biology and medicine. This has been prototypically outlined here for the case of nuclear astrophysics, as discussed before; another prominent field of application, not detailed here, could be laser-driven ion-bunch tumor radiotherapy [171–173]. Full exploitation of the specific properties of the novel laser-driven secondary sources (including, besides electrons and ions, also neutrons and positrons) will require to consider them not just as a challenge to or potential replacement of conventional sources, but as a true complementary enlargement of the experimental portfolio. Next-generation high-power laser facilities, capable of providing full access to optimized laser-acceleration mechanisms, are already on their way to become operational in the near future, a prominent example in Europe will be the Extreme Light Infrastructure (ELI) with its three pillars: ELI-Beam lines (near Prague/Czech Republic), ELI-Nuclear Physics (near Bucharest/Romania) and ELI-ALPS (Attosecond Light Pulse Source, near Szeged/Hungary). Thus a bright future for laser-driven, particle-beam based science and applications appears at the horizon.

References

1. H. Daido, M. Nishiuchi, A.S. Pirozhkov, Rep. Prog. Phys. **75**, 056401 (2012)
2. A. Macchi, M. Borghesi, M. Passoni, Rev. Mod. Phys. **85**, 751 (2013)
3. National Ignition Facility (LLNL, Livermore, CA). Available at: https://lasers.llnl.gov/
4. Le Laser Mégajoule (CEA, France). Available at: http://www-lmj.cea.fr/en/lmj/index.htm
5. H.T. Kim et al., Phys. Rev. Lett. **111**, 165002 (2013)
6. W.P. Leemans et al., Phys. Rev. Lett. **113**, 245002 (2014)
7. M.D. Perry, G. Mourou, Science **264**, 917 (1994)
8. G. Mourou, T. Tajima, S. Bulanov, Rev. Mod. Phys. **78**, 309 (2006)
9. M. Dunne, Nat. Phys. **2**, 2 (2006)

10. J.-P. Chambaret, G. Cheriaux, J. Collier, R. Dabu, P. Dombi, A.M. Dunne, K. Ertel, P. Georges, J. Hebling, J. Hein, C. Hernandez-Gomez, C. Hooker, S. Karsch, G. Korn, F. Krausz, C. Le Blanc, Z.S. Major, F. Mathieu, T. Metzger, G. Mourou, P. Nickles, K. Osvay, B. Rus, W. Sandner, G. Szabó, D. Ursescu, K. Varjú, Proc. SPIE **7721**, 77211D (2010)
11. T. Maiman, Nature **187**, 493 (1960)
12. D. Strickland, G. Mourou, Opt. Commun. **56**, 219 (1985)
13. A. Dubietis, G. Jonusauskas, A. Piskarskas, Opt. Commun. **88**, 437 (1992)
14. T. Tajima, J.M. Dawson, Phys. Rev. Lett. **43**, 267 (1979)
15. S.J. Gitomer, R.D. Jones, F. Begay, A.W. Wheeler, J.F. Kephart, R. Kristal, Phys. Fluids **29**, 2679 (1986)
16. G. Mourou, T. Tajima, Science **331**, 41 (2011)
17. T. Seggebrock, I. Dornmair, T. Tajima, G. Mourou, F. Grüner, Progr. Theor. Exp. Phys. **2014**, 013A02 (2014)
18. E.L. Clark, K. Krushelnick, J.R. Davies, M. Zepf, M. Tatarakis, F.N. Beg, A. Machacek, P.A. Norreys, M.I.K. Santala, I. Watts, A.E. Dangor, Phys. Rev. Lett. **84**, 670 (2000)
19. A. Maksimchuk, S. Gu, K. Flippo, D. Umstadter, Phys. Rev. Lett. **84**, 4108 (2000)
20. R.A. Snavely, M.H. Key, S.P. Hatchett, T.E. Cowan, M. Roth, T.W. Phillips, M.A. Stoyer, E.A. Henry, T.C. Sangster, M.S. Singh, S.C. Wilks, A. MacKinnon, A. Offenberger, D.M. Pennington, K. Yasuike, A.B. Langdon, B.F. Lasinski, J. Johnson, M.D. Perry, E.M. Campbell, Phys. Rev. Lett. **85**, 2945 (2000)
21. L.V. Keldysh, Zh. Eksp. Teor. Fiz. **47**, 1945 (1964.) (in Russian)
22. V.S. Popov, Phys. Usp. **47**, 855 (2004)
23. S.V. Bulanov, I.N. Inovenkov, V.I. Kirsanov, N.M. Naumova, A.S. Sakharov, Phys. Fluids B **4**, 1935 (1992)
24. S.V. Bulanov, N.M. Naumova, F. Pegoraro, Phys. Plasmas **1**, 745 (1994)
25. F. Pegoraro, T.Z. Esirkepov, S.V. Bulanov, Phys. Lett. A **347**, 133 (2005)
26. A. Pukhov, J. Meyer-ter-Vehn, Phys. Rev. Lett. **76**, 3975 (1996)
27. T.D. Arber, K. Bennett, C.S. Brady, A. Lawrence-Douglas, M.G. Ramsay, N.J. Sircombe, P. Gillies, R.G. Evans, H. Schmitz, A.R. Bell, C.P. Ridgers, Plasma Phys. Contr. Fusion **57**, 113001 (2015)
28. E. Esarey et al., Rev. Mod. Phys. **81**, 129 (2009)
29. A. Modena, Z. Najmudin, A.E. Dangor, C.E. Clayton, K.A. Marsh, C. Joshi, V. Malka, C.B. Darrow, C. Danson, D. Neely, F.N. Walsh, Nature **377**, 606 (1995)
30. K. Nakajima, D. Fisher, T. Kawakubo, H. Nakanishi, A. Ogata, Y. Kato, Y. Kitagawa, R. Kodama, K. Mima, H. Shiraga, K. Suzuki, K. Yamakawa, T. Zhang, Y. Sakawa, T. Shoji, Y. Nishida, N. Yugami, M. Downer, T. Tajima, Phys. Rev. Lett. **74**, 4428 (1995)
31. D. Umstadter, S.-Y. Chen, A. Maksimchuk, G. Mourou, R. Wagner, Science **273**, 472 (1996)
32. A. Ting, C.I. Moore, K. Krushelnick, C. Manka, E. Esarey, P. Sprangle, R. Hubbard, H.R. Burris, R. Fischer, M. Baine, Phys. Plasmas **4**, 1889 (1997)
33. C. Gahn, G.D. Tsakiris, A. Pukhov, J. Meyer-ter-Vehn, G. Pretzler, P. Thirolf, D. Habs, K.J. Witte, Phys. Rev. Lett. **83**, 4772 (1999)
34. W.P. Leemans, P. Catravas, E. Esarey, C.G.R. Geddes, C. Toth, R. Trines, C.B. Schroeder, B.A. Shadwick, J. van Tilborg, J. Faure, Phys. Rev. Lett. **89**, 174802 (2002)
35. V. Malka, S. Fritzler, E. Lefebvre, M.-M. Aleonard, F. Burgy, J.-P. Chambaret, J.-F. Chemin, K. Krushelnick, G. Malka, S.P.D. Mangles, Z. Najmudin, M. Pittman, J.-P. Rousseau, J.-N. Scheurer, B. Walton, A.E. Dangor, Science **298**, 1596 (2002)
36. C.G.R. Geddes, C. Toth, J. van Tilborg, E. Esarey, C.B. Schroeder, D. Bruhwiler, C. Nieter, J. Cary, W.P. Leemans, Nature **431**, 538 (2004)
37. S.P.D. Mangles, C.D. Murphy, Z. Najmudin, A.G.R. Thomas, J.L. Collier, A.E. Dangor, E.J. Divall, P.S. Foster, J.G. Gallacher, C.J. Hooker, D.A. Jaroszynski, A.J. Langley, W.B. Mori, P.A. Norreys, F.S. Tsung, R. Viskup, B.R. Walton, K. Krushelnick, Nature **431**, 535 (2004)
38. J. Faure, Y. Glinec, A. Pukhov, S. Kiselev, S. Gordienko, E. Lefebvre, J.-P. Rousseau, F. Burgy, V. Malka, Nature **431**, 541 (2004)
39. T. Katsouleas, Nature **431**, 515 (2004)

40. A. Pukhov, J. Meyer-ter-Vehn, Appl. Phys. B Lasers Opt. **74**, 355 (2002)
41. E. Miura, K. Koyama, S. Kato, N. Saito, M. Adachi, Y. Kawada, T. Nakamura, M. Tanimoto, Appl. Phys. Lett. **86**, 251501 (2005)
42. A. Yamazaki, H. Kotaki, I. Daito, M. Kando, S.V. Bulanov, T.Z. Esirkepov, S. Kondo, S. Kanazawa, T. Homma, K. Nakajima, Y. Oishi, T. Nayuki, T. Fujii, K. Nemoto, Phys. Plasmas **12**, 093101 (2005)
43. B. Hidding, K.-U. Amthor, B. Liesfeld, H. Schwoerer, S. Karsch, M. Geissler, L. Veisz, K. Schmid, J.G. Gallacher, S.P. Jamison, D. Jaroszynski, G. Pretzler, R. Sauerbrey, Phys. Rev. Lett. **96**, 105004 (2006)
44. K. Krushelnick, Z. Najmudin, S.P.D. Mangles, A.G.R. Thomas, M.S. Wei, B. Walton, A. Gopal, E.L. Clark, A.E. Dangor, S. Fritzler, C.D. Murphy, P.A. Norreys, W.B. Mori, J. Gallacher, D. Jaroszynski, R. Viskup, Phys. Plasmas **12**, 056711 (2005)
45. M. Geissler, J. Schreiber, J. Meyer-ter-Vehn, New J. Phys. **8**, 186 (2006)
46. A.I. Akhieser, R.V. Polovin, JETP **3**, 696 (1956)
47. S. Gordienko, A. Pukhov, Phys. Plasmas **12**, 043109 (2005)
48. A. Pukhov, S. Gordienko, Phil. Trans. R. Soc. A **364**, 623 (2006)
49. C.G.R. Geddes, C. Toth, J. van Tilborg, E. Esarey, C.B. Schroeder, J. Cary, W.P. Leemans, Phys. Rev. Lett. **95**, 145002 (2005)
50. D.J. Spence, S.M. Hooker, Phys. Rev. E **63**, 015401 (2000)
51. A. Butler, D.J. Spence, S.M. Hooker, Phys. Rev. Lett. **89**, 185003 (2002)
52. T.P. Rowlands-Rees, C. Kamperidis, S. Kneip, A.J. Gonsalves, S.P.D. Mangles, J.G. Gallacher, E. Brunetti, T. Ibbotson, C.D. Murphy, P.S. Foster, M.J.V. Streeter, F. Budde, P.A. Norreys, D.A. Jaroszynski, K. Krushelnick, Z. Najmudin, S.M. Hooker, Phys. Rev. Lett. **100**, 105005 (2008)
53. A.F. Lifschitz, J. Faure, V. Malka, P. Mora, Phys. Plasmas **12**, 093104 (2005)
54. S.Y. Kalmykov, L.M. Gorbunov, P. Mora, G. Shvets, Phys. Plasmas **13**, 113102 (2006)
55. V. Malka, A. Lifschitz, J. Faure, Y. Glinec, Phys. Rev. ST Accel. Beams **9**, 091301 (2006)
56. W. Lu, M. Tzoufras, C. Joshi, F.S. Tsung, W.B. Mori, J. Vieira, R.A. Fonseca, L.O. Silva, Phys. Rev. ST Accel. Beams **10**, 061301 (2007)
57. E. Cormier-Michel, C.G.R. Geddes, E. Esarey, C.B. Schroeder, D.L. Bruhwiler, K. Paul, B. Cowan, W.P. Leemans, AIP Conf. Proc. **1089**, 315 (2009)
58. W.P. Leemans, B. Nagler, A.J. Gonsalves, C. Tóth, K. Nakamura, C.G.R. Geddes, E. Esarey, C.B. Schroeder, S.M. Hooker, Nat. Phys. **2**, 696 (2006)
59. K. Nakamura, B. Nagler, C. Tóth, C.G.R. Geddes, C.B. Schroeder, E. Esarey, W.P. Leemans, A.J. Gonsalves, S.M. Hooker, Phys. Plasmas **14**, 056708 (2007)
60. S. Karsch, J. Osterhoff, A. Popp, T.P. Rowlands-Rees, Z. Major, M. Fuchs, B. Marx, R. Hörlein, K. Schmid, L. Veisz, S. Becker, U. Schramm, B. Hidding, G. Pretzler, D. Habs, F. Gruner, F. Krausz, S.M. Hooker, New J. Phys. **9**, 415 (2007)
61. X. Wang, R. Zgadzaj, N. Fazel, Z. Li, S.A. Yi, X. Zhang, W. Henderson, Y.-Y. Chang, R. Korzekwa, H.-E. Tsai, C.-H. Pai, H. Quevedo, G. Dyer, E. Gaul, M. Martinez, A.C. Bernstein, T. Borger, M. Spinks, M. Donovan, V. Khudik, G. Shvets, T. Ditmire, M.C. Downer, Nat. Commun. **4**, 1988 (2013)
62. A.J. Gonsalves, K. Nakamura, C. Lin, D. Panasenko, S. Shiraishi, T. Sokollik, C. Benedetti, C.B. Schroeder, C.G.R. Geddes, J. van Tilborg, J. Osterhoff, E. Esarey, C. Tóth, W.P. Leemans, Nat. Phys. **7**, 862 (2011)
63. W.P. Leemans, R. Duarte, E. Esarey, S. Fournier, C.G.R. Geddes, D. Lockhart, C.B. Schroeder, Cs. Tóth, J.-L. Vay, S. Zimmermann, in *Advanced Accelerator Concepts*, ed. by S.H. Gold, G. S. Nusinovich. vol. 1299 (AIP, New York, 2010), p. 3
64. V.I. Veksler, Proc. CERN Symp. of High Energy Accelerators and Pion Physics (Geneva, Switzerland) vol 1, (1956), p. 80
65. F. Mako, T. Tajima, Phys. Fluids **27**, 1815 (1984)
66. T.Z. Esirkepov, Y. Sentoku, K. Mima, K. Nishihara, F. Califano, F. Pegoraro, N.M. Naumova, S.V. Bulanov, Y. Ueshima, T.V. Liseikina, V.A. Vshivkov, Y. Kato, JETP Lett. **70**, 82 (1999)

67. A.G. Zhidkov, A. Sasaki, T. Tajima, T. Auguste, P. D'Olivera, S. Hulin, P. Monot, A.Y. Faenov, T.A. Pikuz, I.Y. Skobelev, Phys. Rev. **E60**, 3273 (1999)
68. A. Zhidkov, A. Sasaki, Phys. Plasmas **7**, 1341 (2000)
69. S.P. Hatchett, C.G. Brown, T.E. Cowan, E.A. Henry, J.S. Johnson, M.H. Key, Phys. Plasmas **7**, 2076 (2000)
70. K. Krushelnick, E.L. Clark, M. Zepf, J.R. Davies, F.N. Beg, A. Machacek, M.I.K. Santala, M. Tatarakis, I. Watts, P.A. Norreys, A.E. Dangor, Phys. Plasmas **7**, 2055 (2000)
71. A. Maksimchuk, K. Flippo, H. Krause, G. Mourou, K. Nemoto, D. Shultz, D. Umstadter, R. Vane, V.Y. Bychenkov, G.I. Dudnikova, V.F. Kovalev, K. Mima, V.N. Novikov, Y. Sentoku, S.V. Tolokonni Kov, Plasma Phys. Rep. **30**, 473 (2004)
72. Y. Murakami, Y. Kitagawa, Y. Sentoku, M. Mori, R. Kodama, K.A. Tanaka, K. Mima, T. Yamanaka, Phys. Plamas **8**, 4138 (2001)
73. A.J. Mackinnon, M. Borghesi, S. Hatchett, M.H. Key, P.K. Patel, H. Campbell, A. Schiavi, R. Snavely, S.C. Wilks, O. Willi, Phys. Rev. Lett. **86**, 1769 (2001)
74. A.J. Mackinnon, Y. Sentoku, P.K. Patel, D.W. Price, S. Hatchett, M.H. Key, C. Andersen, R. Snavely, R.R. Freeman, Phys. Rev. Lett. **88**, 215006 (2002)
75. K. Nemoto, A. Maksimchuk, S. Banerjee, K. Flippo, G. Mourou, D. Umstadter, V.Y. Bychenkov, Appl. Phys. Lett. **78**, 595 (2001)
76. I. Spencer, K.W.D. Ledingham, R.P. Singhal, T. McCanny, P. McKenna, E.L. Clark, K. Krushelnick, M. Zepf, F.N. Beg, M. Tatarakis, A.E. Dangor, P.A. Norreys, R.J. Clarke, R.M. Allott, I.N. Ross, Nucl. Instrum. Methods. Phy. Res. B **183**, 449 (2001)
77. I. Spencer, K.W.D. Ledingham, P. McKenna, T. McCanny, R.P. Singhal, P.S. Foster, D. Neely, A.J. Langley, E.J. Divall, C.J. Hooker, R.J. Clarke, P.A. Norreys, E.L. Clark, K. Krushelnick, J.R. Da Vies, Phys. Rev. E **67**, 046402 (2003)
78. M. Zepf, E.L. Clark, K. Krushelnick, F.N. Beg, C. Escoda, A.E. Dangor, M.I.K. Santala, M. Tatarakis, I.F. Watts, P.A. Norreys, R.J. Clarke, J.R. Davies, M.A. Sinclair, R.D. Edwards, T.J. Goldsack, I. Spencer, K.W.D. Ledingham, Phys. Plasmas **8**, 2323 (2001)
79. M. Hegelich, S. Karsch, G. Pretzler, D. Habs, K. Witte, W. Guenther, M. Allen, A. Blazevic, J. Fuchs, J.C. Gauthier, M. Geissel, P. Audebert, T. Cowan, M. Roth, Phys. Rev. Lett. **89**, 085002 (2002)
80. B.M. Hegelich, B. Albright, P. Audebert, A. Blazevic, E. Brambrink, J. Cobble, T. Cowan, J. Fuchs, J.C. Gauthier, C. Gautier, M. Geissel, D. Habs, R. Johnson, S. Karsch, A. Kemp, S. Letzring, M. Roth, U. Schramm, J. Schreiber, K.J. Witte, J.C. Fernández, Phys. Plasmas **12**, 056314 (2005)
81. M. Roth, A. Blazevic, M. Geissel, T. Schlegel, T.E. Cowan, M. Allen, J.-C. Gauthier, P. Audebert, J. Fuchs, J. Meyer-ter-Vehn, M. Hegelich, S. Karsch, A. Pukhov, Phys. Rev. ST Accel. Beams **5**, 061301 (2002)
82. M. Roth, E. Brambrink, P. Audebert, M. Basko, A. Blazevic, R. Clarke, J. Cobble, T.E. Cowan, J. Fernandez, J. Fuchs, M. Hegelich, K. Ledingham, B.G. Logan, D. Neely, H. Ruhl, M. Schollmeier, Plasma Phys. Control. Fusion **47**, B841 (2005)
83. S. Fritzler, V. Malka, G. Grillon, J.P. Rousseau, F. Burgy, E. Lefebvre, E. d'Humières, P. McKenna, K.W.D. Ledingham, Appl. Phys. Lett. **83**, 3039 (2003)
84. J. Fuchs, T.E. Cowan, P. Audebert, H. Ruhl, L. Gremillet, A. Kemp, M. Allen, A. Blazevic, J.-C. Gauthier, M. Geissel, M. Hegelich, S. Karsch, P. Parks, M. Roth, Y. Sentoku, R. Stephens, E.M. Campbell, Phys. Rev. Lett. **91**, 255002 (2003)
85. J. Fuchs, Y. Sentoku, S. Karsch, J. Cobble, P. Audebert, A. Kemp, A. Nikroo, P. Antici, E. Brambrink, A. Blazevic, E.M. Campbell, J.C. Fernández, J.-C. Gauthier, M. Geissel, M. Hegelich, H. Pépin, H. Popescu, N. Renard-LeGalloudec, M. Roth, J. Schreiber, R. Stephens, T.E. Cowan, Phys. Rev. Lett. **94**, 045004 (2005)
86. K. Matsukado, T. Esirkepov, K. Kinoshita, H. Daido, T. Utsumi, Z. Li, A. Fukumi, Y. Hayashi, S. Orimo, M. Nishiuchi, S.V. Bulanov, T. Tajima, A. Noda, Y. Iwashita, T. Shirai, T. Takeuchi, S. Nakamura, A. Yamazaki, M. Ikegami, T. Mihara, A. Morita, M. Uesaka, K. Yoshii, T. Watanabe, T. Hosokai, A. Zhidkov, A. Ogata, Y. Wada, T. Kubota, Phys. Rev. Lett. **91**, 215001 (2003)

87. P. McKenna, K.W.D. Ledingham, T. McCanny, R.P. Singhal, I. Spencer, E.L. Clark, F.N. Beg, K. Krushelnick, M.S. Wei, J. Galy, J. Magill, R.J. Clarke, K.L. Lancaster, P.A. Norreys, K. Spohr, R. Chapman, Appl. Phys. Lett. **83**, 2763 (2003)
88. P. McKenna, K.W.D. Ledingham, J.M. Yang, L. Robson, T. McCanny, S. Shimizu, R.J. Clarke, D. Neely, K. Spohr, R. Chapman, R.P. Singhal, K. Krushelnick, M.S. Wei, P.A. Norreys, Phys. Rev. E **70**, 036405 (2004)
89. P. McKenna, K.W.D. Ledingham, S. Shimizu, J.M. Yang, L. Robson, T. McCanny, J. Galy, J. Magill, R.J. Clarke, D. Neely, P.A. Norreys, R.P. Singhal, K. Krushelnick, M.S. Wei, Phys. Rev. Lett. **94**, 084801 (2005)
90. P.K. Patel, A.J. Mackinnon, M.H. Key, T.E. Cowan, M.E. Foord, M. Allen, D.F. Price, H. Ruhl, P.T. Springer, R. Stephens, Phys. Rev. Lett. **91**, 125004 (2003)
91. M. Borghesi, A.J. Mackinnon, D.H. Campbell, D.G. Hicks, S. Kar, P.K. Patel, D. Price, L. Romagnani, A. Schiavi, O. Willi, Phys. Rev. Lett. **92**, 055003 (2004)
92. T.E. Cowan, J. Fuchs, H. Ruhl, A. Kemp, P. Audebert, M. Roth, R. Stephens, I. Barton, A. Blazevic, E. Brambrink, J. Cobble, J. Fernández, J.-C. Gauthier, M. Geissel, M. Hegelich, J. Kaae, S. Karsch, G.P. Le Sage, S. Letzring, M. Manclossi, S. Meyroneinc, A. Newkirk, H. Pépin, N. Renard-LeGalloudec, Phys. Rev. Lett. **92**, 204801 (2004)
93. M. Kaluza, J. Schreiber, M.I.K. Santala, G.D. Tsakiris, K. Eidmann, J. Meyer-ter-Vehn, K.J. Witte, Phys. Rev. Lett. **93**, 045003 (2004)
94. H. Ruhl, T. Cowan, J. Fuchs, Phys. Plasmas **11**, L17 (2004)
95. J. Schreiber, M. Kaluza, F. Grüner, U. Schramm, B.M. Hegelich, J. Cobble, M. Geissler, E. Brambrink, J. Fuchs, P. Audebert, D. Habs, K. Witte, Appl. Phys. B Lasers Opt. **79**, 1041 (2004)
96. Y. Oishi, T. Nayuki, T. Fujii, Y. Takizawa, X. Wang, T. Yamazaki, K. Nemoto, T. Kayoiji, T. Sekiya, K. Horioka, Y. Okano, Y. Hironaka, K.G. Nakamura, K. Kondo, A.A. Andreev, Phys. Plasmas **12**, 073102 (2005)
97. S. Ter-Avetisyan, M. Schnürer, P.V. Nickles, J. Phys. D **38**, 863 (2005)
98. M. Borghesi, J. Fuchs, S.V. Bulanov, A.J. Mackinnon, P.K. Patel, M. Roth, Fusion Sci. Technol. **49**, 412 (2006)
99. K. Krushelnick, E.L. Clark, R. Allott, F.N. Beg, C.N. Danson, A. Machacek, V. Malka, Z. Najmudin, D. Neely, P.A. Norreys, M.R. Salvati, M.I.K. Santala, M. Tatarakis, I. Watts, M. Zepf, A.E. Dangor, IEEE Trans. Plasma Sci. **28**, 1184 (2000)
100. A. Noda, S. Nakamura, Y. Iwashita, S. Sakabe, M. Hashida, T. Shirai, S. Shimizu, H. Tongu, H. Ito, H. Souda, A. Yamazaki, M. Tanabe, H. Daido, M. Mori, M. Kado, A. Sagisaka, K. Ogura, M. Nishiuchi, S. Orimo, Y. Hayashi, A. Yogo, S. Bulanov, T. Esirkepov, A. Nagashima, T. Kimura, T. Tajima, T. Takeuchi, K. Matsukado, A. Fukumi, Z. Li, Laser Phys. **16**, 47–653 (2006)
101. P. Antici, M. Fazi, A. Lombardi, M. Migliorati, L. Palumbo, P. Audebert, J. Fuchs, J. Appl. Phys. **104**, 124901 (2008)
102. G.M. Dyer, A.C. Bernstein, B.I. Cho, J. Osterholz, W. Grigsby, A. Dalton, R. Shepherd, Y. Ping, H. Chen, K. Widmann, T. Ditmire, Phys. Rev. Lett. **101**, 015002 (2008)
103. M. Murakami, Y. Hishikawa, S. Miyajima, Y. Okazaki, K.L. Sutherland, M. Abe, S.V. Bulanov, H. Daido, T.Z. Esirkepov, J. Koga, M. Yamagiwa, T. Tajima, AIP Conf. Proc. **1024**, 275 (2008)
104. P.R. Bolton, T. Hori, H. Kiriyama, M. Mori, H. Sakaki, K. Sutherland, M. Suzuki, J. Wu, A. Yogo, Nucl. Instrum. Methods. Phy. Res. A **620**, 71 (2010)
105. M. Roth, T.E. Cowan, M.H. Key, S.P. Hatchett, C. Brown, W. Fountain, J. Johnson, D.M. Pennington, R.A. Snavely, S.C. Wilks, K. Yasuike, H. Ruhl, F. Pegoraro, S.V. Bulanov, E.M. Campbell, M.D. Perry, H. Powell, Phys. Rev. Lett. **86**, 436 (2001)
106. V.Y. Bychenkov, W. Rozmus, A. Maksimchuk, D. Umstadter, C.E. Capjack, Plasma Phys. Rep. **27**, 1017–1020 (2001)

107. K.W.D. Ledingham, P. McKenna, T. McCanny, S. Shimizu, J.M. Yang, L. Robson, J. Zweit, J.M. Gillies, J. Bailey, G.N. Chimon, R.J. Clarke, D. Neely, P.A. Norreys, J.L. Collier, R.P. Singhal, M.S. Wei, S.P.D. Mangles, P. Nilson, K. Krushelnick, M. Zepf, J. Phys. D **37**, 2341 (2004)
108. W.L. Kruer, K. Estabrook, Phys. Fluids **28**, 430 (1985)
109. A. Pukhov, Phys. Rev. Lett. **86**, 3562 (2001)
110. S.C. Wilks, A.B. Langdon, T.E. Cowan, M. Roth, M. Singh, S. Hatchett, M.H. Key, D. Pennington, A. MacKinnon, R.A. Snavely, Phys. Plasmas **8**, 542 (2001)
111. A. Henig, Ph.D. Thesis, LMU Munich, 2010
112. A.P.L. Robinson, P. Gibbon, M. Zepf, S. Kar, R.G. Evans, C. Bellei, Plasma Phys. Control. Fusion **51**, 024004 (2009)
113. A. Macchi, F. Cattani, T.V. Liseykina, F. Cornolti, Phys. Rev. Lett. **94**, 165003 (2005)
114. T. Esirkepov, M. Borghesi, S.V. Bulanov, G. Mourou, T. Tajima, Phys. Rev. Lett. **92**, 175003 (2004)
115. O. Klimo, J. Psikal, J. Limpouch, V.T. Tikhonchuk, Phys. Rev. ST Accl. Beams **11**, 031301 (2008)
116. A.P.L. Robinson, M. Zepf, S. Kar, R.G. Evans, C. Bellei, New J. Phys. **10**, 013021 (2008)
117. S.G. Rykovanov, J. Schreiber, J. Meyer-ter-Vehn, C. Bellei, A. Henig, H.C. Wu, M. Geissler, New J. Phys. **10**, 113005 (2008)
118. V.A. VShivkov, N.M. Naumova, F. Pegoraro, S.V. Bulanov, Phys. Plasmas **5**, 2727–2741 (1998)
119. D. Ursescu, G. Cheriaux, P. Audebert, M. Kalashnikov, T. Toncian, M. Cerchez, M. Kaluza, G. Paulus, G. Priebe, R. Dabu, M.O. Cernaianu, M. Dinescu, T. Asavei, I. Dancus, L. Neagu, A. Boianu, C. Hooker, C. Barty, C. Haefner, Rom. Rep. Phys. **68**, s11–s36 (2016)
120. S.C. Wilks, W.L. Kruer, M. Tabak, A.B. Langdon, Phys. Rev. Lett. **69**, 1383 (1992)
121. X. Zhang, B. Shen, X. Li, Z. Jin, F. Wang, Phys. Plasmas **14**, 073101 (2007)
122. X. Zhang, B. Shen, X. Li, Z. Jin, F. Wang, M. Wen, Phys. Plasmas **14**, 123108 (2007)
123. A. Henig, S. Steinke, M. Schnürer, T. Sokollik, R. Hörlein, D. Kiefer, D. Jung, J. Schreiber, B.M. Hegelich, X.Q. Yan, J. Meyer-ter-Vehn, T. Tajima, P.V. Nickles, W. Sandner, D. Habs, Phys. Rev. Lett. **103**, 245003 (2009)
124. T. Tajima, D. Habs, X. Yan, Rev. Accel. Sci. Technol. **2**, 221 (2009)
125. A. Macchi, S. Veghini, F. Pegoraro, Phys. Rev. Lett. **103**, 085003 (2009)
126. A. Macchi, A. Sgattoni, S. Sinigardi, M. Borghesi, M. Passoni, Plasma Phys. Control. Fusion **56**, 039501 (2014)
127. J. Schreiber, P.R. Bolton, K. Parodi, Rev. Sci. Instrum. **87**, 071101 (2016)
128. J. Schreiber, F. Bell, Z. Najmudin, High Power Laser Sci. Eng. **2**, e41 (2014)
129. F. Wagner, O. Deppert, C. Brabetz, P. Fiala, A. Kleinschmidt, P. Poth, V.A. Schanz, A. Tebartz, B. Zielbauer, M. Roth, T. Stöhlker, V. Bagnoud, Phys. Rev. Lett. **116**, 205002 (2016)
130. D. Jung, L. Yin, B.J. Albright, D.C. Gautier, S. Letzring, B. Dromey, M. Yeung, R.H. Hörlein, R. Shah, S. Palaniyappan, K. Allinger, J. Schreiber, K.J. Bowers, H.-C. Wu, J.C. Fernandez, D. Habs, B.M. Hegelich, New J. Phys. **15**, 023007 (2013)
131. J.H. Bin, W.J. Ma, K. Allinger, H.Y. Wang, D. Kiefer, S. Reinhardt, P. Hilz, K. Khrennikov, S. Karsch, X.Q. Yan, F. Krausz, T. Tajima, D. Habs, J. Schreiber, Phys. Plasma **20**, 073113 (2013)
132. K. Zeil, S.D. Kraft, S. Bock, M. Bussamnn, T.E. Cowan, T. Kluge, J. Metzkes, T. Richter, R. Sauerbrey, U. Schramm, New J. Phys. **12**, 045015 (2010)
133. K. Ogura, M. Nishiuchi, A.S. Pirozkhov, T. Tanimoto, A. Sagisaka, T.Z. Esirkepov, M. Kando, T. Shizuma, T. Hayakawa, H. Kiriyama, T. Shimomura, S. Kondo, S. Kanazawa, Y. Nakai, H. Sasao, Y. Fukuda, H. Sakaki, M. Kanasaki, A. Yogo, S.V. Bulanov, P.R. Bolton, K. Kondo, Opt. Lett. **37**, 2868 (2012)
134. I.J. Kim, K.H. Pae, C.M. Kim, H.T. Kim, J.H. Sung, S.K. Lee, T.J. Yu, I.W. Choi, C.-L. Lee, K.H. Nam, P.V. Nickles, T.M. Jeong, J. Lee, Phys. Rev. Lett. **1**, 165003 (2013)
135. J.S. Green, A.P.L. Robinson, N. Booth, D.C. Carroll, R.J. Dance, R.J. Gray, D.A. Maclellan, P. McKenna, C.D. Murphy, D. Rusby, L. Wilson, Appl. Phys. Lett. **104**, 214101 (2014)

136. D. Jung, Dissertation, LMU München, 2012
137. H.Y. Wang, C. Lin, F.L. Zheng, Y.R. Lu, Z.Y. Guo, X.T. He, J.F. Chen, X.Q. Yan, Phys. Plasmas **18**, 093105 (2011)
138. B. Quiao, S. Kar, M. Geissler, P. Gibbon, M. Zepf, M. Borghesi, Phys. Rev. Lett. **108**, 115002 (2012)
139. A. Sgattoni, S. Sinigardi, A. Macchi, Appl. Phys. Lett. **105**, 084105 (2014)
140. X.Q. Yan, H.C. Wu, Z.M. Zheng, J.E. Chen, J. Meyer-ter-Vehn, Phys. Rev. Lett. **103**, 135001 (2009)
141. F. Negoita, M. Roth, P.G. Thirolf, S. Tudisco, F. Hannachi, S. Moustaizis, I. Pomerantz, P. McKenna, J. Fuchs, K. Spohr, G. Acbas, A. Anzalone, P. Auderbert, S. Balascuta, F. Cappuzzello, M.O. Cernaianu, S. Chen, I. Dancus, R. Freeman, H. Geissel, P. Ghenuche, L. Gizzi, F. Gobet, G. Gosselin, M. Gugiu, D. Higginson, E. d'Humieres, C. Ivan, D. Jaroszynski, S. Kar, L. Lamia, V. Leca, L. Neagu, G. Lanzalone, V. Meot, S.R. Mirfayzi, I.O. Mitu, P. Morel, C. Murphy, C. Petcu, H. Petrascu, C. Petrone, P. Raczka, M. Risca, F. Rotaru, J.J. Santos, D. Schumacher, D. Stutman, M. Tarisien, M. Tataru, B. Tatulea, I.C.E. Turcu, M. Versteegen, D. Ursescu, S. Gales, N.V. Zamfir, Rom. Rep. Phys. **68**, s37–144 (2016)
142. M. Arnould, S. Goriely, K. Takahashi, Phys. Rep. **450**, 97 (2007)
143. I.V. Panov, H.-T. Janka, Astron. Astrophys. **494**, 829 (2009)
144. H.-T. Janka, K. Langanke, A. Marek, G. Martinez-Pinedo, B. Muller, Phys. Rep. **442**, 38 (2007)
145. J.J. Cowan, F.-K. Thielemann, Phys. Today **57**, 47 (2004)
146. C. Freiburghaus, S. Rosswog, F.-K. Thielemann, Astrophys. J. **525**, L121 (1999)
147. R. Surman, G.C. McLaughlin, Astrophys. J. **679**, L117 (2008)
148. K. Farouqi, C. Freiburghaus, K.-L. Kratz, B. Pfeiffer, T. Rauscher, F.-K. Thielemann, Nucl. Phys. A **758**, 631c (2005)
149. D. Habs, P.G. Thirolf, M. Gross, K. Allinger, J. Bin, A. Henig, D. Kiefer, W. Ma, J. Schreiber, Appl. Phys. B Lasers Opt. **103**, 471 (2011)
150. FAIR, An International Accelerator Facility for Beams of Ions and Antiprotons, Baseline Technical Report, (GSI, Kolkata, 2006), http://www.gsi.defairreportsbtr.html
151. SPIRAL II, Detailed Design Study - APD Report, GANIL (2005), http://pro.ganil-spiral2.eu/spiral2/what-is-spiral2/apd
152. S. Baruah, G. Audi, K. Blaum, M. Dworschak, S. George, C. Guénaut, U. Hager, F. Herfurth, A. Herlert, A. Kellerbauer, H.-J. Kluge, D. Lunney, H. Schatz, L. Schweikhard, C. Yazidjian, Phys. Rev. Lett. **101**, 262501 (2008)
153. M. Dworschak, G. Audi, K. Blaum, P. Delahaye, S. George, U. Hager, F. Herfurth, A. Herlert, A. Kellerbauer, H.-J. Kluge, D. Lunney, L. Schweikhard, C. Yazidjian, Phys. Rev. Lett. **100**, 072501 (2008)
154. I. Dillmann, K.-L. Kratz, A. Wöhr, O. Arndt, B.A. Brown, P. Hoff, M. Hjorth-Jensen, U. Köster, A.N. Ostrowski, B. Pfeiffer, D. Seweryniak, J. Shergur, W.B. Walters, Phys. Rev. Lett. **91**, 162503 (2003)
155. K. Blaum, Phys. Rep. **425**, 1 (2006)
156. P. Moller, J.R. Nix, W.D. Myers, W.J. Swiatecki, At. Data Nucl. Data Tables **59**, 185 (1995)
157. P.G. Thirolf, D. Habs, Prog. Part. Nucl. Phys. **49**, 325 (2002)
158. V. Metag, Nukleonica **20**, 789 (1975)
159. R. Vandenbosch, J.R. Huizenga, *Nuclear Fission* (Academic Press, New York, 1973)
160. A. Gavron, Phys. Rev. C **21**, 230 (1980)
161. O.B. Tarasov, D. Bazin, Nucl. Instrum. Methods. Phy. Res. B **204**, 174 (2003)
162. A.B. Quint, W. Reisdorf, K.-H. Schmidt, P. Armbruster, F.P. Heßberger, S. Hofmann, J. Keller, G. Münzenberg, H. Stelzer, H.-G. Clerc, W. Morawek, C.-C. Sahm, Z. Phys. A **346**, 119 (1993)
163. W. Morawek, D. Ackermann, T. Brohm, H.-G. Clerc, U. Gollerthan, E. Hanelt, M. Horz, W. Schwab, B. Voss, K.-H. Schmidt, F.P. Heßberger, Z. Phys. A **341**, 75 (1991)
164. D. Habs, M. Gross, P.G. Thirolf, P. Böni, Appl. Phys. B Lasers Opt. **103**, 485 (2011)

165. NRC Rare Isotope Science Assessment Committee (RISAC) Report (National Academies Press, Washington, DC, 2007)
166. T. Kurtukian-Nieto, J. Benlliure, K.H. Schmidt, Nucl. Instrum. Methods. Phy. Res. A **589**, 472 (2008)
167. Particle Data Group, Phys. Rev. D **66**, 010001 (2002)
168. S. Segre, *Nuclei and Particles*, 2nd edn. (W.A. Benjamin, London, 1977)
169. S. Ichimaru, *Basic Principles of Plasma Physics: A Statistical Approach* (Benjamin, Reading, 1973)
170. H.-C. Wu, T. Tajima, D. Habs, A.W. Chao, J. Meyer-ter-Vehn, Phys. Rev. ST Accel. Beams **13**, 101303 (2010)
171. U. Linz, J. Alonso, Phys. Rev. ST Accel. Beams **10**, 094801 (2007)
172. J. Bin, K. Allinger, W. Assmann, G. Dollinger, G.A. Drexler, A.A. Friedl, D. Habs, P. Hilz, R. Hoerlein, N. Humble, S. Karsch, K. Khrennikov, D. Kiefer, F. Krausz, W. Ma, D. Michalski, M. Molls, S. Raith, S. Reinhardt, B. Röper, T.E. Schmid, T. Tajima, J. Wenz, O. Zlobinskaya, J. Schreiber, J.J. Wilkens, Appl. Phys. Lett. **101**, 243701 (2012)
173. K. Zeil, M. Baumann, E. Beyreuther, T. Burris-Mog, T.E. Cowan, W. Enghardt, L. Karsch, S.D. Kraft, L. Laschinsky, J. Metzkes, D. Naumburger, M. Oppelt, C. Richter, R. Sauerbrey, M. Schürer, U. Schramm, J. Pawelke, Appl. Phys. B Lasers Opt. **110**, 437–444 (2013)

Printed in the United States
By Bookmasters